SELLING THE TRUE TIME

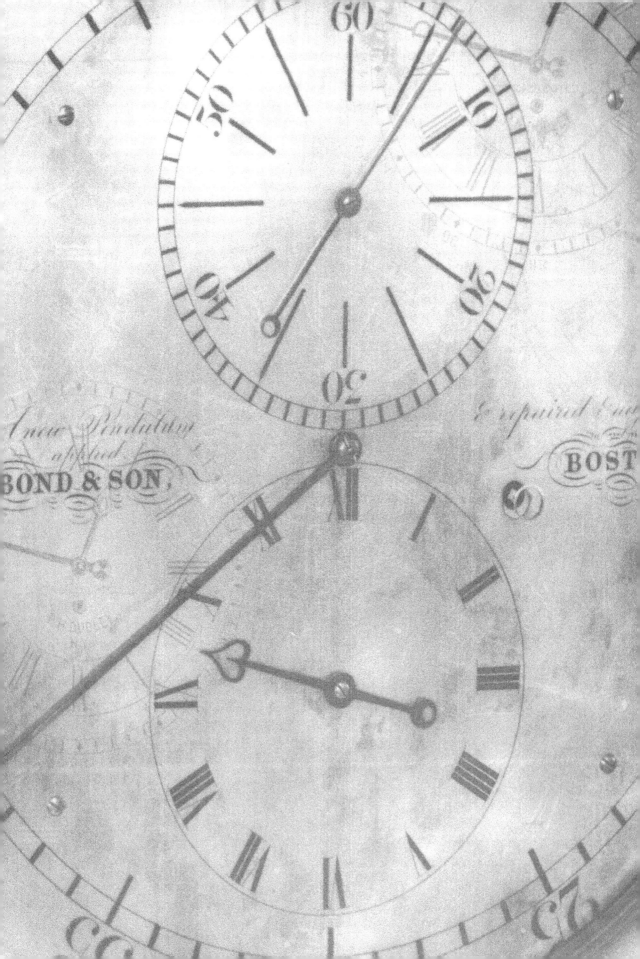

SELLING THE TRUE TIME

Nineteenth-Century Timekeeping in America

IAN R. BARTKY

STANFORD UNIVERSITY PRESS

STANFORD, CALIFORNIA 2000

Stanford University Press

Stanford, California

© 2000 by the Board of Trustees of the

Leland Stanford Junior University

Printed in the United States of America

Printed on acid-free, recycled paper

Library of Congress Cataloging-in-Publication Data

Bartky, Ian R.

    Selling the true time : nineteenth-century timekeeping in America /

Ian R. Bartky.

      p.   cm.

    Includes bibliographical references and index.

    ISBN 0-8047-3874-2 (cloth : alk. paper)

    I. Time—Systems and standards—United States.   I. Title.

QB210.U5 B37   2000

389'.17'09—dc21   99-086739

Designed by Janet Wood

Typeset by James P. Brommer in 11/14 Garamond

Original printing 2000

Last figure below indicates year of this printing:

09 08 07 06 05 04 03 02 01 00

Contents

Illustrations

Preface

Although I was completely unaware of it at the time, my first brush with public timekeeping issues came on Monday morning, January 7, 1974.

Christmas vacation was over and it was time for our children to go to school. I opened the front door and was astonished to see that it was still dark. So we held David and Anne back from their walk to Bethesda Elementary for perhaps twenty minutes, exchanged puzzled glances, and then went about getting ourselves ready for work. Mine was on Capitol Hill; I was a National Bureau of Standards employee, one of that year's departmental Science Fellows providing technical support to a subcommittee of the House Commerce Committee.

At work I learned that in mid-December, Congress had passed emergency legislation, a response to the country's oil embargo crisis. For the next two years all clocks would be advanced one hour—an experiment designed to conserve energy. The law had just taken effect, at 2 A.M. on Sunday. Like most Americans, I had paid no attention to the consequences of this shift in civil time.

My sunrise on that Monday morning was at 8:27—the latest that anyone living in Washington had ever seen. Across the United States sunrises varied from 7:37, in the Florida panhandle, to 9:48, in northwestern North Dakota; everyone everywhere was experiencing the latest sunrise ever recorded in their region. But I knew nothing then about these remarkable spans in sunrise (and sunset) times—having scarcely more than an awareness of the one-hour spacing between time zones and the fact that there are more hours of daylight in summer than in winter.

Asked in July to review the Department of Transportation's interim report on the effects of the country's experience with year-round daylight-saving time, I learned of the many protests directed to members of Congress by parents worried about the safety of their children walking to school or waiting for the bus in unanticipated darkness. Their letters started arriving in congressional offices in January, but by spring—with its earlier sunrises by the clock—the volume of mail dropped. It dawned on me that their concerns would resurface around the end of October, increase in volume to a peak in December, and continue throughout the entire month of February.

The interim report before me indicated no overall energy savings linked to advancing the nation's clocks an hour. Moreover, it contained evidence of an increase in fatalities among schoolchildren during the weekday mornings of the two winter months just past. In my mind, given these results, the country's year-round daylight-saving time experiment had to be altered.

To everyone's great relief, the House subcommittee that I was advising held hearings in mid-August. Then, in October, a Congress rushing toward adjournment and last-minute reelection campaigning voted to drop advanced time for the winter of 1974–75. I returned to NBS, now enmeshed in a subject that coupled public policy issues with well-understood astronomical phenomena, America's geography, and demographic data.

Time passed. An NBS team reviewed the final report on the effects of daylight-saving time for the subcommittee and testified at its hearing. With Elizabeth Harrison, a lawyer on the Commerce Committee's permanent staff, I published an article in *Scientific American* in which we addressed the impossibility of gaining more light in the evening in one region without extending morning darkness in another. Underlying our remarks was a concern that any changes in daylight-saving time periods should do minimum damage to the country's system of uniform time. In my view, Congress forgot its 1967 goal, expressed in legislation, to have everyone everywhere keep the same time; the passage of emergency legislation in 1973 certainly conflicted with that intent.

Eighteen months later a second event challenged my understanding of the history of American timekeeping. I already knew that the first national legislation involving civil time standards could be traced to the public's desire for daylight-saving time during World War I. I was also aware that the zone system that Congress had established then could be traced to the railroads' adoption of Standard Railway Time in 1883. Now, preparing a talk for the U.S. Naval Observatory's 150th anniversary symposium, I was finding other evidence of the federal government's involvement in public timekeeping—some dating back as early as the middle of the nineteenth century. Even more intriguing, I was seeing late-nineteenth-century statements by Samuel Langley, secretary of the Smithsonian Institution, denigrating this early, Naval Observatory time service, and also claiming that his observatory near Pittsburgh had been the first to provide a stable source of public time signals. Later I would learn that some historians asserted Harvard College Observatory's priority in the matter. And, equally remarkable, for many years both institutions were paid to supply time to the public.

Perplexed by the conflicting priority claims, I hedged one or two of my remarks at the Naval Observatory's celebration. Then, in my spare time, I began mining the annual reports of observatories and the histories of railroads. Early on I was able to resolve a number of the astronomers' conflicting statements. However, a framework for understanding American timekeeping did not come so easily. Not until I was able to devote all of my time to the subject did the missing pieces fall into place. This book is the result.

Acknowledgments

During our final years of research at libraries and archives, partial financial support came from two Dudley/Pollock Awards given by the trustees of the Dudley Observatory, and from a grant by the National Science Foundation, SBR-9411696. To those anonymous outside reviewers who supported my proposals to those institutions, thank you. I am also indebted to Ralph A. Alpher, the Dudley Observatory's administrator, for his continuous encouragement.

In the two decades spent on this effort, I was helped by many people. Librarians Brenda Corbin and Gregory Shelton at the Naval Observatory not only guided me through that institution's superb nineteenth-century astronomy and science collections but also spent enormous amounts of time locating materials at libraries throughout the country. Eileen Doudna, former librarian for the National Association of Watch and Clock Collectors, located numerous trade journal references. Early on, Ellen Halteman of the California State Railroad Museum's library steered me to the collection of railroad rule books, thus giving my analysis of railroad timekeeping a firm basis. Marietta Nelson at the National Institute of Standards and Technology (which will always be NBS to me!) opened the library's timekeeping files. At the Library of Congress, Constance Carter and Ruth Freitag of the Science and Technology Division, and Richard Sharp in the Business and Industry Division, were extremely helpful in explaining the various bibliographic listings of science, business, and government documents. Finally, our now-close friend Christine Bain, at the New York State Library in Albany, located the state government documents dealing with early longitudes and time services, and also opened many doors for us there.

Prior to this research effort, my professional world had never intersected the archival one. First contact came with Sharon Gibbs, then at the Polar and Scientific Archives branch of the National Archives. Her successful search for materials became the foundation for several articles on time balls. Later, as Sharon Thibodeau, she encouraged my expanding archival efforts. Marjorie Ciarlante, also at the National Archives, gave me much guidance on making efficient use of the many science collections.

As my study of American timekeeping expanded, archivists at other institutions contributed. Rebecca Abromitis, University Archives, University of Pittsburgh, located and presorted a substantial record of Allegheny Observatory time services; because of her efforts, we performed what would have been over a week of digging and review in less than two days. Others who smoothed our path included Michele Aldrich, former archivist at the American Association for the Advancement of Science; William R. Massa, Jr., Sterling Memorial Library, Yale University; Dorothy Czarnik, Illinois Regional Archives Depository, Northeastern Illinois University; Patrick Quinn, Northwestern University; and Eric Hillemann, Carleton College. At the urging of Bernard Schermetzler, University of Wisconsin-Madison Archives, we reviewed the Washburn Observatory's archives, thereby gaining an understanding of the growth of commercial railroad timekeeping. The efforts of archivist Nancy Langford and the assistance of Rita Spenser at the Dudley Observatory made it possible for us to develop an analysis of antebellum astronomers' involvement in public timekeeping.

The Harvard University Archives contains a wealth of observatory information germane to our interests. We are most grateful to Clark Elliott, Patrice Donoghue, and Gabrielle Green for making our weeks there so profitable. And I cannot praise too highly Dorothy Schaumberg at the Mary Lee Shane Archives of the Lick Observatory, University of California, Santa Cruz. Without her exhaustive efforts, Edward S. Holden's actions in university and commercial time services at both Washburn and Lick Observatories would have remained obscure.

At a very early stage, Carlene Stephens of the National Museum of American History and I worked together uncovering the history of the railroads' 1883 adoption of Standard Railway Time. My research needs segued to railroad timekeeping and then to the control of the trainmen's timekeepers, and thus led to several informative meetings with Dana Blackwell of Connecticut, an expert on watches and electrical clocks. During our Cambridge sojourn, my wife and I also interacted with curator Will Andrewes, Harvard University Collection of Historical Scientific Instruments, and assistant Martha Richardson, who opened their files and facilities. We count ourselves fortunate that Barbara Sladek, Department of the State, Hartford, Connecticut, took an interest in our inquiries regarding companies associated with the electrical distribution of time. Without her detailed review of incorporation records, we would have failed in our quest.

Throughout the last half-dozen years I received strong and steady encouragement from National Air and Space Museum historians David DeVorkin and Robert W. Smith (now at the University of Alberta, Edmonton). Dorrit Hoffleit of Yale's Astronomy Department gave unstintingly of her time when we investigated the nuances of New Haven timekeeping. Marc Rothenberg, editor of the Joseph Henry Papers, and assistant editor Kathleen Dorman dazzled me with their knowledge and interpretation of nineteenth-century American science and technology, and gave most welcome guidance to the correspondence and library of the Smithsonian Institution's first secretary.

Historian of astronomy Steven Dick, my collaborator on the study of the world's early time balls, assisted me in numerous ways. One was sharing his draft history, with supporting references, of the U.S. Naval Observatory. Just as important, Steve gave years of careful attention to my "almost-finished" stories of timekeeping.

On reaching the writing and editing phases, I received excellent advice and frank comments from my longtime friend Allan Lefcowitz, founding director of the Writer's Center in Bethesda. Poet, writer, editor, and now good friend Lane Jenning read all my materials and suggested changes in wording to strengthen the text. To both these men I am extremely grateful.

During the years spent on research for this book, I ignored the changes taking place in word-processing technology. As a result, my notes and drafts existed only as impossible-to-transfer files linked to two obsolete computers. Without my brother Scott Bartky's advice and long-distance guidance, the subsequent copying of altered files to neighbor Jane Dorfman's newer computer, then a second file conversion, and one final transfer onto a current machine, all my words would still be inside a PC-XT. For this essential help I am most grateful to them both.

Throughout this frustrating process of file conversion, I railed against so-called advances in personal-computer technology, which seemed to shift more of the manuscript-processing burden onto the author. But the ability to manipulate images on a computer screen—to remove the foxing from a late-eighteenth-century engraving, or insert professional lettering on line drawings, to cite but two examples—has restored my faith in the machine. Even so, it took hours of computer-software assistance from David H. Morse and uncounted days by my son David working on the files to lead me back to the fold.

I also wish to acknowledge the support and encouragement of my daughter Anne Bartky Kunzman and her husband, Douglas.

At Stanford University Press, I had the great good fortune to receive continuous guidance and support from director Norris Pope. My editor, Anna Eberhard Friedlander, deserves to be congratulated for nurturing my manuscript through the myriad steps needed for it to emerge as a published work. And Janet Wood's elegant text design enhanced its content.

As a child, and even as a young man, I would often sit puzzled after reading an acknowledgment in which some specialist author thanked his wife. No longer. Without Elizabeth Bartky's unswerving support, including working at my side in libraries and archives, this study of American timekeeping would not have been possible. I dedicate this book to her.

Abbreviations

AAAS	American Association for the Advancement of Science
ACSM	American Congress on Surveying and Mapping
AIS	Archives of Industrial Society, Hillman Library, University of Pittsburgh
AT&T	American Telephone and Telegraph Corporation
BAAS	British Association for the Advancement of Science
CAS	Chicago Astronomical Society
DOA	Dudley Observatory Archives
HCO Annals	*Annals of the Astronomical Observatory of Harvard College*
HCO-HUA	Harvard College Observatory Records
HUA	Harvard University Archives
HUCHSI	Harvard University Collection of Historical Scientific Instruments
IEEE	Institute of Electrical and Electronics Engineers
JC&HR	*Jewelers Circular (and Horological Review)*
JHU	Eisenhower Library, Johns Hopkins University
LC	Manuscripts Division, Library of Congress
LO-UC	Mary Lea Shane Archives of the Lick Observatory, University of California, Santa Cruz
LR	Letters received
LS	Letters sent
LS-M	Miscellaneous letters sent
LS-S	Letters sent to the secretary of the navy and to navy officers
NA	National Archives and Records Administration
NAS	National Academy of Sciences
NAWCC	National Association of Watch and Clock Collectors

NBS	National Bureau of Standards
NYPL	New York Public Library
PAMS	*Proceedings of the American Metrological Society*
P&W	Providence & Worcester Railroad
Proc.	*Proceedings*
RAS	Royal Astronomical Society
RG	Record Group
SNET	Southern New England Telephone Company
STC	Standard Time Company, New Haven, Connecticut
Travelers' Official Guide	
	Travelers' Official Guide of the Railway and Steam Navigation Lines of the United States and Canada
UAV	(Harvard) University Archives
U.K.	United Kingdom of Great Britain and Ireland
USCS-NA	U.S. Coast Survey Records, National Archives, RG 23, Microfilm M642
USNO	U.S. Naval Observatory
USNO-NA	U.S. Naval Observatory Records, National Archives, RG 78
WO-WIS	Washburn Observatory Records, University of Wisconsin-Madison Archives
YUA	Yale University Archives

SELLING THE TRUE TIME

It is not therefore a business venture with its attendant risks for an observatory to "sell time." It is rather the utilization of a product which will otherwise be unused.[1]

—Leonard Waldo, 1883

INTRODUCTION

At noon on Sunday, 18 November 1883, North American railroads created the modern era of public timekeeping. They discarded the forty-nine different times by which they had been running freight and passenger trains and replaced them with five new ones. Known collectively as Standard Railway Time, these operating times differed from each other by exact hours across the continent and were indexed to the Royal Observatory at Greenwich's meridian. They were the first elements in what would become the worldwide system of civil time zones that we now call Standard Time.

The process by which time awareness changed in the United States—from local, to regional, and then to national time—provides the background for this book.[2] Americans today take Standard Time for granted, and it is no exaggeration to say that it has become our culture's time—the common standard for business and government transactions around the world. But 116 years ago the need for a uniform system of timekeeping was a matter of debate among time-givers. Surprisingly few Americans voiced opposition after the change, yet where these protesters lived demonstrated a subtle link between civil timekeeping and the daily cycle of light and darkness that all of us live by. Even today, on the brink of the twenty-first century, the Greenwich-based system of civil timekeeping is still not universally observed.[3]

America's railroads adopted uniform time in 1883, not out of need but to forestall federal intervention. In 1882 Congress had passed legislation that promised to restrict the choice of railroad operating times. Another bill, which would have forced the inclusion of the city of Washington's time in all passenger-train schedules, had failed to win passage; but the bill was almost certain to be reintroduced in the next Congress. This pressure on legislators came largely from lobbying efforts by American astronomers. Some of these scientists saw national legislation as a route to uniform time for the entire country: having that kind of system would make it easier to conduct research in geophysics, such as studies of tornadoes, thunderstorms, earthquakes, terrestrial magnetism, and the like. Their ranks also included observatory directors who sold time signals to cities and railway companies, and who were then at the peak of their influence on public timekeeping.

The railroads, by adopting their own Standard Railway Time, virtually ended congressional interest in public timekeeping for the next three and a half decades. No national legislation emerged until 1918, when "an act to save daylight and to provide standard time for the United States" became law.

Astronomers first entered the realm of public timekeeping around mid-century. Events smoothed their entry. In 1853 two terrible train wrecks caused by errors in timekeeping alarmed the nation. Seeking to calm public fears, several railroad companies hired astronomers to supply them with accurate time. Using newly developed research equipment, astronomers could demonstrate unequivocally that observatory time was more precise than the time calculated by city jewelers and clockmakers, who had plied the trade for decades. But a close examination of the records reveals that public relations, not public safety, prompted most railroad companies to seek out astronomers as purveyors of time. Within the scientific community, too, there is evidence of mixed motives. Some astronomers saw time services as a way to gain funds for their cash-starved observatories; for others, the goal was direct personal gain. Yet all claimed a disinterested desire to advance pure science while providing the service.

Ironically, even though advanced technologies gave astronomers an edge in the 1850s, most observatory directors ignored later changes in timekeeping technologies once they had established their own enterprises. Believing in their own product, which was accurate time, these astronomers argued that accuracy was absolutely essential in all time-related activities. Yet, for many customers consistency in timekeeping—making sure that all clocks display the same time—was the important consideration. The commercial world responded to this need. As a result, by the late 1880s new timekeeping and time-distribution devices were replacing observatory time services. Frozen out of the commercial world of time services, American astronomers downgraded their interest in public timekeeping.

A few studies have chronicled timekeeping during the nineteenth century at a single institution, drawing largely on annual reports and the archival materials the directors left behind. But only by examining observatories as a group, within a context that includes both timekeeping technologies and users, can we assess the significance of time services in the development of American astronomy.

Reviewing observatory timekeeping as a whole reveals both the importance of the U.S. Coast Survey in establishing the country's first time services and the contributions made by several long-neglected inventors to America's world reputation as a scientific and commercial power. Examining the associated legislative record exposes the stages by which the U.S. Naval Observatory came to dominate public timekeeping, and illuminates the real achievements of those relatively few astronomers whose efforts—often through words more than deeds—helped make America the first country to adopt a system of uniform time spanning an entire continent.

A handful of key American astronomers participated in public time services during the nineteenth century. At the peak, which came in the 1880s, a dozen and a half institutions were distributing time. No previous study has described the beginnings of the era, detailed the astronomers' actual influence on civil timekeeping, or shown why the country's new technologies displaced observatory directors' dreams of expanded business opportunities.

Considering the rapid growth of the country's railroads and telegraphs in the years following the Civil War, one looks in vain for parallel growth in the number of private observatories selling time signals to those communications links. In fact, the astronomers' influence on public timekeeping was not based on their time services but rather on their publicity campaigns to promote greater time uniformity. Some of these proselytizers demonstrated great skill in lobbying for their cause; others revealed astonishing naiveté when confronted by those—including their own colleagues—opposed to their ideas. How American scientists, initially focused on establishing precise latitudes and longitudes and coordinating data collected from multiple observation points, managed to promote time uniformity and the nationwide adoption of a Greenwich-based time system, offers an object lesson in how science, government, and private interest can interact. It is a lesson that remains important even in our era, in which the task of establishing one's position anywhere on the earth's surface has become routine thanks to Global Positioning System satellites that broadcast the local time of the now-disestablished Royal Observatory at Greenwich.

Once the railroads inaugurated their system of operating times, many American cities and towns embraced it. Via ordinance and statute, a fundamentally different way of reckoning civil time spread rapidly throughout the country. Superficially, the transformation in civil timekeeping was simply a continuation of railroading's influence on the American public. Indeed, one informed observer even likened the changeover to "a noiseless revolution," for most Americans simply altered their clocks and went on with their lives.[4] But here and there a few people objected, and for more than thirty years they prevented the further spread of Greenwich-indexed timekeeping in this country. Their reasons for opposing what was destined to be the nation's civil time were virtually identical to the ones underlying the protests that now roll in two times a year, when Americans advance or retard their clocks in response to daylight-saving time.[5]

This country's near-complete acceptance of uniform time would not have occurred without the extraordinary efforts and remarkable good fortune of a small number of nineteenth-century individuals, sometimes vocally defending their public service, often pursuing their own personal agendas, and largely unknown or ignored by the general public. Uncovering the facts concerning opposition to, and support for, national time in nineteenth-century America sheds light on a momentous process in which government agencies, scientific institutions, and private businesses all played roles that would be impossible in America—or the world—today.

PART I.   EMPLOYING TIME (1801–1856)

1. TRUE TIME AND PLACE

A clock is a measuring device; we use it to quantify time. We want to know either duration—how long some process lasted—or the exact moment of an event—which we fix by taking the value displayed on the clock's dial and joining it to the day, month, and year.

However we use a clock, we must be certain that it keeps time well; we learn this by comparing our timekeeper against some standard. In the first decades of the nineteenth century, the choice of a standard was quite limited. Those Americans wealthy enough to own a well-made clock, usually one imported from England, and who did not reside in one of the country's cities had only the sun or the stars at their disposal.

But the verification task was relatively simple, and little or no understanding of astronomy was needed. Almanacs contained the necessary tables, and instructions for their use had been available ever since the introduction of pendulum-regulated clocks.[1] In 1683, for example, London clockmaker Thomas Tompion published "A Table of the Equation of Days, Shewing How much a good *Pendulum Watch* ought to be faster or slower than a true *Sun-Dial*, every Day in the Year." Along with his table, Tompion instructed the clock's owner to "set the *Watch* to so much faster or slower than the time by the Sun, according to the Table for the Day of the Month, when you set it; and if the *Watch* go true, the difference of it from the Sun any day afterward will be the same with the table."[2]

Some clockmakers had owners check their timepieces against the stars; a label printed in 1789 describes this process in detail. After directing that the clock be leveled properly and the back of its case screwed to the wall, the "Directions shewing how to set up and regulate an eight-day clock" continue:[3]

> To regulate the Clock if not at Time, being set to Time with a good Sun-Dial at Twelve O'Clock. Then observe any particular fixed Star you know, and mark down the precise Time to a second shown by the Clock, when the fixed Star vanishes behind the Door or Chimney, or to pass any hole, made for the purpose of seeing through. Then make this observation Ten Nights after, and mark the time on the Clock, as above, to a Second, when the Star passes the mark; then the Clock ought to be 39 minutes, 19 seconds behind the Star in passing the Mark; if not so much behind, the Pendulum must be screwed down a little [i.e., lengthened]; if more behind, it must be screwed up in proportion as you see the difference require.

This comparison between stellar (sidereal) time and the clock's indications over ten days was quite precise, and simple to perform—always assuming clear night skies.

For the owner of a clock living in one of America's growing cities, checking it was no problem at all. To obtain the correct time, he merely consulted a local clockmaker or jeweler. Some cities also had official timekeepers whose responsibility was to provide the exact time. One of them was David Rittenhouse, a giant of early American astronomy and instrument-making. In 1781 he erected a small observatory in Philadelphia, using public funds. Until his death in 1796, "the Rittenhouse observatory provided Philadelphia with its standard time."[4]

With all this advice and so many resources, a person owning a well-made clock could regulate it and then probably keep it in agreement with local time to within at least a minute or so a week. Indeed, indications are that few Americans needed time more precise than that—or even as precise. Not until the coming of railroads would people be concerned about more precise timekeeping.

Those who made and sold timepieces, however, those who constructed the almanac tables that clock owners used, and those who were the region's timegivers of necessity were interested in accurate timekeeping. As specialists, they employed specific terminology to designate the various times with which they dealt. Few other Americans—only surveyors, ocean navigators, and gentlemen amateurs in astronomy—were conversant with these nuances of meaning.

Basic to measuring time by the sun or stars is the concept of a *meridian*: an imaginary north-south line drawn across the sky from horizon to zenith to horizon. Observers made time determinations using special telescopes—either a transit instrument or a meridian circle—mounted so that they point

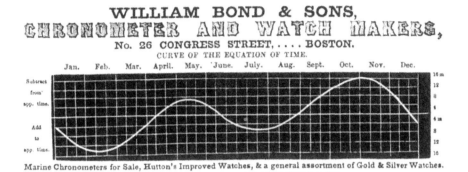

Fig. 1.1. Graphical depiction of the equation of time. From an advertisement in *The Boston Directory for the Year* 1851 (Boston: George Adams, 1851). Library of Congress collection.

only along the band of sky containing the meridian arc. When the sun (or a chosen star) crosses the meridian, the observer notes the time. If it differs from the value given in the table of transits, he corrects his timekeeper.

Time by the sun and time by the clock are different, however. Time based on the sun's passage, which can be measured by a sundial, is called *apparent solar time.* When the center of the sun crosses the meridian, that moment is called apparent noon. Apparent noon varies slightly from day to day when compared with a running clock: these differences occur because the earth spins on a tilted axis as it moves in its elliptical orbit around the sun.

Until late in the eighteenth century, a country's *civil time*—the time displayed on church and public clocks—was based on apparent time. But improvements in mechanical clocks, which tick off uniform intervals, eventually changed this way of reckoning civil time.[5] One key advance occurred in 1656, when Dutch mathematical physicist and horologist Christiaan Huygens incorporated a pendulum into a clock to control its going. Uniformity of timekeeping improved dramatically as a result—from an error of many minutes per day to one of only a few seconds.

To a good approximation, how fast or slow a pendulum clock runs depends only upon the pendulum's length. For a public timekeeper, the goal is to have its display agree more or less closely with the sun. Consequently, one strives to have the yearly sum of the daily differences between the actual sun's time of transit and noon as shown by that clock equal zero. Once such an adjustment has been made, the clock is said to be keeping *mean solar time,* or *mean time.*

The difference between apparent time and mean time is called the *equation of time.* The equation's daily values are the "sun fast-sun slow" entries in the astronomical tables placed in the public's almanacs. Plotting these values as a curve shows immediately that mean time and apparent time agree four times every year, and that the difference between the two can be as much as sixteen minutes.[6]

In the nineteenth century, specialists and others often used the term *true time* to denote the time displayed by a mean-time clock regulated to a specific meridian.[7] For example, in 1853 Harvard chemist Eben N. Horsford proposed a time ball by which ocean navigators would have "true Boston time." In 1859 a Cleveland jewelry and clockmaking firm wrote that "all the clocks, & rail Road time are dependent on us for the true time in this city." And an 1877 editorial in the *Times* (London) described the telegraphic distribution of "true Greenwich time."

Finally, there is *local time*. As used throughout this book, the term local time is the mean solar time at a particular place; it is synonymous with "true time." Like apparent time and mean time, local time varies with east-west position. Taking Washington as an example, any place due north or south of it (that is, on the same meridian) has the same local time. A clock displaying the local time of Baltimore—which lies slightly to the east of Washington—will always be later, by one minute and thirty-two seconds, than one set to Washington's local time. A clock set to Chicago time will always be forty-two minutes and twenty-eight seconds earlier than Washington time.

The local time difference between two places is a measure of the east-west distance between them, for a clock set to mean time mimics the earth's near-uniform rotation. Twenty-four hours on the clock is equivalent to one complete revolution (360 degrees) of the earth, so that four minutes in time is equivalent to one degree of rotation, or one degree of longitude.[8]

As a precision measuring instrument a clock can be used to answer two related questions: "Where is the place I want to reach?" and "Where am I now?" Knowing one's location on the surface of the earth is essentially the same as knowing the latitude and longitude coordinates of that place: degrees north or south of the equator, degrees east or west of some agreed-upon meridian. By the mid-1700s, finding one's latitude anywhere was a relatively simple matter. But determining the longitude was not.

Positions on land were becoming easier and easier to learn, however, for explorers and surveyors had long been engaged in a worldwide activity that was leading to the preparation of ever more accurate maps. Whenever the longitude of a selected place was wanted, these specialists set up a pendulum timekeeper and carefully adjusted it to local time.

After the clock was adjusted, the senior observer watched the moons of Jupiter through a high-power telescope. He called out the exact instant that one of these satellites was hidden by the planet, or emerged from behind it, and his assistant noted the time. Afterward, one of them would look up the

Fig. 1.2. Portable observatory with long-case astronomical clock used in determining longitudes by Jovian satellite and lunar observations. (The engraving has been enhanced by computer.) From Wales and Bayly, *Original Astronomical Observations,* plate II. U.S. Naval Observatory Library.

time of the same event in a (predictive) table prepared by an observatory in Europe and then calculate the difference between the printed value and the observed one, as recorded by the assistant. This difference was the site's longitude—usually reported in hours, minutes, and seconds from the observatory's meridian. (For example, the longitude difference between Washington and Chicago is given as 42^m28^s.)

Longitude at sea could not be determined in this way, however. A pendulum clock keeps time badly on a rolling and pitching ship, and the continual heaving makes it impossible to keep a powerful telescope trained on a small target like Jupiter, much less one of its moons. Of course coastal chart-making was advancing, but finding one's longitude while under sail remained problematic. Moreover, the estimated positions of islands and reefs on sailing charts were often in error by hundreds of sea-miles.

In addition, any timepiece drifts away from its initially set value, and how much it changes—the clock's "error"—has to be factored into all calculations. On land, checking a clock against solar or stellar transits for several days gives the amount of its drift, and thus the timekeeper's daily change: its "rate," a value given in seconds/day, gaining (or losing). In between these check periods, the timekeeper is assumed to be operating stably (uniformly), and its error is determined by projecting this rate. A clock's rate cannot be determined while at sea.

An Englishman, John Harrison, won lasting fame for developing an ocean-going clock that would operate properly, always mirroring the earth's rotation. Harrison's timekeeper looked much like an overgrown watch. During the 1760s, this device—now labeled "H.4" and arguably the most famous time-piece in the world—underwent sea trials, voyages between England and the Caribbean lasting more than six weeks. For these trials it was set to the local time of Portsmouth and its rate carefully measured. The second trial showed that, after using the rate to project the timekeeper's accumulated error, H.4 replicated the astronomically-determined longitude difference between Barbados and Portsmouth within 39.1 seconds (time)—corresponding to a position error of less than ten nautical miles. A reliable method for finding the longitude at sea had at last been demonstrated.[9]

Ironically, Harrison's development came at a time when astronomers had collected enough data on star positions and the complex motion of the moon around the earth to predict the moon's position against the stellar background—a celestial clock, as it were—thus making "longitude by lunars" possible.[10] In the later years of the eighteenth century and beyond, both seagoing clocks, soon called "marine chronometers," and lunars were used to determine longitudes at sea. But the observations and the complexities of data reduction for the latter method required observing and mathematical skills of a high order. Using a chronometer was far easier, and so as these portable precision clocks became less expensive and their reliability improved, ocean navigators came to use them almost exclusively.

By the first quarter of the nineteenth century nautical chandlers were adding these timekeepers to their inventories of charts and instruments. Chronometer firms also sprang up, offering repair and rating services, and, soon after, rentals. In the United States, such firms could be found in every major port and many minor ones.[11]

Nautical chandlers Benjamin and Samuel Demilt erected an observatory in New York City for the rating of marine chronometers. Their services began in the mid-1790s and continued throughout the brothers' forty-year career. Another early chronometer firm was Wm. Bond & Son of Boston. Philadelphians William H. R. Riggs and Isaiah Lukens also rated chronometers, and John Bliss, later an important New York chronometer manufacturer, sold and maintained chronometers in New Orleans. Another small rating observatory, famous in the history of astronomy, was located on Nantucket Island. There Maria Mitchell, the country's first woman astronomer, assisted her father in rating the chronometers taken on multiyear voyages by Nantucket's whaling ships.[12]

Marine chronometer reliability improved so much that European astronomers began sending groups of them between their respective observatories, using them to determine observatory longitudes. Among those who took part in these expeditions was English chronometer-maker Edward J. Dent, who in 1839 determined the longitude between the Royal Observatory at Greenwich and the port of New York—a widely reported technical feat.

The great barrier to using chronometers to measure transatlantic longitudes was the duration of the ocean voyage. Leaving New York and sailing eastward, scheduled packet ships took three weeks to reach Liverpool; sailing westward against the prevailing winds, the trip from Liverpool to New York could take five to eight weeks. A tiny misjudgment of a chronometer's rate at the outset of the western leg would be so magnified by the time the ship arrived in New York, far more than the same journey eastward, that a longitude calculated by averaging the two results was suspect. It was of scant value to chart-makers, or even to dealers wishing to show off the quality of their marine timekeepers.

This impediment vanished in 1838 with the inauguration of steamship service across the Atlantic. Now a passage west or east could be reliably expected to last no more than two weeks.[13] In July of 1839 Dent placed four of his firm's chronometers on board the *British Queen*, bound on its maiden voyage to New York. That fall he reported the results of two round trips, and the following spring his New York agent, George W. Blunt (whose firm was this country's most important supplier of charts and navigation instruments), added results from the subsequent voyages. The ensemble of chronometer-based longitude values between London and New York agreed with astronomical determinations to within a second or two.[14]

Dent and Blunt were elated, the former writing that "all objections founded on the idea that the motion of a steam-vessel would affect injuriously the more delicate movement of the chronometer, and taint the results, must now fall to the ground." Further, beamed the chronometer-maker, "the instruments were sent out unattended by any savant, and brought home their own report."[15]

When the newly established Cunard Line began transatlantic steamer service, the opportunity to accumulate results (thereby increasing one's confidence in the resulting longitude value) shifted from New York to Boston. In 1843 the U.S. Coast Survey began what became a twelve-year effort to compare longitudes by chronometers with values from astronomy—a scientific endeavor subordinate to its primary mission of charting America's coasts and adjacent waters.

Many decades earlier President Thomas Jefferson proposed a coastal survey, and in 1807 Congress approved his request. Ferdinand Rudolph Hassler, a Swiss-born immigrant, was chosen to undertake the effort and was given approval to execute it within the framework of a trigonometric survey. It was a lengthy process requiring years to complete, but it was the most accurate way to proceed.[16,17]

Many countries (among them Denmark, France, Great Britain, and various German states) were already conducting trigonometric surveys, their field efforts supported by many of the world's greatest astronomers and mathematicians. In contrast, the United States lacked precision surveying instruments, and only a tiny handful of individuals had the knowledge of surveying, astronomy, and mathematics needed to conduct such a program. Fortunately, Hassler was one who did. Looking back, one is awed by Thomas Jefferson's prescience. His bold decision to approve a trigonometric survey of the U.S. coastline did much to advance the level of American science.

The first decades of the enterprise were rocky. Hassler was no politician, and he had few skills in dealing with people. His unswerving focus on his task, without interference or compromise, angered many. Battles erupted in Congress over costs, slow progress, and Hassler's intransigence. These led ultimately to his resignation, to years of delay in properly charting the coast, and to bureaucratic maneuvering that included transferring the Coast Survey from the Treasury Department to the Navy Department and back again.

Hassler eventually returned as superintendent, and after the 1830s some progress on the coastal charts was made. But many still considered the pace too slow, and maneuvering for control and direction of the agency continued. Finally in 1843 a frustrated Congress directed that a plan be developed for "reorganizing the mode of execution of the survey."

In response to this legislation, a board of experts was established. Its members included several of Hassler's assistants, among them Edmund Blunt of the New York firm of chart suppliers. The board developed a plan, recommending that the Coast Survey continue its work on "the scientific methods proposed by F. R. Hassler." President John Tyler approved the plan, and the charting of America's coastline and its waters continued at the highest technical levels.[18]

Included in the plan was a provision asserting that it was now necessary "that the difference of longitude between some main points of the survey and the meridians of any or all of the European observatories, be ascertained." The board of experts considered this matter sufficiently important to recommend that it be started immediately.[19]

Enlarging the agency's legislatively mandated program was a portent of the future. The meridian of New York was the Coast Survey's first "zero" of longitude. However, the intent was to shift to the meridian of Washington as trigonometric work advanced down the Atlantic coast. The relation between

Fig. 1.3. Marine chronometer typical of the mid-nineteenth century and after. This particular timekeeper, purchased by the Admiralty in 1841 and assigned to many hydrographic surveying expeditions, remained in active government service until 1922. Courtesy of Charles Frodsham & Co. Ltd., London.

the Coast Survey's base meridian and those of other countries now had to be determined—and with a high degree of certainty.

Those who inserted the longitude-difference requirement into the plan did so for several reasons. American astronomers—a modest but growing group—had produced many values of longitude differences, all of them based on astronomical observations. It was time to analyze their results within a single framework.[20] Dent's use of the *British Queen's* officers to transport chronometers between England and America showed that yet another technology for acquiring quality values had been demonstrated.

Moreover, the "Grand Chronometrical Expedition" was getting underway. This was a major Russian technical effort led by Friedrich G. W. Struve, one of the world's leading astronomers and geodesists. Using multiple transports of several dozen chronometers, Struve was determining a precise value for the longitude between the royal observatories at Altona (near Hamburg) and Pulkova (outside St. Petersburg). The U.S. Coast Survey's experts may have felt that America could—and should—do the same.[21]

Late in 1843 Superintendent Hassler died. His legacy was an organization steeped in science, its staff committed to realizing results at the highest attainable level of quality.[22] Immediately after his death, his successor, Alexander Dallas Bache, expanded the Coast Survey's transatlantic longitude program, which involved both astronomy and chronometer transports. Extremely well connected socially and politically, Bache did not alter the agency in any basic way. But his skills in dealing with his superiors at Treasury and those in Congress who controlled the agency's appropriations were awesome. Coast Survey funds increased enormously, its programs expanded, and progress accelerated.[23]

Bache contracted the observational effort to astronomers in Cambridge, Massachusetts, and Philadelphia. Observations already made by astronomer Elias Loomis at Hudson Observatory in Ohio, ones from the Blunts' observatory in Brooklyn, and values from sites in the city of Washington, were added to the growing body of data. A little later, Bache included current observations from William and Maria Mitchell on Nantucket Island and from Lewis Gibbes in Charleston, South Carolina.

For the chronometer transports, Bache hired William C. Bond, director of Harvard College Observatory (often referred to as the "Cambridge Observatory" in this era). However, Wm. Bond & Son, the family clockmaking and chronometer firm in Boston, had responsibility for acquiring the necessary data. The firm started with values generated from the rating of chronometers carried on the Cunard steamships arriving at the port. It soon became clear that acquiring meaningful longitude differences required special efforts. Dent's earlier assertion regarding the simplicity of the process was not borne out by the evidence, especially in these years when chronometer technology was being pushed to its limit. Accordingly, and under Coast Survey sponsorship, Bond organized an effort along the lines of Struve's 1843 expedition, with two Boston-Liverpool expeditions taking place during 1849 and 1851.

Large numbers of chronometers were carefully rated in Boston, placed on board Cunard steamships, monitored en route by one of the firm's employees, and, on arrival, taken to the Liverpool Observatory.[24] After being rated there, the chronometers were transported by rail to Greenwich, rated again at the Royal Observatory, and returned to Liverpool for the voyage back to Boston.

Despite careful attention to detail and the use of a very large number of

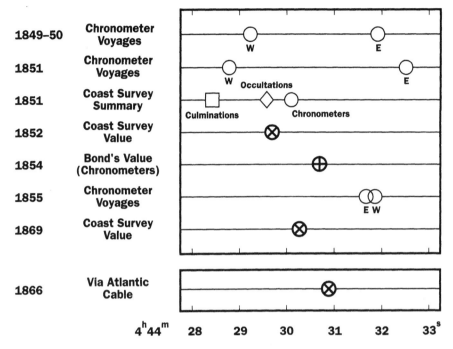

Fig. 1.4. U.S. Coast Survey–sponsored determinations of the longitude difference between Harvard College Observatory in Cambridge, Massachusetts, and the Royal Observatory at Greenwich, England. The values are in hours, minutes, and seconds, and reflect corrections by Coast Survey assistant C. A. Schott.

chronometers to achieve high confidence in the resulting value, the two expeditions yielded disappointing results. The longitude calculated from the eastward transports differed from the average obtained from the westward ones by three seconds—well outside their individual, estimated precisions. Moreover, the longitude arrived at by combining these two averages differed from two astronomically derived values, both of which were also considered quite precise but which differed markedly from each other. Thus, no choice for the "correct" longitude was possible, and averaging the three discordant values offered little improvement over what was already known.[25]

In 1855 another chronometer expedition was mounted in an attempt to resolve the east-west discrepancy. This time the results for east and west voyages agreed, but the resulting longitude by chronometer still disagreed with the astronomical determinations.

A dozen years had now been spent on a task ancillary to the Coast Survey's primary purpose, and a new, proven technology for determining longitude differences was waiting in the wings. Bache terminated Bond's efforts, softening the blow with the remark, "If there were not so fair a prospect of a telegraphic connection with Europe, it might be desirable to continue these researches." Bache discarded the results derived from the first two expeditions, terming them "preparatory." He also stated that the longitude difference between Greenwich and Cambridge that the Coast Survey was using (temporarily) for its program was probably in error by as much as two seconds—which corresponded to a positional error of about a half a mile.[26]

Only the failure of the first transatlantic cable gave any long-term utility to the chronometer-transport results.[27] Although a ship's position at sea could be determined only to within ten nautical miles, the chronometer remained the key navigational aid for the world's shipping for the next hundred years.[28] But in the United States, by midcentury the era of chronometers for determining precision longitudes on land had ended—thanks to the telegraph.

2. RUNNING ON TIME

By the late 1830s, Europeans and Americans could already feel the enormous social changes wrought by the railroads. Some even spoke of rail travel as the "annihilation of space and time." From Paris the exiled poet and writer Heinrich Heine sighed, "I can smell the German linden trees; the North Sea's breakers are rolling against my door. Space is killed by the railways, and we are left with time alone."[1]

Less obvious was the fact that railroad time was entirely different from the public's time. Before the dawn of railroading, east-west position and time were tightly linked. At the end of any journey by horse or carriage a traveler simply set his timepiece ahead or behind to conform with local time. But railroad companies maintained a single time all along their route, no matter how far west or east their rails ran. Thus railroad travelers throughout their journey had to consider city times and railroad times simultaneously. Space had indeed been bested, but multiple times had sprung up in its stead.

This impact of railroad timekeeping on the general public underlies our story. Eventually, of course, city time and railroad time became one. But how railroads set their schedules, and how astronomers and others responded to their special needs, shaped most developments in American timekeeping for the next sixty years.

Railroad timekeeping was simple in concept: use one time everywhere. To schedule traffic, a superintendent first took the time of a city or town along the rail route as his company's operating time. Then he drew up a timetable for the entire system by adding each train's running time to its departure time from a given station. No local time adjustments were required.

These pioneers, however, went even further. Each defined his road's operating time in terms of a particular clock and established rules governing the timekeeper and its use. For example, an 1835 rule of the Boston & Providence Railroad instructed train conductors to "daily compare their time with the time at the Depots." A Baltimore & Susquehanna rule directed all conductors and engine men to get their "time from this [Baltimore] office." And an 1843 rule issued by the Baltimore & Ohio stated that "the clock at the Pratt street depot in Baltimore shall be taken to be the standard time, and all conductors of passenger trains before departing from Baltimore with their trains, are required to have their watches regulated and compared with that clock." And not only were trainmen's pocket watches compared with the designated clock; conductors also were instructed "to see that all other stations which they pass conform to the standard time," and report any differences to headquarters.

Rail operations require a consistent system of time, not necessarily an accurate one. So although a company's master clock gradually drifted from its initial time—as all clocks do—every timepiece throughout the system was adjusted daily to remain in synchrony with the time displayed on the company standard.[2]

In this regard railroad timekeeping differs from timekeeping on board a ship. When an ocean navigator calculates his vessel's longitude, he has to project the chronometer's rate over days and weeks, sometimes even months. Consequently, this timekeeper is never reset, for doing so would cause it to drift irregularly and invalidate his projection of its time. Nevertheless, timekeeping on ship and train are alike in one respect: both are fundamental to safe operations.

In the United States, running on time was a matter of supreme importance, for most of the country's rail links were single-track lines on which trains ran in both directions. In addition to the rules for acquiring and keeping correct time—that is, time in conformity with the company's standard—railroads issued rules detailing the precise steps to be taken by conductors, engine drivers, and other employees once they discovered that a train was late—in railroad parlance, when it was "out of time" or "running wild."[3]

Given the excellent quality of pendulum clocks in this early era, rules mandating the use of one operating time throughout a rail system made time at the stations and depots consistent to within a minute or so. A road's time was probably consistent to within two or three minutes everywhere, including the time carried on trainmen's watches as they ran the cars between stations. Moreover, daily time checks enabled company officials to spot poorly operating timekeepers and have them adjusted or repaired.

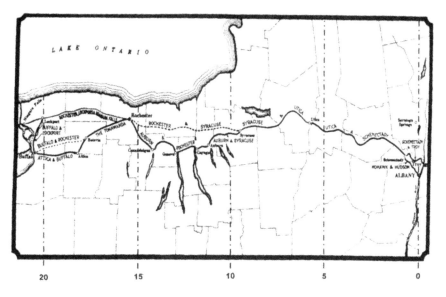

Fig. 2.1. Map showing local-time differences in terms of Albany time. Base map is from Hungerford, *Men and Iron*, facing page 88.

Obviously, any railway company can resolve its timekeeping within its own boundaries. But what happens when two or more roads in the same operating region run on different times?[4] This issue was probably addressed in January 1843, when the eight companies forming the 326-mile route from the Hudson River to Buffalo, New York, met in convention. Traffic on this east-west migration route, one of the nation's earliest rail links, had grown enormously and was projected to increase even more. However, services for through-passengers were woefully inadequate.[5]

When each railroad opened, its operators selected a convenient city time— Albany, Schenectady, Utica, (probably) Syracuse, or Rochester—resulting in a time difference between the Albany and Buffalo terminals of twenty minutes and thirty-two seconds. With so many operating times and the fact that each line scheduled its trains independently, through-passengers could easily miss connections and be stranded. At this 1843 railroad convention, possibly the first ever held, company executives referred the matter to their operating officials, directing them "to arrange the arrivals and departures on the intermediate points on the line and other matters in connexion therewith."[6]

How the roads' superintendents responded is a matter of conjecture. With rather modest train speeds and a gap of four city blocks between rail terminals in Rochester (necessitating a hackney coach ride for connecting passengers), they may simply have altered the schedules so that connecting trains did not leave until all through-passengers were on board. They may not have even considered reducing the number of operating times, for the time differences at adjacent rail terminals were small—ranging from thirty-two seconds to slightly over six minutes.

Nonetheless, the pressure to adopt fewer operating times built rapidly as train speeds increased and traffic volume grew. The gap between the Rochester stations was bridged by rails, and through-passenger cars became the norm. By the early 1850s only Schenectady and Rochester times were in use, and with the 1853 consolidation of the rail lines as the New York Central Railroad, officials adopted a single operating time: the "Standard of Time" given by the "clock in [the] Depot at Albany, which is 21 minutes faster than Buffalo time."[7] Time uniformity had arrived—at least on the "water-level route."

While railroad men in central New York were consolidating operations, those in New England also tried to rid their region of multiple operating times and varying practices. To do so, they invoked the same mechanism: an association of engineering experts. But their "one-time-for-all" solution was fatally flawed.

In New England, scores of railway companies provided passenger services; many were "short lines" linking a few small towns with some larger road. The region's twenty-four-minute breadth approximated the Albany-Buffalo time difference, while its area approached that of England. Boston, two hundred railroad miles and ten travel hours east of Albany, had seven terminals, with ten separate railroads running from them. Worcester, forty-five miles west of "the Hub" and with a local time nearly three minutes earlier, was also a major rail center, served by six rail companies.

In the spring of 1848 nineteen railroad pioneers founded the New England Association of Railroad Superintendents. A powerful organization whose members included the operating officials of the region's major railways, its objects were to increase and diffuse knowledge "upon scientific and practical subjects connected with rail roads" and to promote "harmony among the rail road companies of New England." Gaining new members via nomination and election, by October of the following year its ranks included representatives from thirty of the region's seventy railroads.[8]

At the association's February 1849 meeting, newly elected member I. W. Stowell, superintendent of the Worcester & Nashua Railroad, proposed that the "subject of a standard time for rail roads" be considered. A committee of three was appointed, and in April of that year its report was read. Members present voted that it "be recommitted to see what arrangements can be made for a standard time keeper."

Time passed. In October the committee recommended "the time of a meridian, 2 minutes later than the meridian of Boston," also described as "the time of a meridian about 30 miles west of Boston," as the roads' operating time standard. After discussion, the thirteen members attending the meeting voted "that this association recommend to all the rail road companies in New England the adoption (for this standard) of a time two minutes after the true time at Boston as given by Wm. Bond & Sons [*sic*], 26 Congress Street, and that on and after the 5th of November, all station clocks, conductors' watches and all time tables and trains should be regulated accordingly."

Since membership was voluntary, the association's officers could do no more than announce the standard in its official guide book, recommend it, and list those railroads that agreed to run their trains by it. As of December 1849, only eleven roads, a minority of members, had done so. Clearly, the uniform time standard was being resisted.[9]

The *Records* of the New England Association are silent as to why the superintendent of the Worcester & Nashua even introduced the subject of uniform time. The time difference between his own road's terminals was only one minute and twenty-four seconds, an amount that should have been of no consequence for operations. Although the meridian adopted by the association lies about halfway between Worcester and Nashua, this does not explain why a majority of those present at the October meeting should have selected an arbitrary time standard different from that of Boston, the region's commercial center. Perhaps the meridian was a compromise, chosen after proponents argued that local passengers in other parts of New England would not accept Boston time.

The meridian two minutes west of Boston proved a poor choice. Before two decades had passed, the city's own meridian defined the operating time for most of New England's transportation companies.[10] But with conservatism a strong force in the industry, some of those roads that had chosen the "wrong" time in 1849 stuck by it. As a result, uniform railroad time in New England was delayed for thirty-four years.[11]

As puzzling as the association's choice of a standard time was their selection of Wm. Bond & Son's master clock as the embodiment of their time standard. No evidence has surfaced to suggest that William C. Bond, son of the firm's founder, was involved with any railroad prior to 1849.[12] That market was already being served by Simon Willard, Jr., son of one of Boston's most famous and respected clockmakers. His firm's astronomical regulator, built around 1832 and "tested by daily transit observations," was the source of "standard time for all the railroads in New England," if we are to believe his grandson. Moreover, this particular regulator was described as having been "for forty years the standard of time for all New England."[13]

Wm. Bond & Son was one of forty or so watch- and clockmakers and dealers in Boston.[14] What little Bond wrote about his dealings with the association relate to enhancing public awareness of Harvard College's observatory, a

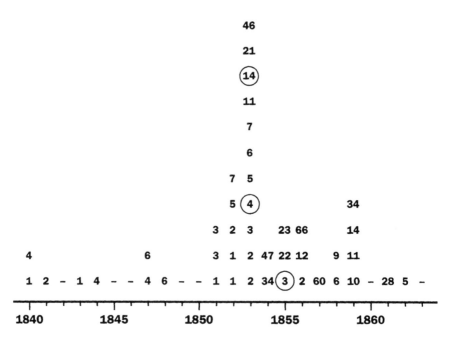

Fig. 2.2. Histogram showing major North American railroad accidents, 1840–63. The values give the number of fatalities, with the circled ones indicating those accidents involving faulty timekeeping. With the exception of the circled 1855 value, all data are from Robert B. Shaw's appendix summarizing railroad accidents, 1831–1977, in his *History of Railroad Accidents.*

subject of scant interest to the railway managers. They had merely fixed the source of their railway time by a business arrangement with a private firm, much as they hired a publisher to print their monthly guides.[15] On the other hand, even though Bond was just one among many able to determine the city's true time, his contracts with the government surveys certainly set him apart from other Boston clock- and watchmakers. And his years of careful astronomical observations at various sites were known to many of Boston's leading citizens. While astronomical precision is of little importance to the actual needs of railroad timekeeping, Bond's reputation may have influenced association members as they made their choice among the city's clockmakers.

Most likely, Bond's involvement with the New England Association was strictly business-related. The new service gave his firm, apparently for the first time, access to the railroad clock and watch market. Inaugurating a railroad time service in late 1849 proved to be a very astute move. Four years later his firm's clock and watch sales soared when proper timekeeping became critical for New England railroads.

The year 1853 is often cited as the point in time at which the excellent safety record of America's railroads vanished. Before then, even the worst accident

Employing Time

had never killed more than a half-dozen people. But starting in January 1853, newspapers began to report a seemingly endless string of terrible operating failures, several of which killed a dozen or more passengers. The public reacted first with horror and then with fear. To them, it appeared that America's rail system was fatally flawed, and railroad owners didn't seem to care.

The role of timekeeping in railroad safety was dramatically demonstrated on 9 August 1853, when two Camden & Amboy trains struck head on, killing four people. Poor timekeeping was implicated, for one conductor's company-supplied watch was two and a half minutes slow. And although excessive speed certainly contributed to the severity of the collision, the underlying cause of the accident was the failure of the train's conductor and engine driver to compare their watches with the station clock and with each other at departure. Both requirements were embodied in the Camden & Amboy's operating rules, thus both men violated these rules. Ignoring facts, the press excoriated company management officials.[16]

On 12 August a terrible collision on the Providence & Worcester brought railroad timekeeping once again under the public's scrutiny. At 7:20 A.M., a P&W train left Providence, Rhode Island, bound for Worcester, Massachusetts. Meanwhile, the P&W's crowded excursion train, running south toward Providence, arrived late at Valley Falls, a town just north of the city. Checking his time, this train's neophyte conductor concluded that he still had four minutes in which to reach the double-track section of rail, three-quarters of a mile away. He signaled his train to move on down a section of curving track.

Out of sight around the bend the up-train waited while its engineer observed the five minutes allotted for the excursion train to cross the switch onto the double track. Five minutes passed. A guard raised the signal ball, giving the up-train the right to move through the switch. The conductor called, "All aboard!" The engine driver waited a minute or so and then started the train forward, passing through the switch to the single track.

The two engines struck less than a minute later. The impact of the collision was so great that some of the excursion train's wooden passenger cars telescoped, ramming and tearing through adjacent ones. In all, fourteen passengers were killed and twenty-three were injured.[17]

The inquest began the following day. Testimony brought out the fact that the watch of the excursion train's conductor was over a minute and a half slow compared with the company's operating time. Further, the conductor had counted on making up lost time in his rush to the double track below the switch. In their report, the coroner's jury placed the blame for the tragedy squarely on him. He was arrested and charged with manslaughter.

Jurors faulted the railroad company as well. In their view, officials had placed an inexperienced person in charge of a train and had not provided him

Fig. 2.3. Collision, 12 August 1853, near Valley Falls, Rhode Island, on the single track of the Providence & Worcester R.R.; fourteen people were killed. The daguerreotype is the first one taken of a railroad disaster. Courtesy of George Eastman House, Rochester, New York.

with an accurate watch. They heaped additional blame on the conductor and engineer of the waiting train for not having started up at "the very second the signal was given." Had they done so, jurors argued, the train "might have reached . . . a point on the road where it would have been in sight from Valley Falls, and this great calamity probably would have been prevented."[18]

In fact, the remedial action that these citizens were urging—always start up at once—would have increased the frequency of collisions on the Providence & Worcester. Apparently the jury simply wanted its report to cover every party associated with the accident. Such behavior demonstrates why an inquiry by nonexperts prone to hasty judgments can be only a first step in the process of determining cause and culpability.[19]

The P&W's directors met while the inquest was underway. They voted to dismiss both the conductor and the engineer of the excursion train, as well as their master of transportation, and to double-track this stretch of the route. In addition, they placed a signalman at the Valley Falls side of the curve.[20]

Many Providence citizens considered these actions insufficient and complained that the company was not moving with dispatch. Feelings ran high;

local newspapers announced a public meeting of concerned citizens, with the mayor attending.[21] Seeking to moderate what must have been a lynch-mob atmosphere, the Railroad Commissioners for the State of Rhode Island announced that a special meeting would be held in Providence. Their purpose was to investigate the causes of the accident, a responsibility that at that time exceeded the agency's authority. Nevertheless, on 25 August commission members began examining company employees and others under oath.[22]

The investigation lasted three days. The commissioners found that company officials had never even checked to ensure that trainmen were using accurate watches, much less providing watches for them. They also learned that the road planned to continue running trains on the old schedule, with no changes contemplated, during the weeks it would take to lay double tracks.

Although these findings exposed several troubling issues, the most compelling one was in the company's timetable. The Railroad Commissioners found that at some stations, a train arrived one minute before the opposing one. Such close scheduling meant that an engineer had only a very short while to move his train to a siding. Further, this finding implied that a variation of only one minute in the time of a conductor's watch would lead to a collision. Although many P&W trainmen compensated for the inherent variability of their watches by waiting a few extra minutes at particular "choke points" before proceeding on, the practice was not universal and, of course, was not one of the company's operating rules.

When the hearings concluded, the Railroad Commissioners adjourned to prepare a report for the General Assembly that was read to that legislative body in mid-September. Local newspapers described this report as "strong in its condemnation of the management of the Providence and Worcester Railroad."[23] Responding to it, the Rhode Island legislators directed the preparation of, and then immediately passed, an act that drastically increased the liability of common carriers in cases where any death resulted from their carelessness or negligence.[24]

Although financial penalties for negligence were increased, and rightly so, surely the critical event was the Railroad Commissioners' unauthorized investigation. A fundamental, unexpected flaw in the country's rail safety system had now been uncovered and highlighted. Almost at once, American railroads began to modify and extend their operating rules. Some focused on time-keeper variations. For example, early in 1854 the New York & New Haven Railroad announced that now it required "a further allowance of three minutes . . . [to be] made by trains from the east, for any possible variation of watches, an allowance on which trains from the west have no right to encroach." In Louisville, Kentucky, a local railway directed that when any train scheduled to meet another at a station arrived to find the opposing train late, it must "wait until five minutes after the time fixed for leaving the station" before proceeding on, and stay five minutes behind its schedule until the other train was met. The company remarked that "the five minutes in this case are intended to guard against a possible difference of watches."[25]

This relaxation of an underlying requirement for highly precise, perfectly maintained timepieces was an enormous step forward in railroad safety. Not the least of its advantages was cheapness, for it cost almost nothing to alter passenger and freight train schedules by modest amounts. Many American railroads adopted this strategy of compensating for any errors in their employees' watches. Even today allowing for possible errors in measuring devices is a sophisticated consideration in systems design.

A second, albeit more expensive, approach for improving safety on single-track lines was also widely instituted: a requirement that "each Conductor must possess a good, accurate, and reliable Time Keeper." This requirement is seen repeatedly in rule books issued after the Valley Falls collision. Presumably companies had some means of ensuring compliance; otherwise, adding the rule would not have improved safety at all. Unfortunately, historical records on this point are sparse. Of the hundreds of railroads operating in the 1850s and after, only one company's records exist in detail sufficient to judge the consequences of such a rule: the Boston & Providence Railroad.

Pioneer railroader W. Raymond Lee had been the road's superintendent ever since it opened for traffic in 1835. The railroad was well run; only one accident early on marred its safety record. Over the years expanding business caused the Boston & Providence to lay a section of double track into the heart of Providence, a section also shared by Providence & Worcester trains. Indeed, it was on the P&W's single track leading to this section that the disaster at Valley Falls occurred.

Superintendent Lee was certainly aware of the inquest, for one of his conductors testified there. He may even have attended the Railroad Commissioners' investigation himself. In any event, just three working days after the conclusion of these hearings, he issued a company circular: "Standard Time."

Lee specified that his railroad's operating time was based on Wm. Bond & Son's standard timekeeper in Boston, and he described how time from that clock was transferred to the road's central and station clocks. Lee also spelled out the responsibilities of train conductors with regard to acquiring the correct time.

The superintendent then detailed an inspection process covering the watches being used by the road's employees. First, all conductors had to submit their timepieces to Wm. Bond & Son. If found satisfactory after examination, the firm would issue a written "certificate of reliability, which will be handed to the Superintendent." Moreover, the firm would continue its

Boston and Providence Railroad.

STANDARD TIME.

1. STANDARD TIME is two minutes later than BOND & SONS' clock, No. 17 Congress street, Boston.
2. The inside clocks, Boston and Providence stations, will be regulated by Standard Time.
3. The Ticket Clerk, Boston station, and the Ticket Clerk, Providence station, are charged with the duty of regulating Station Time. The former will daily compare it with Standard Time, and the latter will daily compare it with Conductor's time; and the agreement of any two Conductors upon a variation in Station Time shall justify him in changing it.
4. Conductors will compare their watches with Standard time in the following order.

> MONDAY,..........Conductor of Steamboat Train.
> TUESDAY,..........   "   Accom'n Train No. 1
> WEDNESDAY,......   "   "   2
> THURSDAY,........   "   "   3
> FRIDAY,............   "   Dedham Train 1
> SATURDAY,........   "   "   2

5. All Conductors of Passenger and Freight trains will compare their time with Station time, Boston and Providence, every day, and report any variations to Superintendent of Transportation.
6. A record will be made by the Ticket Clerk, or in his absence, by the Baggage Master, of the comparisons required by Art. 5, to which they will certify by their signature or initials.
7. Conductors will submit their watches to Bond & Sons, 17 Congress street, Boston, for examination, and procure from them a certificate of reliability, which will be handed to the Superintendent.
8. Conductors will report to Messrs Bond any irregularity in the movements of their watches, and they will clean, repair and regulate them, at the expense of the Corporation, furnishing Conductors with reliable watches in the interim.

> W. RAYMOND LEE, SUP'T.

BOSTON, AUGUST 31ST, 1853.

See vote of the Directors of all the R R Companys which leave Boston = Nov² 5/49

Fig. 2.4. Superintendent W. Raymond Lee's 31 August 1853 rules associated with the Boston & Providence R.R. operating time. Courtesy of the Collection of Historical Scientific Instruments, Harvard University.

inspection and maintenance service by cleaning, repairing, and regulating watches used on the Boston & Providence, all service "at the expense of the Corporation." By focusing on the continuous reliability of the watches carried by his trainmen, Lee inaugurated a new era in railway safety.[26]

These complementary requirements—one compensating for less-than-perfect timekeepers, the other keeping a continuous eye on railroad time-keeping's weakest link—were significant additions to the companies' operating rules. The first became the now-famous "five minutes [allowance] for watches," which in 1887 was included in the first industrywide standard code of operating rules. The second paved the way for the industry's adoption, also in 1887, of watch inspection programs.

Adding these timekeeping requirements to its rules gave every railroad a safe operating system based on time. Despite the enormous growth of rail transportation in the United States and vastly higher speeds in the ensuing decades, this revised system of timekeeping was quite effective. Between late 1853 and 1900, no more than two timekeeping blunders resulted in passenger fatalities.[27]

Less important for actual rail safety, but all-important to reassuring a frightened public, was the accuracy of the railroads' time. In the aftermath of two highly publicized time-related collisions, New York–centered railroad companies hired a university professor. One such company was the New York & Erie (and subsidiaries), whose passenger trains had been involved in five mishaps over the past seventeen months, resulting in the deaths of fourteen passengers. In September of 1853 Erie officials reassured patrons that faulty timekeeping was not possible on their line, for "Mr. [Richard H.] Bull, the astronomer," had been retained "to ascertain the correct time from the heavens each day, and to regulate the company's chronometers," including the primary one, made by Edward Dent of England.[28]

In mid-1854, responding to a public now sensitized to the link between timekeeping and rail safety, Simon Willard, Jr.'s firm advertised English clock- and watchmaker Charles Frodsham's new timepieces, terming their enhanced reliability and stability as "peculiarly adapted to Railroad purposes." In New York, Ohio inventor Alexander Hall, who was seeking investment capital, announced an electrically-driven and -controlled clock system "for maintaining uniform time at all stations . . . in order to prevent accidents, a number of collisions having occurred by the variations of clocks at different stations, and the time kept by the conductors." Nothing much came of this announcement, however, for the Ohioan's focus was on an issue of modest import in this era of high-quality pendulum regulators.[29]

The general public was quite unaware that the critical factor in timekeeping for rail operations was the daily consistency of trainmen's watches. Most astronomers would not have known this either; indeed, some saw in these two disasters a market for selling accurate time, a market that they could easily dominate. After a decade of strenuous effort, American scientists had well in hand a telegraph-based time-distribution technology. This technology, of enormous value to astronomy and geodesy, would soon be touted as a critical component in public timekeeping as well.[30]

This [1851 U.S. Coast Survey] report exhibits the information . . . from all accessible sources . . . and it is not probable that a more impartial statement of the case will ever be obtained.[1]

—William C. Bond, 1856

All who were engaged in perfecting the application of the new method were more anxious to perfect it than they were to get credit for themselves.[2]

—Simon Newcomb, 1897

3. TELEGRAPHING TIME, MAKING HISTORY

Determining longitude differences via the electric telegraph was American astronomy's first innovation, introduced in 1844 and rapidly perfected. Soon after midcentury, European observatory directors began to use this now-mature technology, dubbing its equipment, procedures, and observational techniques "the American method for determining transits."

America's astronomers were delighted by this appellation, for they were struggling to overthrow the world view, long-prevalent and largely justified, that their current status in science was at best mediocre.[3] Being only human, these innovators also vied for credit within their own community, advancing their respective claims in influential newspapers, scientific journals, and federal agency reports.

The group that ultimately won what became a bitter priority battle did so by preparing a report in which key steps in the technology's development were deliberately obscured. In particular, much confusion still surrounds one late-1848 invention that undergirds the "American method." The official record also tells us little about the invention's effect on observatory timekeeping—an application that first gave astronomers the means to supersede all other time-givers.[4]

Time-at-a-distance was good theater at the first public telegraph offices. In 1844 in England, operators of the just-extended twenty-mile line running alongside the Great Western's Paddington-to-Slough railway entertained a skeptical, paying public by sending time-of-day inquiries and calling out the replies. A New Year's greeting from Paddington sent some seconds into 1845 was received at Slough in 1844—a widely remarked-on oddity caused by the local time difference between the two stations.

On Friday, 24 May 1844, the American government's forty-mile line along the Baltimore & Ohio Railroad's right-of-way from Baltimore to Washington opened for trials. After transmitting an Old Testament comment, "What hath God wrought!" operators sent additional messages to awe and entertain those assembled at the Capitol. "What is your time?" was the day's fifth message. From that moment on, Americans were captivated by the near-instantaneous transmission of time to distant places.[5]

Less than three weeks after the inauguration of American telegraphy, a scientific experiment involving time signals took place.[6] Samuel F. B. Morse, superintending the government's line, had given U.S. Navy officer Charles Wilkes permission to use the facility. Commander Wilkes, famous for his leadership of the U.S. Exploring Expedition and a friend of the inventor, began his trials on 10 June.

For the experiment, two mean-time chronometers were used, one regulated to the time of the meridian passing through Baltimore's Battle Monument Square, the other regulated to that of the meridian bisecting the Capitol dome. Then via an exchange of signals over Morse's line, Commander Wilkes compared the time displayed on one clock with that on the other. The time difference, of course, was the longitude between the two places. Commander Wilkes thus became the first person to employ the telegraph to measure a longitude difference.[7]

In his note of thanks to Morse, Commander Wilkes, a recognized expert in the conduct of triangulations and soundings for reliable coastal charts, noted that the results left him "well satisfied that your Telegraph offers the means for determining *meridian distances* [longitude differences] more accurately than was before within the power of instruments and observers." To those scientists familiar with the rigors of careful surveying, his brief report documented that precise values could be obtained in just a few days.[8]

These field trials made evident the striking advantages to be gained by telegraphing time, rather than carrying a marine chronometer from one location to another. But to capitalize on the new measurement technology, wires linking places had to be strung and maintained, telegraph operators had to be trained, and, critical to any hopes, funds to do all that had to be found.

Fortunately for science, instantaneous communication between distant places was emerging as a national desire. Just as fortunately, many entrepreneurs saw the founding of telegraph companies as opportunities that promised excellent profits. Consequently, as one historian noted, by the "autumn of 1845 telegraph builders were constructing lines on virtually all of the important routes."[9] Without this commercial expansion, American astronomers would not have become preeminent in this new way of determining longitudes. Wilkes had demonstrated the technology's promise; now others elsewhere could refine it.

The U.S. Coast Survey turned its attention to the electric telegraph soon after this era of rapid growth began. On 25 November 1845 a request went to failed businessman and gentleman astronomer Sears C. Walker of Philadelphia, directing him to make the necessary arrangements with the telegraph patentees. Walker was the perfect choice. For fifteen years he had worked and dreamed "to give to the world the most perfect determination . . . of the longitude of a point on the earth's surface . . . [ninety degrees] from the meridians of Europe."[10] In addition, his skills at reducing and analyzing astronomical data were impressive. In the United States he had few equals—if any.

Cooperation between two government agencies, and specifically between their recently appointed superintendents, was critical to success. The two were U.S. Navy lieutenant Matthew Fontaine Maury, in charge of what would eventually be called the U.S. Naval Observatory, and the Coast Survey's Alexander Dallas Bache.[11] The Naval Observatory had astronomical equipment and trained observers—and its location was prime—while the Coast Survey had funds, a coastal longitude mission, and long experience with precision measurements. Early on the two agency heads agreed to mount a joint endeavor; however, a crack in their agreement arose. Assisting in its propagation was Sears Walker.

In February of 1846, Walker, desperate for employment, had joined the Naval Observatory but continued to serve Bache as an unofficial, paid advisor on longitude matters. Walker resigned his observatory position in March 1847, complaining that he had not received full credit for his outstanding efforts in astronomy. He took with him enormous contempt for the Naval Observatory's superintendent. Within a few weeks, Walker received an appointment to the Coast Survey as assistant for telegraph operations.[12]

At the time Walker resigned, only a promising but unsuccessful longitude trial between Washington and Philadelphia had been effected. The first real success came in August, when longitude differences were measured via the commercial telegraph lines between Washington, Philadelphia, and Jersey City. Although the Naval Observatory and the Coast Survey collaborated, Lieutenant Maury took part with great reluctance. There were no more joint efforts.[13]

That two years of intense effort produced only a pitifully small number of successes was due neither to frequent breaks in the commercial telegraph lines nor to poor observing conditions. The true cause lay in the nature of the method itself. To gain a reliable value required exchanging numerous signals to compare the separated sidereal clocks. This task must then be followed by masses of observations of star crossings. Relating a stellar event to the precise time challenged both an astronomer's eye and ear. Simultaneously, he had to hear the telegraphed signal and estimate that fraction of a second during which a star's moving image was precisely aligned in the transit's crosshairs.

An alternative measuring technique—coincidence of clock beats—was just as exhausting. The observer-operator had to tap out seconds on a telegraph key synchronously with his local timekeeper for up to fifteen minutes at a stretch. Even then, doing so was likely to produce no more than two or three coincidences between these pulsed transmissions and the beats of the astronomical clock at the receiving observatory—an inadequate number for high-precision results; consequently, many trials were needed.

The method's ultimate precision depended upon years of observing experience, experience of a kind that only a few American astronomers possessed. And those few were busy throughout most of the year teaching students or running a family business. In 1847, for example, the astronomers critical to Walker's longitude program were Elias Loomis at New York University and Otis E. Kendall at Philadelphia High School. (William C. Bond joined Walker's effort in mid-1848, becoming the fourth highly skilled observer available.)

All participants in these very first longitude efforts must have wished for some automatic means to transmit the required time signals. Loomis himself wrote that he communicated this need in August 1847, just after the first real success with the telegraphic method.[14] However, the historical record generally starts with William C. Bond.

In August of 1848, after the New York–Boston longitude campaign was concluded, Walker reported that Bond had "contrived a mechanical method of making the astronomical clock strike its beats at all the telegraph stations in the circuit when these are receiving signals." The Coast Survey assistant added that Loomis was pleased with the idea, and that the cost of modifying an astronomical clock was about fifty dollars.

In September, Walker reported that Bond's "Telegraph Clock" would not be ready for the longitude campaign at the Cincinnati Observatory.[15] On his arrival there, Walker informed at least two scientists of the need for a transmitting clock; he also told them of Bond's proposed solution. At this moment Walker was unaware that suitable European transmitting clocks already existed. Like most American astronomers, he was quite unfamiliar with the technical details of electricity-based devices, telegraphic ones in particular. Superintendent Bache knew equally little. Indeed, some years earlier the Coast Survey had rejected the proposal of its own instrument-maker, Joseph Saxton, for an electrical switch attached to a clock. The device would have met the program's requirements, but both Bache and Walker feared that it would degrade the timekeeper's performance.[16]

One scientist who responded with concrete ideas was Ormsby MacKnight Mitchel, director of the Cincinnati Observatory. But distracted by his impending survey of a new railroad route from Cincinnati to St. Louis and by the need to promote the sale of stock in the fledgling transportation company, Mitchel tried what Walker would describe as "one or two experiments of a very obvious and simple nature." Nonetheless, these few experiments became the basis for Mitchel's priority-of-invention claim, a claim that seems untenable.[17]

The other respondent was John Locke, a medical doctor and lecturer in chemistry at the Medical College of Ohio. Well known in the scientific community, Locke had published extensively on geomagnetic observations, an area of interest he shared with his colleague and friend, Alexander Dallas Bache. Locke was also skilled in the design and construction of pendulum clocks and apparatus to demonstrate electrical principles and effects.

For Walker, Locke was a godsend. Together the two began a program of device design and construction, with Walker supplying requirements and Locke building the equipment. This collaboration began in earnest upon the completion of the Coast Survey's Cincinnati longitude campaign—a day or so after Mitchel departed to take up his railroad survey duties.

An existing astronomical clock, designed and built by Locke, was fitted with a circuit-breaking device and tested on 17 November 1848. Its going caused an electrical circuit linked to the recording pen of a Morse telegraph register to open momentarily, thus printing a line of dashes on the register's moving paper tape. Trials demonstrated that the attachment did not alter the timekeeper's constancy of rate to any great extent. Walker and Locke were elated. At this first blossoming of success, they publicly acknowledged each other's technical acumen.

Almost at once (the next day, according to Locke), the enormous advantages of connecting a closed electrical circuit to the Morse register's pen, instead of the usual open, battery-conserving one, became apparent: an observer need only break the closed circuit by means of a second, normally closed switch—a telegraph key—to record events on the same moving surface.

This method for combining time signals and events in the same record altered forever many scientific endeavors. A line exactly one or two seconds long now contained a short break indicating the moment of the event. At leisure, the resulting segments could be measured with a precision scale and the moment determined to within a few hundredths of a second. The astronomer's estimation of the time of an event—made "on the fly," and precise to no more than a few tenths of a second—was now obsolete.

Locke took credit for this invention and its execution; all evidence, including Walker's initial statements to Bache and the newspapers, support his contention. Nevertheless, this key development would not have occurred when it did without close interactions between both scientists, for Walker provided Locke with an understanding of the parameters associated with stellar transit observations.

Yet Walker had asked for a transmitting clock. Instead, he got a recording apparatus, one that could be used to compare star transits or distant astronomical clocks. Locke himself articulated this difference, writing that his invention *"consists in making a clock or other timekeeper by electrical means to generate a time scale, on which are registered by points, dots, or otherwise, epochs of events, and of course those epochs may be observations of Astronomical phenomena."*[18]

Since Locke was not under contract to the Coast Survey, he was perfectly free to contact others regarding his invention. Late in December he did just that, writing to another friend, Lt. Matthew Fontaine Maury, superintendent of the Naval Observatory. In the letter he included a paper tape from a Morse register to show that measurements to hundredths of a second could, and had been, made. The inventor noted that in addition to its use in electric longitude campaigns, such a clock system "might also be useful in a local observatory as a faithful and convenient register of observations." He proposed that "two such clocks be made by our government, one for your Observatory and the other for the Coast Survey." Although this remark certainly implied contacts with that agency, Locke neglected to mention his just-completed efforts with Walker.[19]

From the inventor's standpoint, it was well that he wrote Lieutenant Maury when he did. Bache had already rushed notice of the invention to the world's leading astronomy journal, and the Coast Survey superintendent was taking credit for urging "on *Mr. Walker* the importance of availing himself of the visible imprint, and of all the precision which is incident to it." With his journal submission, Bache included Walker's not-yet-published report to Congress. Meanwhile, the Coast Survey appropriated Locke's switch design, and Walker went to New York to have a copy built. In short, almost from the start Bache decided to minimize Locke's contribution to the technology's development.[20]

Lieutenant Maury was enormously excited by Locke's news, and he immediately penned a report to Secretary of the Navy John Y. Mason. He emphasized the invention's critical importance to the Naval Observatory, "not only for recording observations" but also "for perfecting the geography of the country."[21] The superintendent was taking the necessary first steps to get equipment funds added to the Navy Department's pending appropriations bill.

In one of those delightful coincidences, Lieutenant Maury's report to the secretary of the navy appeared in Washington's *National Intelligencer* on the same page as a copy of the secretary of the treasury's transmittal to Congress of both Superintendent Bache's and Walker's reports. Placed there by the two barons of government science in order to influence members of Congress, the reports became fuel for the already-smoldering priority-of-invention conflict.

Over the next weeks, Lieutenant Maury lobbied hard. Language was inserted in the Navy Department's appropriations bill providing ten thousand dollars to pay for the construction and erection of a recording system at the Naval Observatory, the government securing unrestricted use of Locke's invention in exchange.[22] The bill passed the House early in February, the Senate in March, and it was signed into law soon after.

Superintendent Maury had fought and won a skirmish. As the bill was moving through Congress, Cincinnati Observatory director Ormsby Mitchel, seeking "justice," battled publicly and privately with his old enemy, John Locke. Trying to derail the Naval Observatory's appropriation—Locke's reward—Mitchel allied himself with Bache and registered a stiff complaint with an influential Ohio senator.[23]

Like so many scientists of his day, Locke was no stranger to priority controversies.[24] Isolated in Cincinnati, he read "conspiracy" in the events swirling around the appropriations process. So he had Bache's official correspondence searched by a House Ways and Means Committee staffer to ensure that true copies of the correspondence to his (former) friend were being prepared. Responding to this pressure, Bache ordered the cessation of all longitude trials using the Coast Survey's unauthorized copy of Locke's device. Bache also gagged Walker, who was justifying his actions (according to Locke) by accusing the inventor of betraying the secrets of the Coast Survey to Lieutenant Maury at the Naval Observatory—thereby depriving his agency of its honors.[25]

Even gentle William C. Bond entered the fray. After complaining bitterly for weeks, Bond broke his public silence at news of the enactment of the bill. In a letter to an editor he asserted his own priority, giving the history of his electrical clock proposal and its subsequent approval by the Coast Survey. Rhetorically he asked, "Must I ask leave of Dr. Locke to use my own invention" in order to fulfill "a contract made long before he, Dr. L., had turned his attention to the subject?" Bond had put himself into a no-win situation. His complaint was taken up at once by Boston and New York newspapers, whose editors demanded an explanation from Locke.[26]

Locke asserted that his invention was "a new *combination* of things already known," and that "so far from being identical with Mr. Bond's, and a plagiarism or piracy from it, it does not include his invention, or even the smallest part of it, as a constituent element." Locke poured salt into the astronomer's open wounds by pointing out that Bond's timekeeper had not been built, much less tested. He noted that Carl Steinheil, Charles Wheatstone, and Alexander Bain had all developed such clocks years before. Then turning to his own invention, Locke expressed great surprise that Bond was complaining publicly about the chronograph. "I sent him a letter . . . [with] a specimen of the printed fillet, as I did to several other scientific gentlemen, and even requested his co-operation in the perfecting of some of the subordinate parts."[27]

This ended the matter—at least in public. Newspaper editors moved on to livelier controversies. Bond, of course, never accepted Locke's interpretation, even though his transmitting clock was not completed for almost a year, and even though he considered his time-and-event recorder a development quite separate from such circuit-breaking timekeepers. In later remarks to a close friend, Bond scoffed at Locke's "reinvention" of his own circuit-interrupting device. And Bond's son George later added his own words of contempt, writing: "One thing I am certain of is, that we never heard the names of Prof. Mitchel & Dr. Locke in connection with this subject, till after the whole ground had been gone over at Cambridge."[28]

The Bonds' perception of their priority claims did not, however, include the rapid development of the instrumentation immediately after the bill became law. Now agencies had the right to use Locke's invention without paying a fee. The Coast Survey rushed to build equipment, installing a circuit-breaker on its primary timekeeper. Saxton, the Coast Survey's gifted instrument-maker, designed and built a chronograph—its recording surface a rotating cylinder—to replace the Morse register and its linear paper tape. Optical instrument–maker William Wurdemann prepared a new set of transit-telescope diaphragms, with forty-five fiducial lines instead of the usual five. (Only with a chronograph could so many transits be observed.)

With no fanfare, the Coast Survey also established an observing site at the north-east edge of Capitol Hill, where it placed a transit of the first order. By mid-July, Seaton Station was in operation, becoming at once the de facto "zero" for the country's expanding electric-longitude net.[29] Over the next weeks Walker recorded a variety of stellar and clock data, documenting the system's greatly improved precision of results even in the hands of inexperienced observers. He completed this trial of brand-new equipment just four days before speaking at the annual meeting of the American Association for the Advancement of Science (AAAS).

By August 1849 the Coast Survey thus had a technical triumph well in hand, one that had taken only three years to accomplish. Unquestionably, Bache and Walker were loath to share success with anyone except "the officers proper of the Coast Survey, or of commissioned officers and civilians acting temporarily as assistants." So in his AAAS talk Walker mentioned Locke's name twice: last in a long list of inventors and proposers of circuit-breakers for clocks, and parenthetically elsewhere. He even denigrated the significance of the circuit-breaker—this an essential component of the system—by noting that it was "not an American invention."

With regard to Locke's invention, the essence of the chronograph (the simultaneous recording of time and event), Walker asserted that it had been made earlier "in an evanescent form, on the *auditory nerve* of the ear"—a most misleading assertion. And though the assistant for telegraph operations did allude to his mid-December report to Congress (which cited Locke's work), he announced to his audience that he would not discuss "the respective claims of Americans for priority or superior excellence of inventions and suggestions, believing that it will be becoming for all of us, to look to the great work that has been accomplished, by our united efforts, rather than to the single share of each." And so a story of a single-agency triumph, cloaked in the guise of a united effort, became the Coast Survey's position for the next forty-eight years.[30]

Locke completed the Naval Observatory's recording system in late 1849. His transmitting clock, built with a gravity escapement, had a tubular glass pendulum rod and a mercury bob. In theory, if not in practice, this astronomical timekeeper was the state of the art. The chronograph was a paper-wrapped rotating cylinder, with the recording pen moving parallel to the cylinder axis via a rotating threaded rod—analogous to the travel of a lathe's cutting tool. To make recordings, Locke employed a closed electrical circuit, with seconds pulses from the clock and keyed-in events interrupting the spiral line being traced out by the pen.

The two major components of Locke's recording system were constructed at the Boston clock and instrument shop owned by Edward Howard and David P. Davis. Many other artisans and mechanicians also contributed to the system's development, and Locke himself acknowledged a special debt to "the proprietors of the Morse Telegraph." Lieutenant Maury announced the equipment's arrival at the Naval Observatory in the *National Intelligencer* and placed the system into regular operation on 7 December 1849.[31]

In 1850, William C. Bond caught up with, and then surpassed, his fellow instrument developers. All chronographs constructed thus far were plagued with one annoying problem: the printing of time intervals was not completely uniform. Although this was not considered an overwhelming defect, nonuniformity in the printed record complicated and slowed the data reduction process. Indeed, given the masses of data the new measurement system was capable of producing, a solution was needed.

The Bonds solved the problem by increasing the uniformity of rotation of the chronograph's cylindrical recording surface. The first mention of their "spring-governor" control was in mid-February of 1850, when the senior Bond

PLATE XIII

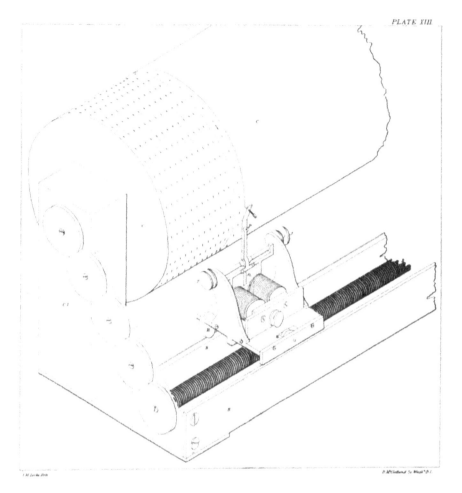

Fig. 3.1. John Locke's drum chronograph, shown tracing out periodic time ticks and stellar events on the same line. (The engraving has been computer-enhanced.) From Locke, *Report of Professor John Locke*, plate XIII. U.S. Naval Observatory Library.

wrote: "George [Bond] has hit upon a plan for regulating the gyratory motion by means of the oscilatory [*sic*], it is now in process of experiment, is simple and I feel confident will succeed."

According to Walker's later history, William C. Bond submitted a model of the chronograph and its controller on 12 April, after which the Coast Survey agreed to fund its completion, even constructing some of the necessary components in its own instrument shop. A near-complete system was taken to New Haven for the annual meeting of the AAAS, modified slightly afterward, and then completed—according to Walker—in November 1850. Having already been awarded a gold medal by the Massachusetts Mechanic Association for their clock-chronograph system, Wm. Bond & Son was making plans to market an excellent product.[32,33]

With this new, more uniform recording equipment in hand, Bond now turned to the Coast Survey's—and his own—most frustrating technical problem: the unaccounted-for disagreement of the chronometric and astrometric values for the longitude difference between the observatories at Greenwich, England, and Cambridge, Massachusetts.

The chronometric value—Bond's contribution to the effort—was actually the sum of three longitudes: Cambridge-Boston, Boston-Liverpool, and Liverpool-Greenwich. Each leg had a degree of uncertainty (error) associated with it. In all high-precision endeavors, the standard procedure is to determine the magnitude of the error, and, if possible, to reduce it.

Under Bond's supervision, the Coast Survey had the Harvard College Observatory linked by telegraph line to Wm. Bond & Son in Boston, a task completed by 21 January 1851.[34] Now all chronometers for the transatlantic longitude expeditions brought to the firm in Boston could be compared there against the master clock at Cambridge via the new recording equipment. By stringing this telegraph wire, one possible source of error was eliminated.

Bond had also been given approval to exhibit the Coast Survey–owned devices in England. Seizing this opportunity, he proposed to Bache that a similar error-reducing process be established at Liverpool Observatory, writing: "I shall try to economise sufficiently to enable Richard [Bond] to expend, if it should be found proper, a sum not exceeding two hundred dollars towards effecting a communication between the Greenwich and Liverpool Observatories by means of the Electric Telegraph."[35]

Indeed, this was an opportune proposal. Within the Coast Survey's assigned budget, the longitude difference between Liverpool and Greenwich—the third in the triad of values—could now be determined by the "American method," using the very best equipment available. No longer would chronometers have to be transported between Liverpool and Greenwich.

The superintendent, however, refused. More correctly, this brilliant manager-bureaucrat responded, "Before taking order in regard to the telegraph connection to which you refer between Liverpool and London I think it right to say to you that Mr. Walker takes the matter very hardly [hard] & considers it as removing from his hands part of the work to which he has devoted himself & in regard to which he had made a special proposition that he be allowed to execute."

Bond was stunned. Replying at once, he informed Bache that he had "rarely been more surprised and grieved than I was at the views in your letter" which, according to Bond, outlined Walker's desire "of reserving this portion of the work for his own personal superintendence." But having no discretion in the matter, Bond told Bache that he would direct his son "to take no part . . . in any determination of differences of Longitude by means of the

Magnetic Telegraph." He followed up his reply to the superintendent with a curt note to Walker. Once again, questions of credit and priority swirled around Walker and the Coast Survey, poisoning a technical endeavor.[36]

Richard Bond had already departed for Liverpool before the senior Bond received a response from Walker. Apparently Walker offered a lengthy explanation designed to mollify the Harvard astronomer. And it seems he succeeded. In a letter to his son dated 14 April 1851, William C. Bond describes it as "written in the kindest and most equitable spirit." "His principal object," Bond goes on, was "to secure [credit for] the american [*sic*] Electric Telegraph system as the result of Coast Survey operations." Bond concludes the letter by admonishing his son "upon all proper occasions [to] secure the share of credit due to the Coast Survey."[37]

For Bond, this episode had a very happy ending. Ten days after the incident, Walker submitted to Bache an "abstract of the progress . . . in the art of determining longitudes by the Electric telegraph," which contained his judgment of Bond's soon-to-be-operating chronograph. "Mr. Bond's machine surpasses in excellence all devices of the kind yet tried in the Coast Survey service," Walker wrote, ending with the prediction that "Mr. Bond's method [of control] is likely to supersede all other methods yet known."[38]

Bond was certainly pleased by this judgment, for in 1852 he placed a copy of Walker's report before his observatory oversight committee, calling it "an historical document relating to a very interesting but much-vexed question." Committee members pronounced it "an exact historical account of the course of invention and improvement in the art of determining longitudes by the electric telegraph." Five years later the astronomer printed Walker's report in the observatory's *Annals*, with his judgment of its quality—the statement quoted at the beginning of this chapter.[39]

Walker's long string of efforts to gain glory for his country, his agency, and himself ended in 1851. Illness forced him to retire in 1852, and he died the following year.[40] The Coast Survey's monopoly on automated recording equipment for astronomy also ended in 1851, when Bavarian astronomer Johann Lamont published a description of the electrically based equipment in use at the Royal Observatory at Munich. Illustrated was the "Galvanic Time-registering Apparatus," the direct result of Walker's first report on Locke's invention, brought to Lamont's attention by the American consul in Leipzig. Astronomers elsewhere soon began building similar equipment.[41]

Though perhaps it was not yet apparent by late 1851, Bond's participation in telegraphic longitude campaigns with the Coast Survey was also over. Already the Coast Survey had shifted its base of operations to Seaton Station. With a set of precision longitude differences between American observatories in hand, its surveyors could fix the location of any place on land simply by wiring that site to the commercial telegraph lines and thereby to any observatory already linked to them. Moreover, with no one able at that moment to reduce the known error in the value of the Cambridge-Greenwich transatlantic longitude difference, any North American longitude had exactly the same accuracy as one determined directly at Cambridge.[42]

Of course Bond knew this. But like all who prepare annual reports, the astronomer "accentuated the positive" in his discussion of the importance of his observatory to the Coast Survey's longitude program. Eager to make history, Bond printed an analysis by Walker, incorrectly terming it the basis by which "Harvard Observatory was adopted . . . as the *zero of longitude of the United States Coast Survey*, in 1851 [emphasis added]."[43] While Bond's friends on the Visiting Committee applauded, the astronomer turned to other agencies for financial support.[44]

The Coast Survey's support of the development of the measurement technology, particularly the recording devices, had an unexpected spinoff. Since determining the time of transit of a star is no different from comparing the ticks of one clock against the time of another, astronomers now had a means for documenting, via the chronograph, the accuracy of their mean-time clock in terms of the observatory's master standard—its sidereal clock, which was checked against the stars. Having wired their observatories into the commercial telegraph lines, Bache had given astronomers the means by which they could distribute mean time—the true time of the observatory—to the general public.[45]

PART II.   DISPENSING LOCAL TIME (1845–1875)

I fixed the clock in the vestibule by the ball today, and it is right by that.[1]

—James J. Gifford, stage carpenter at Ford's Theatre, 14 April 1865

4. INTRODUCING CITY TIME

By the mid-1840s, many people understood that a precise time signal could be sent great distances, but few saw much value in doing so. The boom in telegraphy had no impact on American time distribution; no observatory was even wired to a telegraph office until 1847, and that connection was made by the Coast Survey to further efforts in geodesy. The first public time signal in this era was a visual one.[2]

In October 1844, Lt. Matthew Fontaine Maury became the first superintendent of the Depot of Charts and Instruments. Two months later he received an order from Secretary of the Navy John Y. Mason: "You will be pleased to devise some signal by which the mean time may be made known every day to the inhabitants of Washington," Mason wrote. Lieutenant Maury hastened to comply.[3]

Why Secretary Mason decided to have the U.S. Navy initiate this public service is unknown. Perhaps he saw it as the first step in a plan to extend the depot's functions, for the secretary had considerable authority to do so—if the president agreed and no other department objected. Moreover, U.S. Coast Survey legislation enacted a dozen years before still forbade the erection or maintenance of a permanent observatory. Many supporters of astronomy concluded that this magnificent new facility, with its congressionally approved investment in significant astronomical equipment, circumvented that prohibition; they called the depot the "National Observatory."[4]

No official announcement of the inauguration of this time service has been

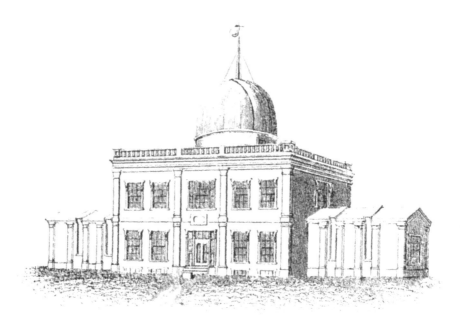

Fig. 4.1. Time balls, on the first Naval Observatory building (above) and on the (currently named) Old Executive Office Building (right), both in Washington. From reproductions in Bartky, "Naval Observatory Time Dissemination."

located. But starting sometime before mid-April 1845, citizens of Washington and neighboring Georgetown could watch every day, at noon, a black India-rubber ball falling from a staff erected on the dome of the observatory building, whose site was on a hill by the Potomac River.[5]

When raised, the ball was 157 feet above the ground (200 feet above high-water mark). Three feet in diameter, it was either "thrown down by hand at the word of command," or "when released [it] drops upon the dome and thence rolls to the roof beneath"—the authors of these differing descriptions, both of which were written some years later, were deliberately suggesting something less than strict accuracy. However, the moment of the time ball's *release* was the important instant: noon, mean time, at the observatory's meridian. For the next forty years at this Foggy Bottom site, and for fifty-one years more atop what is now the Old Executive Office Building, the U.S. Navy announced noon, Washington time, by dropping a time ball.

The superintendent had chosen a nautical time-signal device, the first one of its type having been erected and tested by the Royal Navy at Portsmouth, England, in 1829. The intent of such devices is the same: to provide a precise signal, which, when coupled to the locale's longitude, allows a navigator to determine his chronometer's error (and rate) without bringing it ashore. Lieutenant Maury probably based his selection on a description of the Royal Ob-

servatory at Greenwich's time ball, erected in 1833 at the direction of the Lords of the Admiralty. The time ball at Greenwich was only the third such device ever built; and the Washington installation became the world's twelfth.[6]

The Naval Observatory's signal ball did not serve as an aid to navigation, for few if any vessels plying the Potomac River carried chronometers. Navy officials, however, wrapped it in all the trappings of a navigational device in their budget requests and public documents. Actually, the time ball was an innovation in public timekeeping, giving residents what they lacked: an authoritative source by which all other timekeepers—clocks and watches—could be regulated. Mentioned in city guide books, the Naval Observatory's hilltop location made its noon signal accessible to a host of users.[7]

Time balls eventually became a fairly common way for American observatories to distribute a daily signal. In Britain, the time gun, similar in concept, supplemented such visual devices. But no matter how prominent the ball's drop, or how loud the gun's discharge, these devices share one fundamental limitation: not everyone can see or hear them.[8] Some better way was needed to distribute authoritative time throughout a large city. So it is to Boston that we must turn to chronicle the beginnings of regional timekeeping.

The city's official time was based on the meridian of the State House and was determined by its official timekeeper—one of the Willards, perhaps.[9] However, Wm. Bond & Son had its own time source: Harvard College Observatory, across the Charles River in Cambridge. Observatory time came to the Boston store via a chronometer, this distribution mode being brought about by the transfer of William C. Bond's telescopes to the college's control in 1839.[10] For the Coast Survey's chronometer-transport program of the 1840s, the firm's master clock was set to observatory time—earlier than Boston time by sixteen seconds. When, in 1849, the several area railroads adopted a two-minute offset from the "true time of Boston, as shown by Bond's clock," their station clocks were actually two minutes and sixteen seconds behind Boston time. So within Boston itself there were three true times differing from each other by sixteen seconds to over two minutes.

Most Bostonians were unaffected by any of these differences. Only those who chanced by the Bond establishment would have known that it was keeping observatory time. A few Bostonians rushing for their trains may have wondered why the depot's clock was always about two minutes slow with respect to their own timepieces, but they would have done no more than breathe collective sighs of relief as they sank into their seats. Indeed, Boston's church bells would have drowned out anyone's concerns, for their peals, sounding the start of every hour, spread out over ten minutes—if we may believe the assertions of later uniform-time advocates. Nevertheless, the absurdity of three true times within a tight geographic region would not have been lost on those whose livelihoods depended on keeping time within a second or so.

In 1852 a new electrical technology was introduced by the city government: the municipal fire-alarm telegraph. Once time-giving was linked to it, the centuries-old issue of nonsimultaneous time signals was resolved, and one true time for all of Boston—or any city—became inevitable.[11]

The concept of a municipal telegraph system was first described in a letter to the *Boston Daily Advertiser* entitled "Morse's Telegraph for Fire Alarms." Published on 3 June 1845, the letter was signed only with a "C," but its author is known to have been William F. Channing, a physician and member of a prominent Boston family. Channing proposed that the electric telegraph could be used not only to warn of blazes but also to indicate where they were located. Authorities could then "ring out an alarm, defining the position of the fire, simultaneously on every church bell in the city." Channing envisioned a central location to which the local warnings would arrive, and from which alarms would be distributed to fire fighters and citizens in that area.[12]

While Channing's idea of a wired city was a most innovative application of Morse's telegraph, it went nowhere until 1851.[13] In March of that year he submitted a formal proposal to city authorities and a few weeks later met Moses Farmer, superintendent of the Vermont & Boston Telegraph Company, an enterprise formed to exploit Alexander Bain's electrochemical telegraph.[14] Farmer had just overseen the linking of both the Harvard College Observatory and Wm. Bond & Son to the company's line, and he had recorded the first signals telegraphed directly to the firm. When Channing met Farmer, Harvard College Observatory time, free from any possible transport errors, was available in Boston.[15]

Channing and Farmer joined forces in April. Channing's impressive and comprehensive plan was considered by the city authorities, and, after lobbying by Boston insurance companies and a favorable municipal report, ten thousand dollars was appropriated for the construction and installation of a complete fire-alarm telegraph system, one that extended throughout the city. Farmer, resigning from the telegraph company, was appointed to supervise the construction of the system.

Farmer's simultaneous invention, development, and construction of a city-wide system of call boxes and alarm bells tied to a central location gave him national recognition, and a summary of progress appeared in the *Scientific American*. The editors were not particularly impressed, noting that a similar system was already operating in Berlin. They added that Werner Siemens, the inventor, was attaching electromagnetically driven clock dials to that city's alarm wires. Such a system of dials would provide "uniform and exact time" throughout Berlin, the editors opined. They also reminded their readers that electromagnetic clocks were not new, stating (incorrectly) that Bain had received the world's first patent, and that his clocks had been exhibited in New York three years earlier.[16]

Channing responded publicly to the implications of the *Scientific American* editors' remarks a few months later. He and Farmer had developed their fire-alarm system independently of Siemens, he claimed. Even though he demonstrated scant knowledge of the system in Berlin, he also asserted Farmer's priority of invention, basing it on the inventor's 1848 electromagnetic bell strikers. Conceptualizer Channing ended this lengthy discourse on Boston's fire-alarm telegraph by urging that a master-slave clock system be installed in the city:

> The application of the Telegraph to Fire Alarms is a step in Municipal organization which has become necessary and must lead to others of a higher order. The beautiful chronometer application of the Telegraph, by which a single clock registers its time on a number of dials throughout a city, by the simple magic of the electric circuit, is also one which deserves to be brought into immediate public use.[17]

Fig. 4.2. Boston Fire-Alarm Telegraph Office, showing the first apparatus, with Moses Farmer's automatic circuit-testing clock on the back wall. In *Gleason's Pictorial Drawing-Room Companion* 2 (24 April 1852): 264. Original in author's collection.

In December the Common Council elected Moses Farmer superintendent of the Fire Alarm Office. On 20 April 1852 he placed the city's fire-alarm telegraph into operation, and the first alarm was signaled the next day. Boston's system encompassed over forty miles of wire; forty street boxes, each capable of signaling its identity to a central office; and nineteen strikers mounted next to city and church bells, all on circuits connected to the central office.

Every hour of the day and night operators at the central office tested the continuity of the alarm and signal circuits, one after the other. The device used to perform this absolutely vital function was Farmer's invention. In appearance it resembled a clock, which every hour switched in the circuits for testing.

However, America's first fire-alarm telegraph did not incorporate a time-signal function. Its absence was deliberate: Farmer (and Channing?) had decided to sell time services. Two weeks after the fire-alarm telegraph's acceptance by the municipal authorities, Farmer completed his patent application, "Improvement in Galvanic Clocks," and submitted it to Washington. This invention was an electromagnetically impulsed pendulum, its maintaining power supplied by a battery. (Farmer's pendulum drove the alarm-circuit testing device.) Extolling its potential as a reliable source of time, one Boston writer wrote that "one hundred or [a] thousand clocks all over the city, all ticking at the same instant, and keeping the same time, may be carried by the pendulum." That fall Farmer received the first American patent ever granted for an electric clock.[18]

Despite the inventor's extremely strong patent position, nothing happened in 1852 with respect to marketing this timekeeping innovation. First-ever fire-alarm problems were cropping up and required the superintendent's attention. Farmer's genius was for technical inquiry, and practical application issues in long-distance telegraphy were starting to lead him away from electrical clocks.

In late 1853 Eben N. Horsford, Rumford Professor at Harvard, proposed that time be distributed via the fire-alarm telegraph. Boston's mayor had asked the chemist to develop a public-time plan; no doubt the disastrous railway collision in Providence in mid-August was the catalyst driving his concerns. In *Respecting the Regulation of Timepieces in the City*, Horsford recommended that a time ball be erected on the cupola of the Old State House, its drop controlled by a noon signal from the "electric clock connected with the fire-alarm office." He pointed out that "accurate time is sent in from the Cambridge Observatory to Boston, twice a week, and with the correction due to the difference in longitude, can be employed to regulate the dropping of the ball."

Professor Horsford judged the cost of erecting and connecting a time ball at five hundred dollars and estimated that no fewer than 210,000 inhabitants—those living within a three-mile radius of the State House—would benefit from the time signal. Specific beneficiaries included "State House officials who will have standard time indicated on the electro-magnetic clock," "railway passengers [who] will take the time to every part of New England," and "ship-masters [who] will regulate their chronometers . . . and take with them true Boston time."

Horsford added that his ideas stemmed from the operation of two time balls in England—the well-known one at the Royal Observatory at Greenwich, now dropped by electricity, and one just erected (August 1853) on the Strand by London's Electric Telegraph Company, also controlled by the observatory. (He did not mention the manually dropped time ball in Washington.) He ended his remarks with the opinion that a time-signal device in Boston "will be a fitting accompaniment to the scientific fire-alarm system, which we owe to the conception of Channing, united to the fertile invention of Farmer."

The Harvard professor's plan was referred to the city's committee on fire alarms, summarized in local newspapers, and printed in full a little later.[19] On seeing the summary, William C. Bond penned an extremely negative analysis of Horsford's proposal. The time balls in England were not successes, he asserted, and those Londoners who needed accurate time—railway companies and chronometer makers—used other means. Only a very small fraction of the city's inhabitants could ever watch the ball's drop, Bond continued. And, according to Bond, with the single exception of those who chanced by the remote island of St. Helena in the South Atlantic, no benefits accrued to ocean navigators anywhere in the world from the erection of a time ball. Even at St. Helena, a time ball's value was problematical. Those few mariners anchoring at this tiny island had very little else to do, so they could "afford time to prepare and watch for the signal."

Bond urged instead that the city adopt "the Electric process and [thus] have all our great Clocks regulated by one pendulum, substituting for their present cumbersome machinery one of the various contrivances . . . devised to produce simultaneous motion to an indefinite number of Clocks by the intervention of Electro-magnets." Noting the periodic transmission of time from the observatory to Boston (that is, to his own firm) and elsewhere via telegraphy, Bond asserted that "no other known method will bear comparison with it in point of accuracy or convenience." Further, Bond wrote, "The requisite machinery can be arranged by Mr. Farmer or any one of our ingenious mechanics who has been familiar with the working of the Electric Telegraph."[20]

Bond's low opinion of time balls was distinctly a minority view among astronomers. Yet even though his criticism was never published, clocks running synchronously via electricity became the technology of choice for keeping accurate public time in Boston.

A few weeks after Bond's analysis, the Joint Committee on Telegraphic Fire Alarms began their report with a graceful rejection of Professor Horsford's proposal. Then, relying heavily on Moses Farmer, "our intelligent Superintendent of Fire Alarms," whose "Plan for Regulating the Public Clocks of Boston" was before them, committee members recommended that the city's ten public clocks be connected via telegraph wires to the magnetic clock in the Fire Alarm Office, and that the public clocks be driven by new electromagnetic (that is, Farmer's) movements. This approach would "secure the greatest good for the greatest number," for not just the time ball's "single hour, but *every hour* of the day could be noted with perfect accuracy." And, the committee reiterated, by adopting Farmer's plan, "all sections of the City proper may be furnished with *uniform* and *absolute* time at *all hours*"—at a cost of slightly more than forty-eight hundred dollars.[21]

In his report to the Joint Committee, Farmer had suggested a trial based on the purchase of only one electromagnetic clock movement, estimating the total cost at $910. So on 29 December 1853, the Board of Mayor and Aldermen passed an order, in which the Common Council concurred: "That the Superintendent of Fire Alarms . . . is hereby authorized . . . to cause to be placed on the Old State House, a new bell, with all the machinery and apparatus necessary to test the correctness of the Magnetic Clock in the Fire Alarm Office." A sum of $950 was budgeted, so the test must have taken place, even though no accounts have been located. Without question, this public-time distribution system passed with flying colors, for in contrast to the lengthy development of Boston's fire-alarm telegraph, all necessary devices were already at hand.[22]

This early-1854 feasibility demonstration was a key segment of Farmer's planning. In his remarks to the Joint Committee, he had also outlined how such a system could be extended: "Railroad depots, corporations, and individuals, could easily receive their time from these main clocks, by means of 'local circuits,' without injury to the performance of the Public Clocks." Thus in April, Farmer—allied with clock- and apparatus-maker Edward Howard; Jesse Rowe, superintendent/chief operator of the Boston & Vermont Telegraph Co.; and Channing—petitioned the Commonwealth of Massachusetts "that they may be incorporated by the name of the Boston Electric Time Company, for the purpose of manufacturing and selling electric clocks, and connecting the same by telegraphic lines, for the purpose of securing uniform time, and furnishing electricity to electric clocks and telegraphs; the said company to be located in the City of Boston."[23]

The usual legislative process followed, and on 29 April 1854 the governor approved "an Act to incorporate the Boston Electric Clock Company." An entirely new kind of business—selling uniform time—had been founded in the United States.[24]

Now the system of electromagnetic dials envisioned for the time service required the continuous, exclusive use of wires tied to a central clock. The ability to connect such clusters of private circuits at various points to, say, a citywide system of public-clock wires meant an enormous reduction in construction costs. Given Farmer's position as superintendent, the Boston Electric Clock Company seemed destined to become a most profitable enterprise.

Furthermore, these public and private clocks and dials would display accurate time—not simply be consistent with each other. The obvious source was Wm. Bond & Son. But at that moment the firm's link to observatory time in Cambridge was a combination of private telegraph company lines and wires owned by the U.S. Coast Survey—one section of which passed through the Watertown Arsenal. So in his November 1854 report to the Visiting Committee, Harvard College Observatory director William C. Bond appealed for "a [telegraph] line of our own." Having a more direct route to Boston, the astronomer argued, "would lessen the risk of interruption, as well as the expense of frequent repairs," noting that such a line would be "immediately under our own supervision." Of course Bond did not point out that a dedicated circuit was the only way of supplying continuous time signals to the city. But then, the director really didn't have to articulate this feature to his close friends on the observatory's oversight committee.[25]

Early in 1855 a hitch arose. Farmer, now famous throughout the country for his electrical inventions, was in serious trouble in Boston. Once again the Common Council had elected him superintendent, but the Board of Mayor and Aldermen did not concur. Having received a report of poor fire-alarm service from the Fire Department's engineers, board members tabled Farmer's certification of election. In June, when the city fathers reduced the superintendent's salary by 20 percent, incumbent Farmer was not notified. Upon learning of his reduction in pay two months later, Farmer promptly resigned. The inventor's access to the citywide network of telegraph wires—a wonderful test-bed, indeed—was gone; the nascent Boston Electric Clock Company faded into history. Almost thirty years would pass before another time-services company was incorporated in the United States.[26]

With Farmer out of the picture, Bond continued with his own plans. During 1856 he purchased a section of a telegraph company's line that ran from North Cambridge into Boston, and then had a wire strung from the observatory site through Cambridge and along the Fitchburg Railroad right-of-way to connect to this section.[27] He described the observatory's acquisition and new construction in the fall, claiming "the value of [the observatory] giving certain and accurate time for the regulation of businesses and economical affairs and Railroad movements and the regulation of marine Chronometers"—although these activities were his firm's business ones. Unaccountably, Bond included the "important object of determining differences of longitude of position along our extended sea-board" among his reasons for acquiring the telegraph line. But his contract work for the U.S. Coast Survey was already over. Moreover, his overseers knew that only a few evenings of observations were needed to fix precisely any point in the United States; renting a circuit for this task would be vastly cheaper than owning and maintaining a dedicated line of wire. Bond was presenting the members of the Visiting Committee with a fait accompli. Monies disbursed by Harvard paid for this more direct route to Boston, and the line became observatory property.[28]

Concurrent with this new link to his Boston firm, Bond began experiments with a *normal* [standard] pendulum designed by his son Richard. The pendulum, equipped with contacts for producing electrical pulses, was wired into the observatory's signal-distribution panel. This action suggests that Bond was testing it as the source of continuous time signals to Boston.[29] But the fourteen-foot-long pendulum rod was quite sensitive to temperature changes, and there is no indication that the observatory ever transmitted the pulses generated by its inconsistent swings.[30]

In mid-1858, William C. Bond, who had relinquished his interest in the family firm, wrote that Joseph Stearns, superintendent of the Boston Fire-Alarm Telegraph Office, "has made [an] arrangement, through Messrs. William Bond & Son . . . to have a twelve o'clock signal struck on the different bells of the city." According to the astronomer, observatory time, derived from the signals sent to the Boston store, was previously "carried by Chronometer" to the fire-alarm office.[31] But "Stearns was willing to adopt yet more accurate methods," he wrote, for the superintendent had had a short branch wire strung from the firm's location at Congress Street to the City Building. And with the observatory linked directly to the fire-alarm office, "the mean solar time for the meridian of Boston" was now being determined. "At the instant of mean noon, the circuit is closed by a simple tap on the key . . . which causes all the alarm bells in the City to be struck simultaneously."[32]

Harvard College Observatory's shift of its mean-time signals from scientific needs to public ones put an end to the sixteen-second difference between the firm's true time and Boston's official one. Simple enough to execute, time signals unifying the entire city could have been distributed as early as 1852. Only competing business interests had prevented their introduction.[33]

The Bonds' business interests, however, were now altering Boston's timekeeping. The wire linking Wm. Bond & Son to the Fire Alarm Office meant that the firm was, de facto, city timekeeper. Indeed, Superintendent Stearns's "arrangement" was a business one, so it was Wm. Bond & Son, not Harvard College Observatory, that received compensation—some sort of *quid pro quo* —for its services.[34,35]

This change in Boston timekeeping was noticed early on. The editor of the *Firemen's Advocate* ended a lengthy description of the municipal alarm system with: "But the Fire Alarm is not wholly confined to the giving of fire alarms; at 12 o'clock, each day, one below [blow] is struck upon all the alarms, to announce the true time, as taken at the Cambridge Observatory."

Time-giving became a marketing point in the selling of fire-alarm systems—albeit always a secondary one. Eventually the largest manufacturer of alarms simply placed in its catalogs a notice that the fire alarm "may be used to establish standard time throughout a city. With an instrument placed at any point where correct time is kept, and properly connected with the [municipal] telegraph, the alarm bells may be sounded at mid day, or at any hour desired." The number of municipal telegraph installations soared after the Civil War, and, though the general public was scarcely aware of it, uniform time within cities and towns became the norm—quite independently of the country's astronomical observatories.[36,37]

In 1862 telegraphed time signals from the observatory to Boston ceased when Harvard sold a section of the telegraph line. Although a financial loss resulted, observatory director George P. Bond put the very best face on the sale when he wrote that a connection "was no longer needed for any astronomical work; the longitude operations, for which it was originally designed,

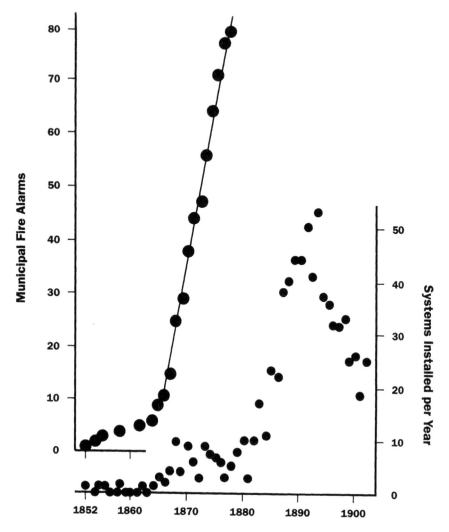

Fig. 4.3. Growth of fire-alarm telegraph systems in the United States, 1852–1902, with 764 installations in 1902. Data are from U.S. Department of Commerce and Labor, Bureau of the Census, *Municipal Electric Fire Alarm and Police Patrol Systems*, table 2.

have long since been concluded."[38] Striking noon on the Boston fire bells continued, nevertheless. And Wm. Bond & Son, its owners now no longer associated with the observatory, may have reverted to chronometer transports or perhaps even paid to receive occasional time signals from Cambridge via some other telegraph line.[39]

Either delivery mode decreased the quality of the city's uniform time. Yet no Boston trains struck head-on, and no citizens missed appointments. In public affairs, consistency—time that is "good enough"—is the desideratum; the high accuracy that astronomers require is of lesser importance.

5. ANTEBELLUM OBSERVATORY TIME SERVICES

The impetus for developing public time services was the magnificent failure by Benjamin Apthorp Gould, Jr., at the Dudley Observatory in Albany. Earlier, astronomers at private observatories had focused on specialized customers; but Gould's aggressive pursuit of funding for his nascent observatory caused them to extend their activities. By the start of the Civil War, three new public time services were operating—two in the state of New York and one in Boston. Like that of the Naval Observatory in Washington, the observatories' true times were local ones; but their signals were distributed by telegraph.

If early practices in England are any guide, American telegraph companies routinely transmitted one particular station's time to their other offices, for the need for uniformity was well understood by all such systems operators.[2] But the time transmitted by these early enterprises was not based on any observatory determinations.[3] Although by midcentury seven astronomical observatories were wired to the country's "lightning lines," none transmitted time with any regularity until 1851, when the Harvard College Observatory was connected to the main line of the Vermont & Boston Telegraph Company.[4]

Soon after this branch line began operating, William C. Bond made an arrangement with Moses Farmer, the telegraph company's superintendent. Just before one o'clock every Monday, Bond connected the observatory's circuit-breaking clock to the company's wires, thereby giving "the beats of the [mean-time] clock at every station all along the line" from Boston north to Burlington, Vermont. The signals were independent of the chronometer-rating ones transmitted to the family firm.[5,6]

Wm. Bond & Son's Boston railroad customers did not receive time directly from the observatory, however, for neither the owners of the city's Morse line nor this Bain (chemical telegraph) system had erected branch lines to the local rail depots.[7] Still, Bond's claim that his Monday signals reached New England railroads rings true. Very likely Harvard College Observatory time was being supplied by the chemical telegraph line's local operators to railroad men, for such transfers were becoming the norm during the 1850s.[8]

One American railroad was already synchronizing its clocks daily via telegraphed time signals. The action was the happy accident of opportunity and interest: the railroad itself owned the telegraph line, and its chief operating official was also a telegraph pioneer. Charles Minot, superintendent of the New York & Erie Railroad, revolutionized American railroading in 1852 when he inaugurated the movement of trains by telegraphed orders. Superintendent of the Boston & Maine prior to coming to New York, he was a founder of the New England Association of Railroad Superintendents, serving on its Committee on Standard Time. Minot, who could also tap out and receive Morse-coded messages, soon convinced the Erie's directors to construct a line of telegraph along the road's right-of-way.[9]

Minot issued his first book of rules for the Erie a few months after the completion of the company's telegraph line. These rules took effect on 31 March 1851. The rule book outlined the Erie's timekeeping, listing a variety of measures for safe, reliable, and efficient railway operations. In one section Minot described his innovation:

> For the convenience of [those who] . . . cannot regulate [their watches] by this Clock [the railroad's standard of time at Piermont, New York], the Clocks at Suffern's, Delaware, Susquehanna, and Honellsville, which are regulated by Telegraph to agree with the Pier Time, shall be taken as *Standards*.

For regulating (synchronizing) these standard clocks, the company's telegraph operator tapped one beat every second, beginning at four minutes before noon. Exactly on the hour, the operator switched to the letter *i* (which in Morse code is represented by two dots), sending the pair of dots every second for one minute. At four minutes past noon, the time-giver relinquished control of the telegraph line. Although in practice no different from the nighttime transmissions of astronomical time for longitude determinations, this daytime process gave station agents the road's operating time within a second or two—when conditions were perfect. Yet even if noise on the line or distractions meant that a local agent missed the exact moment of noon, the string of *i*'s gave him the road's operating time to within a minute.

Easy to institute, the safety value of synchronizing clocks telegraphically lay not so much in better overall consistency as in reducing the burden on the train conductors: no longer were these busy men responsible for checking the clocks at every station. So as railroads began using the telegraph to control their trains' movements, telegraphing time to station agents became the common way to synchronize clocks—this manual signaling continuing well into the twentieth century.[10]

In December 1852, when Gould succeeded Sears Walker as the person responsible for the Coast Survey's longitudes-by-telegraph expeditions, he was unaware of the Erie's pioneering clock-synchronization practices, nor did he know any other details of the road's timekeeping. Residing in Cambridge and interacting frequently with William C. Bond, Gould may have been aware of Wm. Bond & Son's business arrangements with the local railroads, but only in a general way. However, Gould was completely familiar with the periodic time transmissions from the observatory via the chemical telegraph line: the Coast Survey's contract with Bond was still in effect, and the link between Cambridge and Boston had been specifically created to eliminate a possible source of error in the Greenwich-Cambridge longitude determinations.

Living near Boston, Gould certainly knew of the city's pioneering fire-alarm telegraph and would have understood its technical principles. Like all New Englanders, he must have been shocked by the August 1853 time-related collision on the Providence & Worcester Railroad. But public timekeeping was of no particular interest to him until 1855, when he agreed to become astronomer-in-charge of the Dudley Observatory. Although the distribution of mean time was ancillary to his interests, Gould's efforts with regard to this "non-astronomy" activity did much to advance America's timekeeping technologies.[11]

The Dudley Observatory building, completed in 1854, though an imposing structure, was actually an empty shell: no astronomical equipment had ever been purchased for it, much less installed. During the summer of 1855, observatory trustees renewed their attempts to put this very visible white elephant to work. Secretary James H. Armsby traveled to Providence for the American Association for the Advancement of Science's annual meeting. There he met with Alexander Dallas Bache, with Bache's close friend, the eminent mathematician Benjamin Peirce, and with Gould.

They struck a bargain. The trustees agreed to purchase a heliometer—an expensive, specialized telescope capable of measuring stellar separations with high precision. In exchange, the Coast Survey promised to place observers at the site once the heliometer had been installed. The trustees also created a technical advisory body—the Scientific Council—consisting of Albany-born Joseph Henry, the secretary of the Smithsonian Institution; Bache; Peirce; and Gould. Although remaining an employee of the Coast Survey, Gould accepted responsibility for bringing the Dudley Observatory into working operation. His first steps included a trip to Europe to purchase various telescopes and an astronomical timekeeper.[12]

Gould's training, interests, and ongoing responsibilities were in precision measurements, a research environment requiring close attention to a multitude of details. In positional astronomy programs, the need for accuracy and precision was well understood, and a particular measurement framework had been in use for decades. Working at the state of the art meant acquiring the finest equipment that could be built. Any person responsible for an observatory devoted to this branch of astronomy would also try to improve on existing designs. Gould was no exception. Unfortunately, he tried to improve virtually every piece of equipment that the Dudley Observatory needed simultaneously.

Sidereal timekeeping was critical to the anticipated program. Erastus Corning, the founding president of the New York Central Railroad, agreed to pay for the observatory's *normal* (primary standard) clock. Gould ordered the "Corning Astronomical Clock" in October 1855 from Krille, a well-regarded precision clockmaker in Altona, near Hamburg. This timekeeper was designed to be unaffected by changes in barometric pressure: the pendulum and clockwork were to be surrounded by an airtight case. The resulting clock thus had to be "adjusted, regulated, set in motion, stopped and wound by apparatus from the outside." Seemingly unaware of pioneering studies undertaken more than twenty years earlier, Gould termed this innovation "one of the simplest things in the world" and remarked that it was "strange that it was not thought of before."[13]

A sidereal clock with a constancy of rate significantly better than current ones would certainly have advanced the state of the art in the United States—and in Europe. But the clock was never finished, and Gould had to settle for a more conventional timekeeper, one constructed by England's Edward Dent. In the eyes of the trustees, the fact that the Corning Clock was never delivered became one of Gould's many failures.[14]

Early on, the astronomer decided to have this primary timekeeper transmit signals simultaneously to three observing stations within the observatory building. That seemed an innovative way to reduce the huge overhead burden associated with determining clock errors, an essential step in calculating final stellar positions. To design and build this distribution system, Gould engaged Moses Farmer. Prospects for success were excellent. At that moment the inventor was the country's foremost expert on electrical-signaling systems.[15]

Determined to improve chronographs as well, Gould also hired Farmer to construct one that embodied a new control principle. But when installed at the observatory, Farmer's chronograph did not operate as well as the ones fitted with the Bonds' spring-governor control. This technical failure added more

Fig. 5.1. Dudley Observatory master (sidereal) astronomical timekeeper. Originally constructed by Dent of London and later modified (new pendulum, suspension, and dial face) by Wm. Bond & Son, Boston. Courtesy of the Dudley Observatory Archives, Schenectady, New York; photograph courtesy of the New York State Museum, Albany, New York.

fuel to the eventual confrontation that ended Benjamin Gould's career at the Dudley Observatory.[16]

Early in the fall of 1855, the Dudley's trustees, aware that maintaining an observatory was an extremely expensive proposition, asked Gould to consider distributing signals to provide mean time to paying customers.[17] For the trustees, the proposed activity had only one goal: to generate income. For Gould, it became yet another way to move beyond the state of the art.

Gould's model for public timekeeping was the Royal Observatory at Greenwich. He had arrived there at the end of his European buying trip, as the invited guest of Astronomer Royal George Airy. Gould had been overwhelmed when he saw the extent of the observatory's services. The Royal Observatory's decades-old distribution of mean time had expanded in 1852 via the several telegraph lines brought to the site by the South Eastern Railway Company. By 1855 time-giving was a major observatory function, and Airy focused his energies on developing electromechanical devices for use within and outside the observatory. Airy was proud of these accomplishments and often extolled the public value of Greenwich Observatory time. Every hour, signals were sent to the heart of London for subsequent distribution throughout the realm; every second, transmitted pulses controlled a number of clocks in the city. At one o'clock every day a signal from Greenwich automatically released three time balls—one of them located eighty miles away at Deal.[18]

During Gould's six-day sojourn at Greenwich, the astronomer royal personally escorted him to the telegraph company that distributed the observatory's signals, to the railroad stations and clockmakers that received them, and to the company manufacturing the various electrical components. However, Gould, who judged Airy's electromechanical devices ingenious in design, dismissed them as too massive and expensive. "We can improve them greatly while copying what is *not* improvable," he wrote soon after. Then in a classic example of technical naiveté mixed with hubris, Gould added, "Can it not be arranged that we begin on the day of the inauguration to give a signal to the shipping in NYork [*sic*] harbor . . . another on the [state] Capitol . . . & regulate the clocks of the [New York] Central & [New York and] Hudson R. Railroads." (Gould penned these words early in December; the Dudley Observatory's inauguration took place the following August.)

Willing to have the observatory transmit any time desired by would-be consumers, Gould nevertheless asserted that "New York City is so nearly on the meridian of Albany that it ought to use Albany time just as all England uses Greenwich time." For the privilege of receiving the signals, "the New York merchants, or city or Chamber of Commerce, and the two railroads ought each to give the observatory the means of supporting one or two assistants," Gould added, leaving to the observatory's trustees the task of deciding "how this [marketing] can be brought about."

Gould ended his review by confidently stating that "the difficulties of detail are already mastered, and all that remains is to simplify and cheapen the apparatus." In short, Benjamin Gould believed that what had taken the skilled director of one of the world's major observatories years to accomplish could be duplicated via his spare-time oversight of a not-yet-functioning observatory. And all in a mere eight months.[19]

The astronomer's enthusiasm pleased the trustees enormously. Within days of his return to the United States, the sale of time signals was being pursued vigorously. Gould made several estimates of the income that could be gained from various time services. He predicted that it would be one thousand dollars a year from the state government in Albany and another thousand dollars from each of three railroad companies—one of which, the New York Central, was actually headquartered in the capital. Later, however, Gould lowered this estimate to a total of three thousand dollars annually.[20]

A formal proposal for electrically controlling the New York Central's station clocks, carrying both Gould's and Bache's signatures, was submitted. Since the railroad company's board of directors included men on the Dudley Observatory's board of trustees, approval seemed certain. Rushing to complete scores of tasks before leaving for the South and his regular Coast Survey duties, Gould ordered all the necessary devices from Farmer, supplying the inventor with drawings and a host of suggestions for remedying what he referred to as "serious defects" in Airy's equipment.[21]

The electric synchronization of independent timekeepers was an entirely new concept in the United States. Joseph Henry was the only member of the Dudley Observatory's Scientific Council actually able to judge the proposal's technical feasibility. Responding in the matter, the physicist began: "It may at first sight be thought that such extreme accuracy [within a second of time] is not necessary, but since time is the first element of precision in matters of this kind, it is of the highest importance that it should be as exact as the state of science can render it." For Secretary Henry, "matters of this kind" meant railroad safety. In support of Gould's proposal, Henry wrote, "Any plan which tends to lessen the chances of the frightful collisions which within the last year have been so numerous and attended with such melancholy and destructive consequences should be adopted." His judgment came from the heart, and not from a first-hand knowledge of rail operations. Nonetheless, a decided advance in American timekeeping technology was now underway.[22]

In addition to the synchronized clock system, Gould contracted with Farmer for the construction of a time ball and triggering apparatus, to be used in the time service for New York City. Earlier, Gould had estimated that placing a visual signal there required $10,000, an amount that would pay for the purchase and installation of telegraph wires from Albany, construction of the signal devices themselves, and the expenses associated with one year of operations; the last item was projected at $600 per year thereafter. Gould did not include in the estimate the enormous cost of placing supporting poles along the 150-mile route from Albany to the city, for he counted on using the telegraph company's. He proposed acquiring what was still a significant amount—$10,000 in those years was equivalent to more than $190,000 in today's money—via a public appeal to the merchants of New York. However, Bache did not favor this tack, and so an approach to the New York City government was decided on.[23]

During this planning period Gould's vision expanded. One impetus for this was access to the New York–centered railroads—a market much larger than the single line headquartered at Albany. As Gould himself expressed it, there would be "absolutely no limit to the extent of roads & lines to which we might give [that is, sell] the time."[24] Another ambition that now seemed in reach for the Dudley Observatory was a national role—similar to that of the Royal Observatory at Greenwich. As Gould envisioned it, Albany time, supplied by the Dudley Observatory, would establish the true time for all of America.

On 22 January 1856 the Dudley Observatory's proposal—the country's first example of public time based on signals transmitted by an observatory—was presented to Mayor Fernando Wood of New York. Endorsed by the observatory's trustees and the Scientific Council, an appeal for funding began, using England's Royal Observatory as the model of national timekeeping. The proposal's authors bolstered the case for choosing Dudley's service by denigrating the Naval Observatory's manually operated time ball.[25]

The Dudley Observatory offered to drop a time ball in New York and to regulate three of the city government's clocks. The daily controlling signal would be free, and the city's only costs would be new batteries and such to maintain the devices, and to pay someone to arm the signal ball and wind the clocks. Nowhere does the proposal mention that New York was expected to pay for the wires to carry the Albany time signals to it; perhaps talks with Mayor Wood covering this subject had already taken place.

The observatory's scientists mentioned the public's need for correct time— in passing. Their targeted beneficiaries were those involved with the "shipping in the harbor of New York." The proposal emphasized the quality of the observatory's time signal: "accurate to the tenth of a second." This was vastly more precise than the one-second accuracy Henry had supported for railroad timekeeping—and indeed was a level not even needed in the rating of marine chronometers.

Simultaneous with the observatory's proposal to the mayor, someone associated with affairs in Albany approached the editor of the *New York Daily Tribune*. Thus in "Astronomical Time for the Railroads," the newspaper cited a 31 December railroad disaster near Pittsburgh, calling it "only a specimen of a very numerous class of accidents which result from the incorrectness of time-pieces," and a "demonstration of the need of introducing the uniform and absolute accuracy of astronomical time." Noting that the Dudley Observatory had just proposed a fee-for-service distribution of time from its astronomical clock to the New York Central Railroad, the editorial writer voiced the hope "that the railroads in this vicinity will promptly avail themselves of the opportunity, which will now be put in their reach."[26]

Although the *Tribune* had reported earlier that the conductor's watch had been some *fifty* minutes slow at the moment of collision, this editorial made absolutely no mention of other, more effective ways to prevent similar disasters—ones that the country's railroads were already using or incorporating in their operating rules. The myth that accurate time was essential for safe rail operations had already taken hold.

Time finally ran out for Benjamin Gould. He had spent months on Dudley Observatory planning and was far behind in his Coast Survey duties. With the timekeeping proposal now in Mayor Wood's hands, he departed Albany for New Orleans. Stopping in Washington, he received the first indication of a countervailing interest in public and railroad timekeeping in New York. But totally immersed in his official responsibilities, the astronomer could only summarize a newspaper clipping as an alert to those in Albany:

> under "Answers to correspondents", is the reply to query "Is there any authorized time-regulator in this city?" The answer is that two years before, the Common Council authorized the faculty of the N.Y. Univ. to build an obsy on their building, "from which telegraph wires would diverge to every police and railway station, & public clock in the city, giving them true astron[omica]l time the year round". It adds that the purpose not having been carried out there is as yet no authorized time regulator!

Gould dismissed this information with the remark that the Dudley Observatory's agents "will take care of the matter in the Common Council of N.Y."[27]

Exactly when Gould learned that obtaining the city's authorization involved New York University astronomer Richard Bull is not known; the Common Council was already terming the professor its "Timist," this title reflecting Bull's private time services for both the New York & New Haven and the Erie. Even if Gould had been aware of these business and city government interests prior to January of 1856, he probably would not have altered the Dudley Observatory's plans. Gould's contempt for almost every other astronomer in America—an open secret among this small community of scientists—was equaled only by his high regard for his own skills, desires, and goals.[28] A conflict between the two became inevitable.

Accompanied by the mayor's endorsement, the Dudley Observatory's proposal was sent to a committee of the Common Council, where it was completely ignored. In subsequent letters Gould railed against those assigned the task of influencing the council on the observatory's behalf. Bache finally interceded with chronometer dealer George W. Blunt, one of the Dudley Observatory's now-reluctant agents. Still, nothing happened.

In July, Gould wrote that he had given the matter up, and turned to his share of the large and small tasks associated with the observatory's imminent inauguration. But the trustees still faced the absolutely critical task of convincing New Yorkers to contribute funds to an Albany institution, and they appealed to the entire Scientific Council. A marketing prospectus was developed and printed; Bache, Henry, Peirce, and Gould reiterated the great value of accurate time for the port city.[29]

The inauguration of the Dudley Observatory took place on 28 August 1856. In his address on this occasion as astronomer-in-charge, Gould extolled the value of what he termed the soon-to-be-functioning system of observatory time signals for the entire state of New York. He pointed to a running (slave) clock, and informed the huge audience of distinguished citizens and guests that the observatory's mean-time clock—constructed by Farmer and mounted at the still-incomplete facility—was generating the driving pulses. Finally, he announced that this was the "Corning Clock." His ambiguities regarding these timekeepers unleashed one of the most contentious and hilarious issues associated with his tenure at the Dudley Observatory.[30]

Five days later an editorial appeared in the *New York Times* urging Common Council action on the observatory's stalled public time services proposal; the Committee on Arts and Science remained unmoved. A few weeks later Gould informed Astronomer Royal George Airy that the proposal's defeat was caused by local jealousies.[31]

The issue festered. Late in the fall, Bache made an extremely clumsy appeal to the city's nautical and merchant groups. Anxious to secure their financial support, he asserted that New York City was an unsuitable location for an observatory. Claiming that the observatory in Albany was almost functioning, he termed it the closest one able to serve the practical needs of the state's great commercial center—that is, able to provide accurate time for rating the chronometers of oceangoing ships.

This seemingly casual dismissal of all the efforts made—and being made —by prominent New York City educators to establish a public observatory tied to a local university caused an immediate uproar. The *New York Times* issued a rebuke, its editorial writer charging that the Coast Survey superintendent "errs viciously and radically" in urging New Yorkers "to adopt the

Fig. 5.2. Detail from Tompkins Matteson's "Inauguration of the Dudley Observatory, August 1856." On the table is a subsidiary clock dial, with wires running to Moses Farmer's driven pendulum installed in the observatory building a mile away. Seated in the row behind the table (l. to r.) are former New York governor Washington Hunt and three members of the observatory's Scientific Council: Harvard mathematician Benjamin Peirce, U.S. Coast Survey superintendent Alexander Dallas Bache, and his assistant for telegraphic longitudes, Benjamin A. Gould, Jr. Gift of Gen. Amasa J. Parker, collection of the Albany Institute of History & Art, Albany, New York.

Dudley Observatory as a sort of foundling substitute for the offspring our sterility forbids us to hope for."[32]

With funding from New Yorkers now unlikely, the observatory's proponents swung into action, focusing their efforts on the Common Council. Should an appropriation to erect a time ball controlled by signals from Albany be made, then, de facto, the Dudley Observatory would be the port city's observatory. Gould, sputtering against New York's educators, supported the trustees' lobbying effort by calculating the modest difference in longitude between Albany and New York, 150 miles away. Comparing this difference to those at European national observatories hard by their port cities, he showed that the Dudley's siting with respect to New York was similar. Although that relation had nothing to do with the issue, in the astronomer's view these "facts" spoke for themselves.[33]

The trustees in Albany and their supporters in New York failed, however, in this last-ditch effort. Within days of the *New York Times* editorial, Mayor Wood issued an order "to the bell-ringers at the different fire alarms," directing them "to strike the hour of nine every evening upon receiving a signal from Professor Bull, of the New York University, who will announce the hour from transit observations taken at the private observatory in Eleventh-street near Second-avenue."[34]

In his eagerness to aid the city's beleaguered universities by circumventing the Dudley Observatory's proposal, Mayor Wood chose well. The observatory was the one owned by the well-regarded amateur astronomer Lewis Rutherfurd. Located on the Stuyvesant estate in Manhattan and in operation since the late 1840s, it boasted two transit telescopes for time determinations: one mounted late in 1856 when Rutherfurd expanded his own operation, and the other owned by Columbia College. Both were close to the Coast Survey's "Stuyvesant station"—a point on the estate geographically (and geodetically) as well established as any in the country. Corrections amounting to a few hundredths of a second brought observations made at either transit in conformance with mean solar time along the City Hall meridian—had such refinements been necessary, which they were not. Thus, starting in January of 1857, Rutherfurd's Observatory became the first American site to supply uniform time throughout a metropolitan region. For this "great convenience," as the *New York Times* termed it, New Yorkers could thank the Dudley Observatory's trustees and Scientific Council.[35]

Furious at the turn of events, Gould decided to have the last word—a private one. Convinced that he knew the identities of those at New York University responsible for "the exquisite manoeuvre" that "defeat[ed] our time-project," the astronomer penned the following verses set to the meter of "Who Killed Cock-Robin?"

> Who'll toll the nine o'clock bell?
> "I" said Professor Bull
> "Because I can pull—
> I'll toll the nine o'clock bell!"

> Where'll we get the time?
> "Here," said Chancellor Ferris
> "On my University terrace—
> Here's where we'll get the time."

> Who'll make the necessary calculation?
> "I," said Prof. Loomis
> Who so unspeakably *dumm* is—
> "I'll make the necessary calculation."

No doubt Gould was amused by his doggerel; history does not record his feelings when his lines were printed in the newspapers for all to see.[36]

Failure, the outcome of this fifteen-month effort to penetrate New York City's public and railroad time markets, now became the astronomer's personal hallmark. For the next year and a half, he struggled with Dudley Observatory endeavors in astronomy, and his plans for public time services in Albany fell by the wayside. Gould, now observatory director, was dismissed in June of 1858. A month later, as time from Harvard College Observatory began reaching Boston's citizens via the fire-alarm bells, all he could do was grimace. His dream of a "Greenwich-on-the-Hudson" was in total disarray.

Despite his dismissal, the astronomer refused to vacate the site, and so the "public entertainment" documented in James's *Elites in Conflict* began. In January, escorted by police deputies armed with a judge's order, Benjamin Gould finally left the Dudley Observatory. Even though the facility was still not operating, advancing the science of astronomy was certainly not uppermost in the trustees' minds. With the turmoil so visible to Albany's citizens, their own reputations were at stake.

This time fortune smiled on them. Cincinnati Observatory director Ormsby Mitchel, the Dudley's new director, could not leave for Albany immediately, so he convinced Franz Brünnow to accept the position of associate director. Brünnow, already a distinguished, capable astronomer and director of the University of Michigan's Detroit Observatory at Ann Arbor, arrived around May. He plunged into the task of making good on all of Gould's unfilled promises.

In the area of public timekeeping, Brünnow supervised the construction and erection of a time ball on the roof of the capitol in Albany. Tested late in 1859, this public time signal became operational in January 1860.[37] Although the ball itself was not dropped automatically via a clock signal, that scarcely mattered. Its daily descent, with a precision good enough for all public purposes, signaled to Albany citizens that the Dudley Observatory was finally—to use Gould's earlier words—"contributing to the material welfare of the State."

By May a time ball was operating in New York City, too, its noon drop on local time controlled by a signal telegraphed from the Dudley Observatory.[38] It is tempting to conclude that this device embodied the Dudley Observatory's proposal made five years before; indeed, its trustees apparently never gave up on their intention of distributing time to the port city. But in fact the New York time ball can be linked to local business interests; specifically, its erection may have been a counter to J. N. Gamewell's new fire- and police-alarm system, a stratagem designed to keep his sales proposal from being considered by city authorities.[39] In any event, the New York City time ball was abandoned very soon thereafter.

Fig. 5.3. Time ball on the capitol, Albany, January 1860 to mid-1861(?). The ball was dropped by a noon signal telegraphed from the Dudley Observatory. From an original in the collection of the Albany Institute of History & Art.

Mitchel arrived in May to take up his duties. Two months later Brünnow resigned as associate director and returned to the University of Michigan. His heroic efforts, which led to the inauguration of both the observatory's astronomical and time-service programs, were promptly forgotten.

Timekeeping underwent several modifications during Mitchel's sixteen months as director of the Dudley Observatory. Hiring George Washington Hough as his assistant, Mitchel had him modify many of the instruments— among them Farmer's seconds-beating electromagnetic pendulum (now no longer called the Corning Clock). Hough made it capable of sending an electrical pulse automatically once every twenty-four hours. The time ball on the capitol roof was connected to this once-a-day circuit, as were bells in the state senate and assembly chambers.

Concurrently, a tiny time ball was placed in the shop window of one Benjamin Marsh, a prominent Albany jeweler. This, too, was controlled by the observatory's mean-time clock. When the task of arming the capitol's time ball prior to its noontime drop became a burden, Mitchel had the signal device discontinued. Jeweler Marsh's time ball then became the city's public signal, and remained so through 1867.[40]

Mitchel left Albany in August 1861 to serve in the Civil War as a brigadier-general of volunteers from Ohio. Attaining the rank of major-general of volunteers and subsequently placed in command of the Department of the South at Hilton Head, South Carolina, he died of yellow fever on 30 October

1862. Hough became the Dudley Observatory's director. He expanded its time service, transmitting a daily signal to the Albany fire-alarm telegraph office (1868), whose operators signaled the city's time on the fire-alarm bells at 9 A.M. and 9 P.M. The observatory's daily signal was also sent to Western Union's central office. Cities along the New York Central's Albany-Buffalo route received time via the railroad's own once-daily signals, which were based on observatory time transmitted to Marsh's store. Proud of these services (which of course reflected many of Gould's own ideas), Hough wrote: "To all practical purposes, the Dudley Observatory time is the standard for the state" —after the signals were suitably adjusted for differences in longitude. But, sad to say, the observatory's time service failed in its primary mission: generating funds for scientific programs.[41]

Farmer's transmitting clock system remained in service for more than two decades, a remarkable record for the country's earliest electrical time-distribution system.[42] Gould himself remembered this pioneering clock system with great pride. In a biographical notice published just before his death, the now-distinguished and successful astronomer counted both the (undelivered) sidereal timekeeper and the mean-time clock among the most important accomplishments of his career. Still ignoring facts, Gould went so far as to claim that "his" clock (actually built and put into operation by others) was the "clock that gave the time signals to New York."[43]

To the four American observatories providing public time services prior to the Civil War should be added the one established at the University of Michigan in Ann Arbor.

Franz Brünnow returned from Albany in 1860 to resume his duties as director of the university's Detroit Observatory. In the spring of 1861 he had the observatory connected to the commercial telegraph lines. Despite extremely heavy message traffic in the first weeks of the Civil War, Brünnow was able to determine the longitude difference between his site at Ann Arbor and Hamilton College's observatory at Clinton, New York. A further benefit was that he determined the longitude, in terms of the Greenwich meridian, for the Topographic Engineers' base station in Detroit proper. This was the "zero" point for the Survey of the Northern Lakes, and equal in importance to the Coast Survey's own zero point for electric longitudes, Seaton Station in Washington.[44]

Although Brünnow inaugurated public time services while he was at the Dudley Observatory, no evidence exists suggesting that he did so at Ann Arbor, despite having shown keen interest in establishing one for the city of Detroit before he went to Albany.[45] His sojourn there must have made him realize how much time would need to be stolen from research pursuits in order to maintain a public time system. It is more likely that James C. Watson, Detroit Observatory's second director, initiated the time service and in 1864 had the badly deteriorated telegraph wires and poles replaced. The observatory's time service remained operating for decades.[46]

Before the Civil War, American astronomers certainly demonstrated that time-giving was as difficult an undertaking as astronomy itself. And like research in astronomy, creating and maintaining a service required stable funding—or some other sort of subsidy. For these pioneers in network services, Benjamin Franklin's business advice, "Time is money," must have had a very hollow ring.

6. LOBBYING FOR TIME AND NEW TECHNOLOGIES

When the Civil War ended, no more than five astronomical sites were supplying accurate public time on a regular basis: the Naval Observatory in Washington, the Dudley Observatory in Albany, the University of Michigan's Detroit Observatory at Ann Arbor, Rutherfurd's Observatory in New York City—Richard Bull's source for railroad time—and possibly the Hudson Observatory near Cleveland.[1]

New York City's public time was again the province of local jewelers. One firm, Benedict Brothers on Broadway, actually advertised itself as "Keepers of the City Time." In Boston, the city continued to strike noon on "all the alarm bells [to communicate] . . . a uniform and correct time . . . to our citizens," but the Harvard College Observatory was no longer touted as the source of true time.[2]

The cost of purchasing and maintaining the apparatus for transmitting periodic time signals—sending them routinely, in contrast to supplying an occasional time check—must have deterred many observatory directors. However, manpower was the critical issue in this era of one- and two-man institutions. Up to half of a working observatory's most precious resource, its available observing time, had to be devoted to the task of public timekeeping. Every clear night several stars ("clock stars") were observed to determine the sidereal clock's error. After reducing the stellar data, one of the astronomers compared this master timekeeper with the mean-time clock, altering the latter to make it conform to the locale's true time. Every day without fail one of them adjusted the telegraph circuit, and at the proper moment transmitted the contracted-for signal.[3]

Nevertheless, offering time bearing an observatory's imprimatur was one of the very few ways for such institutions to secure operating funds, and during the half-decade following the Civil War, three of the country's new generation of observatory directors tried their hand at selling time. These three were Cleveland Abbe, Truman Safford, and Samuel Langley. All proposed time services that went far beyond the daily transmission of signals. Their planning initiated a transformation in public timekeeping for the United States.

In June 1868 Cleveland Abbe (1838–1916) arrived in the "Queen City of the West" to take up his duties as director of the Cincinnati Observatory.[4] Situated on a low hill overlooking this bustling river port, the observatory was in desperate shape. It lacked an endowment, and thick plumes of smoke from more and more house and factory chimneys were destroying the "seeing" (the steadiness of the atmosphere). More than a decade earlier, Ormsby Mitchel, the observatory's founding director, had warned members of the Cincinnati Astronomical Society that for any long-term program in astronomy, the telescopes must be moved. But Mitchel left in 1860, and in his absence the Cincinnati Observatory drifted into obscurity and neglect.

However, the facility's 12-inch equatorial telescope was still in good condition, and the observatory's very existence remained a source of pride to local citizens. Abbe's task was threefold: to convince civic leaders to champion one of the city's cultural assets, to secure operating funds, and to initiate a meaningful research program.[5]

Within thirty days of his arrival, Abbe drafted a memorial to Cincinnati's Common Council. Claiming that "much inconvenience is felt in the city from the want of standard mean time," the astronomer proposed that Cincinnati Observatory time be provided "as often as may be required." Abbe also informed his overseers that to discharge this important task he would need money for new equipment—a transit and a mean-time clock—and to pay an assistant's salary. But the Cincinnati Observatory's Board of Control could supply no funds, so Abbe's proposal went no further.[6]

The following February the astronomer wrote to the mayor, offering to provide the city with observatory mean time free of charge. Although pleased by Abbe's proposal, municipal authorities moved slowly; there were clearly costs associated with the observatory director's "free" offer. While waiting for their response, Abbe won a victory of sorts: the city's five principal jewelers all agreed to adopt "Observatory mean time as their standard," with the actual comparison of their timekeepers to take place once a month. Heeding the decade-old words of astronomer Elias Loomis, Abbe did not try to compete against the jewelers' business interests. As a result they were now allies in his battle for city financing.[7]

After two years, Abbe's lobbying efforts began to pay off: a special committee of the Common Council was directed to report on the issue of a public standard of time. Committee members solicited Abbe's views. Seizing this opportunity, the astronomer transformed the issue into a municipal need for a "legal standard clock." He already had a model clock system in mind, one that he had seen operating at Pulkova Observatory in Russia.[8]

Abbe's choice was a good one. The Pulkova Observatory's public time system had been constructed by James Ritchie & Son of Edinburgh, embodying the electricity-based clock control principle patented in 1857 by R. L. Jones, a railway station manager in Chester, England. Jones took advantage of the fact that a wire carrying an electric current generates a magnetic field and used that effect to keep pendulum clocks locked in synchrony to within fractions of a second. His invention solved many of the problems associated with electrically based master-slave clock systems and electrically actuated clock synchronizers. The Jones system became one of the best ways to distribute public time continuously in a metropolitan area.[9,10]

Abbe's formal response to the Common Council's inquiry was a model of completeness. He began with a list of current and proposed time services at American observatories and a few foreign ones. He quoted from an article describing how the installation at the Royal Observatory at Greenwich functioned. The article's author decried the city of London's lagging distribution of public time. The implication was clear: Cincinnati could do better.

Abbe attached correspondence from James Ritchie & Son and Astronomer-Royal for Scotland Charles Piazzi Smyth, the director of the Edinburgh Observatory, to document the Jones system's reliability. And he included a carefully reasoned budget. Capital costs were given as twenty-five hundred dollars for the purchase and installation of a Jones-system master clock at the Cincinnati Observatory, a subsidiary clock at the city's post office building, and all equipment needed to transmit noon signals that would activate the city's fire-alarm bells. Annual operating expenses were listed at $1,060, almost all of that being earmarked for the salary of an assistant at the observatory.[11]

Favorably disposed to Abbe's plan, the council's special committee submitted its report and recommended expending public funds on a clock system for Cincinnati. Subsequently, an order was placed with James Ritchie & Son. The clocks were tested at the Edinburgh Observatory and shipped to Cincinnati.[12]

Before the timekeepers arrived, however, Cincinnati Observatory was in crisis again. Late in 1869, the observatory's governing members found themselves unable to raise funds to purchase a new site. They informed Abbe that, for the foreseeable future, the observatory would have to limp along at its present location. Abbe resigned in late 1870 and soon after moved to Washington to take up a senior position with the U.S. Signal Service, which had just been given national responsibility for weather reports and forecasting. His days as a practicing astronomer were over.[13]

The shock of Abbe's resignation, coupled with the complete failure of the Cincinnati Observatory's governing board to establish stable funding, set the stage for the observatory's absorption into the city-supported University of Cincinnati. Once that had occurred, proper funding was provided by means of a property tax—an unusual way of supporting astronomy. The observatory's telescopes were moved to a new location outside the city, needed ancillary equipment was purchased, and a new director was hired. When, after 1875, the Cincinnati Observatory finally began the distribution of mean time for the public, its reliability was all that Abbe had promised.[14]

With the absorption of the Cincinnati Observatory by the university, the Queen City of the West continued its thirty-year love affair with astronomy. In contrast, the Queen of the Lakes did not even embrace astronomy until 1862, when a group of influential business leaders founded the Chicago Astronomical Society. They made no little plans, however, constructing and then equipping the society's observatory with what was then the largest refracting telescope in the world. Located slightly more than three miles south of the city's commercial center—on the grounds of the first University of Chicago and closely affiliated with that institution—Dearborn Observatory began operating early in 1866.

To direct this prestigious facility, the Chicago Astronomical Society selected Truman H. Safford (1836–1901), for many years assistant astronomer at Harvard College Observatory. Upon his arrival in Chicago, Safford inaugurated an observing program, and, as a professor at the university, taught courses in astronomy. A meridian circle—a telescope whose mounting orientation is identical to that of a transit telescope—was purchased in 1868, most of its cost, over seven thousand dollars, being fully paid by a former mayor who was a society member. Plans for using the meridian instrument included cooperative efforts in longitude determination, and, of course, observations for sidereal time. By the end of the year the telescope was in working condition.[15]

In sharp contrast to all other private observatories, there is no evidence that the Chicago Astronomical Society's leaders or director actually wanted to provide the city's public with Dearborn Observatory time. Apparently, the idea was thrust upon them; but they certainly made the most of it.[16]

Since 1847 the municipal government had appointed an official timekeeper. Indeed, in mid-May of 1867, Giles Brothers, a major Chicago jewelry firm, was once again designated the source of city time—at no cost to the city treasury.[17] Almost exactly two years later, however, "A 'Timely' Suggestion" appeared in the *Chicago Tribune*. Signed "J. C. D.," the letter began with the following remarkable assertion: "It is a notable fact in the present history of this enterprising city that its citizens have no regular standard by which to regulate time-pieces." As evidence the writer reported that timekeepers in watch dealers' stores varied "from fifteen to forty-five seconds, and in some instances one or two minutes, so that it is impossible . . . to determine what is the correct time."

Displaying an awareness of the Dearborn Observatory's new meridian instrument, civic booster "J. C. D." proposed that the observatory site be connected by telegraph wire to some building in the central business district. A time ball would be erected there, and a time signal from the observatory would regulate its drop. The writer, very likely the president of the Chicago Board of Trade, ended with a plea for support from the *Tribune*, arguing that his proposal would "further the interests of the 'champion' city."[18]

"A very good suggestion," seconded "Time" a week later. "There is scarcely a jewelry firm or railroad in this city, whose time agrees with any other, and they all . . . disagree with the city time as struck [by hand] from the Court House." Reinforcing the views of "J. C. D.," "Time" argued that this sorry state "should not be [the norm] in a city of the business and importance and public spirit of Chicago. Millions of dollars worth of business is decided by the Court House time, and yet there is sometimes five minutes variation in that time." "Time" ended his letter by asserting that local commerce depended so much upon correct time that "the City authorities should see to it that time is accurate." And so a campaign to have the Dearborn Observatory supply the public's time began.[19]

Safford was certainly aware of the difficulties associated with such an endeavor, but with city time now a public issue, the Dearborn Observatory director had to consider it. No doubt he discussed the situation with those leaders of the Chicago Astronomical Society who were paying his salary; he probably also consulted with Elias Colbert, one of the society's most enthusiastic members and commercial editor of the *Chicago Tribune*. Most likely he did not talk with executives running the Chicago-centered railroads, for in July railway superintendents and others petitioned the city's Common Council for "the establishment of a system whereby the correct time shall be furnished to the people, by means of a telegraphic connection of the city clock with the Dearborn Observatory." This proposal included an appeal for the appointment of a particular watchmaker as City Timekeeper.[20]

After reporting the submission of the railroads' petition, the *Chicago Tribune* printed two editorials. The first dismissed the proposal for an observatory-controlled time ball. The second questioned the petition, which proposed that subsidiary dials be installed at rail depots, with a clock at a central location to keep them in synchrony within fractions of a second and also control the striking of the hours on the Court House bell. Although the editorial writer supported the idea of a standard time for Chicago, he opposed the use of public funds to purchase these devices, asserting that time accurate to within five seconds was sufficient for all business purposes and available at little cost using a twice-weekly signal, by telegraph or chronometer transport, from the Dearborn Observatory. He also declared that the observatory's astronomers were "the most suitable person[s] to entrust with the regulation of the city time"; that is, one of them should be, de facto, City Timekeeper.[21]

The next day the newspaper printed a letter from Safford. The Dearborn Observatory was ready and willing, its director declared, "to undertake the regulation of the principal clocks in the city to uniform standard time." To do so, however, would require some sort of compensation, for he and his assistants—his pupils—were completely engaged in a scientific program and could not be spared even to undertake this worthy public service.

Safford went on to describe the great advantages of electrically regulated clocks and how easy it would be to connect an automatic bell striker to a central public clock. A fair distribution of costs among the interested groups— "the municipality, the general public, and railway and other corporations"— should certainly be feasible, he concluded. And though the observatory's board of directors was unwilling "to provide time for any single interest," Safford reiterated the importance of assuring "uniformity of time throughout the city."[22]

Having laid to rest both competing proposals, all that remained was to ensure proper compensation for the observatory. This matter came before the Common Council a month later when it received Safford's proposal on behalf of the Chicago Astronomical Society. Dearborn Observatory time signals would actually save the city money, the astronomer stated: The society was requesting only one thousand dollars a year for its time services, while the combined annual salaries of the watchmen assigned to strike the hours on the Court House bell totaled eighteen hundred dollars. And no other municipal costs were envisioned, for the fifteen hundred dollars needed for the equipment—a central clock and strikers—would be raised by a private appeal for funds.[23]

Since the Chicago Astronomical Society's proposal had already been endorsed by the city's Board of Fire and Police Commissioners, Common Council approval appeared certain. However, formal submission came too late in the legislative process, and the matter was held over. It was not until mid-April 1870 that the new council's members approved a resolution supporting the funding request, and the Chicago Astronomical Society launched its public appeal for funds to buy equipment.

Giles Brothers, seeing itself about to be superseded as the city's official timekeeper, battled to retain its position. In August it presented a petition, signed by a number of Chicago firms and individuals, offering to place an astronomical-grade timekeeper—one varying by no more than five seconds a month—in the rotunda of the Court House. The clock would be equipped with electromechanical devices to strike the City Hall bell every hour and drive subsidiary dials in city offices. Giles Bothers promised to maintain this two-thousand-dollar timekeeper "free of expense to the city."

The municipal authorities were now faced with a delicate choice: to approve a one-time capital expenditure of two thousand dollars or authorize an annual expenditure of one thousand dollars. The Common Council referred the Giles Brothers' proposal to its Committee on Finance.

Learning of the jewelry firm's proposal during August, Chicago Astronomical Society officials immediately stepped up their canvassing for equipment funds. Simultaneously, the society's secretary informed Chicago's Board of Public Works of their continuing interest in providing Dearborn Observatory time to the city.

Almost at once, the Board of Public Works decided that "it would greatly add to the usefulness of this [time standard] project if there should be erected . . . large dials showing the correct time at all hours." The board then informed the Common Council's Committee on Finance of the negotiations currently underway with the Chicago Astronomical Society. The Committee on Finance responded with a recommendation that Giles Brothers' petition be tabled. The jewelry firm's maneuver had failed.[24]

In November, the Common Council passed an ordinance directing that illuminated clock dials be erected on the cupola of the Court House, at a cost not to exceed two thousand dollars. Chicagoans would have the city's true time all day and all night, but that luxury came with a price: thirty-five hundred dollars in capital costs, and one thousand dollars annually.[25]

During the following ten months, the many actions needed to bring Dearborn Observatory time to the heart of the city were completed. The city's Fire Alarm Telegraph superintendent designed an electrically controlled bell-striking mechanism and then oversaw its construction and placement in the cupola of the Court House. In addition he had wires run to connect the striker directly to his office. A pendulum clock capable of automatically transmitting hour signals was placed there and came under the daily scrutiny of Dearborn Observatory astronomers. It was regulated to conform to the Dearborn Observatory's meridian, approximately four seconds earlier than that of the Court House. Starting in mid-January of 1871, hours were struck on the Court House bell.

In mid-February a sidereal clock made by E. Howard & Company arrived at the Dearborn Observatory and was placed in service. In June the Common Council appropriated three thousand dollars for observatory operating expenses and the city's share of the Court House clock. Also that summer a Howard tower clock purchased by the Chicago Astronomical Society was installed in the cupola of the Court House and set to time. Early in October the Dearborn Observatory received its first payment from the appropriation for providing Chicago time. (Still to come were the four large-diameter, gas-illuminated clock dials that had been ordered from England. As of early October they were on a train bound for Chicago.)[26]

More than two years had passed since "J. C. D." and "Time" had publicly suggested that Dearborn Observatory provide true time for Chicago residents. With this accomplished, the next phase in the Dearborn Observatory's plan was to set up a system of synchronized clocks for Chicago railroad depots and business establishments. Years before, Safford had termed such an installation "most worthy of a great city like Chicago." But the problem now was which technology to choose, for the Jones system of clock control had an American competitor. Presumably, Safford was considering the choices.

The Great Chicago Fire of October 1871 ruined all these plans. Although Dearborn Observatory was completely untouched by the conflagration, the Chicago Astronomical Society was devastated. The considerable assets of its primary benefactor, city banker J. Young Scammon, burned to the ground, and his insurance company was bankrupted by the magnitude of its fire losses.[27] Owing enormous sums, Scammon could no longer pay the director's salary. Safford's career at the Dearborn Observatory was over. (Safford became a wandering astronomer, engaged by the federal government for its Western surveys; eventually, he gained a professorship in astronomy at Williams College.)

In the aftermath of the Great Fire, Colbert became the force driving the observatory's endeavors. This committed amateur astronomer quickly seized upon the only financial opportunity available: the now-suspended time service. Writing articles and editorials in the *Chicago Tribune*, Colbert cajoled and pleaded, reminding readers that Dearborn Observatory remained one of Chicago's great cultural assets. By 1875 the observatory was supplying mean time, via the Western Electric Manufacturing Company and Western Union's wires, to Chicago railroads and businesses, but it received only five hundred dollars per year from Western Electric for its time signals.

When in 1879 astronomer George Washington Hough became the observatory's director, funding was still insufficient to pay him a salary. While Hough revived the research program, Colbert continued writing his editorials and maneuvering behind the scenes. Two years later the long-dormant contract for city time was renewed. Dearborn Observatory began sending a daily signal to the Fire Alarm Telegraph Office, then extended its service to provide continuous transmissions. The authorities paid two thousand dollars a year for this "Time Service of the City of Chicago."

Finally, in January of 1881, the future looked bright. The Dearborn Observatory had survived, and once again the Chicago Astronomical Society could pay an astronomer for his research. Even a program of facility rehabilitation and modernization appeared feasible. Unfortunately, it was a false dawn. Still lacking an endowment, the observatory remained almost entirely dependent on the income from its time service contracts.[28]

Samuel Pierpont Langley (1834–1906), the third astronomer of this period we will deal with, developed perhaps the most famous observatory time service of the nineteenth century. Part of Langley's fame stems from his skills at self-promotion in his history of the era, much of which was written after his appointment as secretary of the Smithsonian Institution in Washington. Langley not only emphasized his accomplishments but also claimed to have had great influence on fellow astronomers who established their own observatory time services. In fact, Langley's influence was minor at best. On the other hand, the astronomer's struggle to inaugurate a time distribution service did spark important developments in American timekeeping technologies.

Langley was appointed to the chair of astronomy and physics at the Western University of Pennsylvania (now the University of Pittsburgh) in 1867. In contrast to his contemporaries Cleveland Abbe and Truman Safford, Langley's professional skills were quite modest. He had served for a few months as an assistant under Joseph Winlock at Harvard College Observatory, leaving in 1866 to take a position at the U.S. Naval Academy. There he was in charge of the student observatory for one academic year. After leaving Annapolis, Langley came to Allegheny, a city directly across the Allegheny River from Pittsburgh and now part of it. There he was expected to devote almost all of his time to teaching—a requirement he apparently found distasteful. Among Langley's other duties was the supervision of the badly deteriorated and almost completely unequipped Allegheny Observatory.[29]

This facility must have seemed as depressing to Langley as Cincinnati Observatory would appear to Abbe six months later. But in the critical matter of funding, Langley was as fortunate as Safford at Chicago's prestigious Dearborn Observatory. William Thaw, an extremely wealthy Pittsburgh businessman, wanted the Allegheny Observatory to prosper. As observatory trustee, Thaw took steps that eventually released Langley from his teaching duties and allowed him to concentrate on research. More important still, Thaw's considerable influence with the Pennsylvania Railroad proved essential. Without it, the Allegheny Observatory's time service could never have started, let alone succeeded.

By the end of his first year at Western University, Langley had garnered sufficient private funds—most of them from Thaw—to repair the observatory building, refurbish its equatorial telescope, and to purchase a transit instrument, a sidereal clock, and a drum chronograph. Using this new equipment, and via a temporary wire strung to the Western Union Telegraph Company's line a mile away, Langley participated in determining the longitude difference between Allegheny Observatory and Harvard College Observatory. (This March 1869 expedition was led by Winlock, Langley's astronomy mentor, who provided the Cambridge observations in a two-week program of exchanges conducted under the aegis of the Coast Survey.)

This series of astronomical observations was probably the first activity for which Langley had an investigator's responsibility, and he may have felt somewhat ill at ease performing his share of the tasks. But in the matter of inaugurating a time service, Langley took splendid—and proper—advantage of his technical naiveté, writing for help and advice to almost everyone in England and America who had knowledge of the electrical distribution of time signals.[30]

Langley began his investigation of time distribution technologies for the observatory by considering both the direct control and the regulation of near and distant clocks. He knew that he must provide multiple times, for his anticipated customer was neither Allegheny nor Pittsburgh but the far-flung lines of the Pennsylvania Railroad.

The first response to his queries came from Edmands and Hamblet in Boston. One of the very few domestic manufacturers of electrical clock systems, this firm practically dominated the American market. In existence since 1862, Edmands and Hamblet manufactured many kinds of electrical equipment: telegraph instruments, James Hamblet's "Electro-magnetic Watch-clock," and the "Electric Plural Time Dial." This last product was a patented system that consisted of a pendulum clock and associated slave dials whose hands advanced via electrical pulses generated by the master clock.[31]

Writing for the firm, Hamblet offered this master-slave system, terming it suitable for a portion of Langley's needs. However, he emphasized the firm's strong interest in developing a new system for "controlling Clocks once in 24 hours [along the lines of] a plan that has been in use in England." Unfamiliar with the details of this English system, Hamblet added, "We will advise you as soon as we can learn anything that seems to be useful and *practicable*."[32]

Hamblet also described timekeeping practices in Boston, where jewelry stores and offices at the railroad depots were connected to the wires of the city's fire-alarm bells. He suggested that if Langley wanted to regulate Pittsburgh's time, the astronomer need only transmit a premonitory signal to the chief operator of the Pittsburgh Fire Alarm Office an exact minute before noon. This individual would then adjust the office clock so that its time would be true when he transmitted the noontime pulse to the city's alarm bells and to the telegraph sounders in the various private firms and offices. But given Langley's focus on railroad timekeeping, these suggestions regarding city timekeeping were of little interest. Moreover, putting them into practice would require lengthy negotiations with the city government.

Continuing his time-distribution study, Langley wrote to Hough at the Dudley Observatory. Hough described his clock system, which transmitted time signals automatically every hour to a jeweler in Albany who passed them to the New York Central Railroad. Hough, whose system had been operating since 1864, offered Langley copies of his plans and electrical diagrams.[33] Here again, Langley showed no interest, for this sort of system would give his own observatory only an indirect influence on railroad timekeeping.

Early in the summer Langley's study took on a sense of urgency. Railway superintendents in Chicago were completing their plans to have Dearborn Observatory time transmitted to various rail depots, including those used by the Pennsylvania. The railroaders' petition came before the Common Council on 12 July; Thaw and Langley were soon aware of the action.

In late August, Hamblet submitted his firm's proposal. His plan was an electrically based system to regulate all station clocks of the Pennsylvania Railroad, including those associated with the rail lines running to Chicago. Envisioned as the primary standard was a mean-time clock at the Allegheny Observatory, wired to a number of secondary clocks located at Pittsburgh train depots. (Since the roads composing the Pennsylvania Railroad used several different operating times, more than one secondary clock was needed.) These clocks were to be powered by weights, but their hands would be under the direct control of the observatory master clock and would advance only once a minute. Periodically, all message traffic on a railroad telegraph line was to be stopped, the observatory-linked secondary clock would be switched into the circuit, and synchronizer-equipped clocks at the local stations would be connected to the telegraph circuit. At a particular minute, the secondary clock would then transmit a pulse to reset the minute hand of all the local clocks.[34]

Of course synchronizers were nothing new in the world of horology, with Steinheil's patent now forty years old. Working in Boston, Hamblet perhaps even knew of the aborted Dudley Observatory plan to regulate railroad clocks with Moses Farmer's clock-control devices. But Hamblet's proposal was far more complex than anything that had been tried before. It was a time-distribution system able to regulate numerous distant clocks, even ones located hundreds of miles away, and to keep them in excellent agreement with the railroad's operating time. (While a Jones-type system could keep clocks in closer synchrony with a primary clock's time, the dedicated telegraph line required made such a system quite uneconomical for regions larger than, say, a city.) And though in essence no different from the manual synchronization of railroad timekeepers inaugurated by Charles Minot on the Erie nearly two decades before, Hamblet's proposal was a decided advance in American timekeeping.

The groundwork for the introduction of this very ambitious system of railroad timekeeping was being laid by observatory trustee William Thaw through his lobbying of the Pennsylvania's operating managers. Indeed, unifying all roads' timekeeping at Pittsburgh remained a viable approach, for Thaw had learned that the Pennsylvania roads running to Chicago had not yet entered into any time agreement with the Chicago-centered companies. In September the Pennsylvania's general superintendent responded to Thaw's proselytizing, inviting the Allegheny Observatory to submit a time-services proposal, with costs.[35]

This invitation—a request, actually—caught Langley off guard. Just back from a field expedition to observe a total eclipse in Kentucky, he was busy writing up a report of his observations; his teaching schedule that fall was a full one as well. And even though he had already sent mean-time signals from the Allegheny Observatory to a Pittsburgh jeweler, the transmissions had been only experimental trials.[36]

The astronomer had no choice, however. It was now or never. The very next day he wrote to Moses Farmer asking for a formal proposal. He also queried the Western Union Telegraph Company on the cost of a dedicated line from the observatory to its Allegheny office so that he could connect to railroad stations in Pittsburgh. He wrote again to Hough.

Busy with other interests, Farmer declined to bid. But he sent detailed diagrams of the required electrical and mechanical circuits. The inventor added that both Edmands and Hamblet and the Boston firm of E. Howard & Co. were capable of building the devices.

Hough offered his plans to Langley once more. He reminded the Allegheny astronomer that the Dudley Observatory's public time service involved only the transmission of hourly signals; clocks beyond the observatory proper were not controlled. For such a (Jones-type) system, Hough directed Langley to the Naval Observatory in Washington.

Langley's other American respondents claimed to have little or no knowledge of electrically controlled time systems, and urged him to write the leading experts: English astronomers George Airy and John Hartnup, and the firm of James Ritchie & Son in Edinburgh.[37] But it was now late October; Langley had run out of time. The astronomer took what he had and began to write. Already he had made one decision: to reject Edmands and Hamblet's untested plan for synchronizing clocks.

Set in type, Langley's "proposal . . . for regulating, from this Observatory, the clocks of the Pennsylvania Central, and other Railroads associated with it" carried a 1 December date and a blank address line. The eight-page pamphlet was little more than a tutorial to which the astronomer attached some tentative ideas for Allegheny Observatory time services; three pages were devoted to the Jones system of clock control, including an illustration copied from Jones's 1857 patent description.[38]

With copies of the pamphlet in print, Langley finally responded to the general superintendent's long-standing invitation to submit a proposal. The observatory director asked for more time, since information from England had not arrived. To show progress, he enclosed a copy of the pamphlet "in advance of the [promised] proposition," as a place-holder for the Allegheny Observatory.[39]

Langley and observatory trustee Thaw had a far better reason to stall than simply lack of information from abroad. In November the astronomer had written to E. Howard & Co. of Boston. The firm replied that "Mr. E. Howard will be in Philadelphia next Saturday and will leave for your place on the following Monday [6 December]. We would like to have a decision deferred till he consults with you. The work must not go out of the Country if it is possible to prevent it."[40]

No other record of this critical interaction has been located; however, the outcome is known. No Jones-system clocks from Great Britain were purchased, even though to do so was the logical choice, considering the dozen years of experience embodied in James Ritchie & Son's systems. Edward Howard must have been most persuasive when he met with the Allegheny Observatory's principals!

One day after penning his letter to Pennsylvania's general superintendent, Langley wrote E. Howard & Co. He asked for a cost estimate for a master clock and secondary ones, to be controlled on the Jones principle. The firm replied at once, also notifying Langley that it was sending a mean-time transmitting clock "on Trial."[41] And thus Edward Howard, eager to penetrate a major market, took a small first step by lending a timekeeper to an observatory that was destined to be coupled to one of the country's largest railway systems. (Almost immediately afterward, Howard took a second step, for in 1870 Hamblet dissolved his partnership with Edmands and joined E. Howard & Co. Over the next half-dozen years Hamblet's electrical expertise guided the design and installation of all the firm's transmitting clock systems.)

The loan of a mean-time transmitting clock allowed Langley to do away with the manual keying of time signals and develop a proper observatory time service. Once installed, the Howard pendulum clock generated seconds pulses continuously in an electrical circuit that went, via a dedicated telegraph wire, into the heart of Pittsburgh. There the line branched to Morse sounders located in city jewelry stores. The time "ticks" were coded in such a manner that the start of the minutes and the hours could be easily identified. For this service, Allegheny Observatory received a total of five hundred dollars annually from these firms.

Negotiations on selling time to the railroads continued, but progress slowed: a new chief operating official had been appointed, and he was busy with more pressing matters. Thaw continued his lobbying. Finally, General Superintendent A. J. Cassatt issued General Order No. 4, which announced that time for Pennsylvania Central Railroad operations would come from the Allegheny Observatory beginning on 1 February 1871. Langley acquired yet more equipment—designed by Hamblet and paid for by Thaw—and by mid-1871 the Allegheny Observatory Time Service was in full operation, with the observatory collecting one thousand dollars annually for its labors.

The time distribution system that the Pennsylvania adopted was a hybrid, based on technologies developed by an earlier generation of electrical engineers and clockmakers: the reception of automatic time signals coupled to the manual setting of distant clocks.[42] Using the observatory's time was a simple matter, with ticks sounding in an office of the Pennsylvania in Pittsburgh, where two railroad telegraphs converged. One of these message lines ran east to Altoona and other railroad stations, terminating in Philadelphia; the other ran west through Columbus, Ohio, and on to Chicago. At specified hours, the railroad's chief telegraph operator connected the observatory line to one of these two railroad telegraph wires, allowing the time ticks to pass. Comparing the telegraphed signals with their own displays, station agents then manually adjusted their local clocks. After five minutes, the chief telegraph operator broke the connection to the observatory's clock, and regular message traffic resumed.[43]

The master clock installation at the Allegheny Observatory reflected none of the complications associated with the Pennsylvania's three operating times: the local times of Philadelphia, Altoona, and Columbus. Necessarily, and of no import in rail operations, these times were rounded to the nearest minute, for the observatory clock, which transmitted seconds continuously, signaled the start of each minute in terms of its own meridian. No secondary clocks were directly synchronized, but part of the installation was what Hamblet and Langley termed a "Journeyman clock"—a time-switch that opened the circuit every hour for a brief period so that the exact second when the hour began could be identified.[44] The overall combination of mechanical and electrical devices was not really a technical advance in time distribution, but Langley always termed the approach "the Allegheny system of electric time signals."[45]

With a time system now operating well, Thaw continued to lobby the railroads, while Langley focused on other possible customers. In 1873 another cluster of the Pennsylvania's railroads agreed to use time from the Allegheny Observatory, and the corporation began paying an additional thousand dollars per year for the service. Also that year the city of Pittsburgh agreed to purchase its official time from the observatory, for one thousand dollars annually.[46]

Langley was proud of his accomplishments, and rightly so. Starting with scarcely any technical knowledge, he had established a timekeeping system that was generating thirty-five hundred dollars in annual receipts—an amount greater than any other American observatory would ever receive.[47] Moreover, because Allegheny's signals were used by one of the country's largest railway companies, they were being propagated across a considerable fraction of the United States: to New York, Philadelphia, and into New Jersey on the east; northward to the shores of the Great Lakes; and to Indianapolis and Chicago in the west. Langley even confided to a colleague many years later that establishing the service "was the first thing I did of any consequence."[48]

Having a paying time service eased Thaw's financial responsibility for the Allegheny Observatory; and Langley was free to take on more research. Partly to keep the astronomer at the university and also to reward him for his income-producing success, Thaw directed that all receipts from the time service be assigned to the observatory, and that the astronomer should personally receive one thousand dollars annually from the gross. Langley, who no longer taught courses, was now the university's highest-paid professor, with an income greater than the salary being paid to the chancellor.[49]

As time passed, Langley's opinion of his accomplishments grew. In 1871 he had noted that "other observatories . . . have before employed not dissimilar means" and that his own part in introducing these signals at the Allegheny Observatory had been "less the contribution of any novel device, than an adaptation of what seemed the best features of plans in use abroad." But in 1884 he was writing that "previous to . . . [1869] time had been sent in occasional instances from American observatories for public use, but in a temporary or casual manner. The Allegheny system . . . is believed to be the parent of the present ones . . . in that it was, so far as is known, the first regular and systematic system of time-distribution to railroads and cities." Langley's oft-repeated statements surely angered some astronomers, and echoes of a long-standing priority controversy were heard at the time of his death.[50,51]

Justifiable pride certainly colored Langley's words. But the changing environment for mean-time signals may have figured in the astronomer's later exaggerations—ones proffered not just to safeguard his share of the time-service's revenues but also to perpetuate the Allegheny Observatory's research program. In the dozen years between the two statements quoted here, three issues had come to the fore: uniform time for the public, competition from non-astronomers, and new electrical technologies for timekeeping.

PART III.   PROMOTING A NATIONAL VIEW OF TIME (1869–1881)

7. ABBE'S ROAD: UNIFORM TIME

By 1870, visionaries were exploring the idea of a single time for civil purposes. For some of them the driving force was money, and a few actually gained an economic benefit from their labors, even though their advocacy tracts had no influence on public timekeeping. Eventually, though, portions of these early concepts were linked to become the country's system of uniform time.

In the first years after the Civil War, almost no one was interested in altering public timekeeping. Local time and uniform time were one and the same in many communities, for the daily ringing of the fire-alarm bells ensured a high level of consistency among clock dials. As in the antebellum era, one might see a town clock that displayed a time different from that of its neighbors, but now that would be cause for repair. Yet in many towns and cities—even those with municipal alarm bells—one timekeeper was never in synchrony: the clock at the railroad station.

Certainly by 1870 this "problem" was the country's norm. Over four hundred railway companies existed, and their superintendents had assigned more than seventy-five different times to control the running of trains. In 1870 the United States had more than fifty thousand miles of railway track, half of it laid in just the previous twelve years. By 1882 work gangs would lay an additional fifty thousand miles, connecting more and more towns and bringing the railroads' times to them.

Reducing the number of railroad times did not seem feasible. As early as 1852 an industry writer proposed that New York City time be telegraphed everywhere, as the "first [sole] meridian of railroad time," citing increased efficiency and safety as near-certain benefits from a consolidation of the companies' operating times.[1] But as no collisions resulted from these scores of overlapping and arbitrary times, no incentive existed for railroads to alter their timekeeping "system."

The public was quite aware of multiple railroad times. In 1857 the editors of *Dinsmore's American Railroad Guide* published a "Comparative Time-Table." Based on longitude differences, the table listed local times of 102 American cities and towns at the moment of noon in Washington. (Supposedly a traveler could use the "Comparative Time-Table" to decipher railway schedules, but actually this information was useless without specific knowledge of company operating times.) The comparison table cost pennies to prepare and never had to be revised, so it was reprinted again and again. Publishers of competing railway guides noted this "useful" filler material and simply added their own listings of local times.

One 1868 guide prefaced its table with this remark: "There is no standard 'Railroad Time' in the United States or Canada, but each railroad company adopts independently the time of its own locality or of that place where its principal office is situated." As for reducing the number of times, the guide's editor suggested that "there is but one remedy, and that a simple one, viz: the adoption of some central point, Washington or Pittsburgh for instance, as the [railroads'] standard." Of course, a schedule of the route to that remedy was not included in the guide, for changing times was not an editor's responsibility.[2]

A few post–Civil War railway guides did include some operating times; however, what got published varied from one railroad company to another. The caution in an 1857 guide remained good advice: "To avoid disappointment"—a missed train!—it was best to "inquire at the hotel by which particular time" a certain rail line operated.[3]

No doubt so much exposure to multiple railroad times made America's late-1860s railway travelers cautious. When they arrived at a station to begin their journey, they probably rushed up to the ticket counter and demanded, "What time does the train leave?" Seasoned travelers would certainly have added, "By which clock?" for these multiple operating times were all displayed.

Years of familiarity, however, bred acquiescence. Surely these long-standing differences in railroad times must have annoyed some passengers. But no evidence suggests very many complaints, much less a ground swell of indignation as the railways grew. Asking for the time solved a traveler's immediate problem.

Nevertheless, suggestions continued for some way to resolve the issue of multiple railroad times. One of these appeared in an 1868 letter printed in *Scientific American*. Written by a Cleveland subscriber who viewed the country's rail operations as a specialist task, the letter proposed that the world's rail-

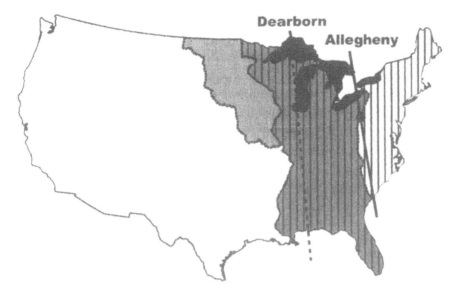

Fig. 7.1. Map showing the boundaries and time meridians of the railroad time zones proposed by Truman Safford (Dearborn Observatory) in August 1869 and by Samuel Langley (Allegheny Observatory) in February 1870.

roads and telegraph companies adopt a time distinct from all local times. "Suppose, then, that Greenwich be the standard time as well as the calculating point for longitude," this unknown prophet continued—thereby posing a specific solution more than a decade before it became fashionable among the country's scientists and engineers.[4]

Astronomers also began to write about multiple operating times. Perhaps the first to do so was Truman Safford of the Dearborn Observatory. Safford noted that Cincinnati, Columbus, Buffalo, and New York City times were being used by the railroad companies operating within a large region of the United States, and with some justice claimed that so many different operating times left people "much puzzled by the railway guides." Safford proposed, "It would be quite practical and very proper to run all trains west of Buffalo, Pittsburgh, or Wheeling, and east of the Missouri river, by the central Chicago time." Not surprisingly, this time happened to be the standard that he hoped to provide to the fifty-six Chicago-centered railroads.[5]

Safford's 1869 suggestion evoked memories of 1848, when New England railroaders voted to adopt a single operating time for running the area's trains. Having come from Cambridge, Safford may even have known that most railways there soon drifted away from that initial meridian and adopted Boston's local time as their operating standard. In any case, the Chicago meridian running through one of the country's centers of commerce must have seemed to

Safford a quite proper and very practical choice. Equally important, Safford's proposed railroad "zone" had natural boundaries: rivers along one side, and the termini of several important rail lines at the other.

The proposal had one downside, however. It encompassed an enormous area, some sixty-eight minutes in breadth.[6] How Safford might have worked with the affected railroad companies to gain an agreement is impossible to predict. All hopes of adopting Chicago's local time as the single operating time throughout the Midwest region ended with the Great Chicago Fire of October 1871.

Samuel Langley of the Allegheny Observatory started his uniform time campaign early in 1870, using both the public and the technical press to advance his ideas. Already in the midst of negotiations with one division of the Pennsylvania Railroad, he urged the adoption of "*one* railway time through the part of the country [that lies] east of the Mississippi."

Langley extended the utility of uniform railway time by arguing that "no reason exists why this [one] time may not be adopted by the whole community, so that the city clocks in New York and Pittsburgh and Chicago shall indicate this same hour." Of course, that hour was to be defined in terms of the meridian of the city of Pittsburgh (actually the Allegheny Observatory's meridian, offset by a few seconds).[7]

Langley's model for uniform time was the Royal Observatory at Greenwich, with its program of telegraphed mean-time signals. But his argument for one time for other civil purposes was based on a misunderstanding of public timekeeping in Great Britain. "Throughout England [and Scotland] the trains on every railway are run by one and the same time," he wrote, and claimed that "the clocks and every body's watch in the island is [*sic*] already set or soon will be to the same hour." The astronomer simply did not know that in 1858 the English courts had ruled that a locale's civil time was its local time, not the Greenwich meridian time that was being telegraphed to it. The "tight little island" was still in the throes of changing to uniform time—and the process would continue for another ten years.[8]

In addition, Langley did not mention that Great Britain's single time zone was only thirty-five minutes in breadth, while his proposal embraced an area ninety-two minutes wide; he would have found it hard to explain away this enormous disparity between his (flawed) English model and American reality. And he felt that adopting a single civil time was an evolutionary process: the area's railroads would initiate the change and slowly the public would follow. Langley did not foresee any great difficulties, asserting that what was a fundamental change for residents in the proposed time zone would lead to no more trouble than what would arise "from having to correct the almanac for the time of sunrise."

In 1870, the idea of one time for all public uses was a radical notion, indeed. Even today, it is difficult to view Langley's proposal as anything more than an innovative sales pitch for the Allegheny Observatory's yet-to-be established railroad time service. And once it was operating, Langley sold his customers what they wanted: Philadelphia, Altoona, and Columbus times.

Langley's position on civil time seems inseparable from his railroad time activities. While the Pennsylvania Railroad's eastern and western divisions debated adopting one operating time to replace the three times then in use, Langley wrote of "some single time for all of the country east of the Mississippi, by which not only the railroads, but cities and the public generally, will regulate themselves." Yet the astronomer may have been something of a visionary. Even when the Pennsylvania's roads decided to use only Philadelphia and Columbus times, he expressed the hope that the two operating times "will ultimately be exchanged for one"—a change unlikely to affect his revenues. Still, the issue of one time for all civil purposes was moot; Langley's customers had made it so.[9]

Both Truman Safford's and Samuel Langley's proposals received favorable notice in the *Railroad Gazette*, an important industry weekly. However, the *Gazette*'s perspective on railroad time was independent of the astronomers' efforts. Instead, its editors proposed using the "local time of some central city, [say] St. Louis" as the one by which to operate the entire country's railroads. Like Langley, these writers also believed that if the railway companies adopted a uniform time, all others involved in public affairs—those setting the official times of cities, say—would have no choice but to follow.[10]

In spite of the *Railroad Gazette*'s ties to the industry, its proposal for one operating time nationwide fared no better than the astronomers' more parochial appeals for regional agreements. The *Gazette*'s April 1870 editorial produced a crop of letters to the editor, but in the ensuing decade no further mention of the subject appeared. It seemed that, publicly at least, those company officials actually responsible for the running of trains were not interested.

Ironically, within six months of the *Railroad Gazette*'s editorial, one railroad official was hard at work derailing the most significant uniform time proposal ever to come before the industry. He was A. J. Cassatt, general superintendent of the Pennsylvania Railroad, and the plan that he opposed was outlined in a pamphlet prepared by Charles F. Dowd, president of Temple Grove Seminary for Women in Saratoga Springs, New York.

Dowd first articulated his uniform time ideas in late 1869, when he spoke to a group of railroad officials meeting in convention in New York. They expressed interest and suggested that he develop a plan. Dowd returned home, began a set of calculations, and in September 1870 published his *System of National Time*.[11] His document deserved the industry's careful attention. In contrast to all other visionaries, Dowd, a schoolteacher, had tested his concepts using actual data: the railways' routes and the locations of their passenger stations.

Dowd began his study of multiple railroad times with the idea of one time for all companies. Invoking the notion that "Longitude and Time are convertible" and boldly asserting that "with respect to time, longitude is as important on the land as upon the sea," Dowd was drawn inexorably to the Washington meridian as the proper basis for railroad time.

Had he stopped there, the educator would deserve no more than a footnote.[12] But after calculating the time difference between Washington and every station served by the country's five hundred rail lines—a task that required looking up and converting the longitude of eight thousand places—Dowd discovered that they varied dramatically from one minute to almost four hours. Corresponding with railroad officials, Dowd became convinced that it was impractical to create a time system in which differences in time greater than an hour were required. Thus he conceived the idea of "hour sections" to span the country, defining within each section a meridian located an integral hour away from "National Time" at Washington; he termed these four meridians *Railway Time*(s)." Using this concept, Dowd turned to the preparation of his document. As part of it he included a map of the United States divided into hour sections, with the time-defining meridians shown.

Ninety-three pages in Dowd's 107-page *System of National Time* was a gazetteer, still in skeleton form. Organizing the gazetteer alphabetically by company, Dowd listed all train stations, assigning two indexes to each one: the hour section in which the station lies and the number of minutes that its corresponding Railway Time was slow or fast of the station's local time. Boston, for example, was contained in the "First Hour Section," so its gazetteer listing was ^{0}Boston^{-24}, Railway Time there being twenty-four minutes slower than Boston time. Chicago, in the "Second Hour Section," was given as ^{1}Chicago^{-18}. Similarly, ^{0}Pittsburg[h]$^{+12}$, ^{1}Omaha^{+16}, and ^{3}San Francisco^{+2} had their own time indexes.

At first glance, this way of presenting information may seem forbidding. Dowd emphasized the special nature of railroad timekeeping, a use distinct from the general public's. His solution for reducing travelers' confusion was to create a list showing every station location in the United States with two times: its local time and its Railway Time.[13]

Actually, Dowd's publication was rather easy to use. At the start of his journey, a traveler would open the gazetteer section to his station. Looking up that station's indexes, he would reset his watch, changing its display from local time to National Time. Now, and throughout his journey, the traveler could compare his watch with the railway's schedule without worrying about

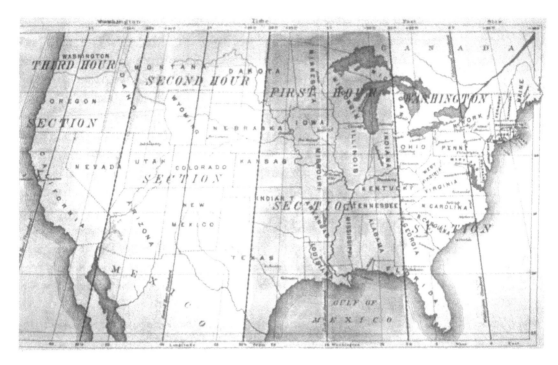

Fig. 7.2. Charles F. Dowd's 1870 map showing the United States divided into four railroad time zones, with the time in each differing by an exact hour from that of its neighbors. The "zero" meridian is the nation's capital.

different operating times—since all companies would be on the same basis. Arriving at his destination, the traveler would consult his gazetteer again, and taking the index numbers for that place, set his watch to the local time.

Dowd's map and gazetteer are compelling evidence that the concept of hour-difference zones was his alone. His railroad time plan was national in scope—another first. Dowd had confronted the realities of railroad time-keeping within a country almost four hours wide.

So far, however, the *System of National Time* was little more than an expansion of the one-page, local-time comparison tables already common in the railroad guides. In order to fill out the pages of his gazetteer with the roads' timetables, Dowd needed the railroad companies to accept his concept of Railway Time(s). Then he would indeed have a "National Railway Time-Table," its entries simpler and easier to use than those of any existing railroad guide. Moreover—or so Dowd thought—if railway superintendents used the National Time system that he had created, the railroad companies "would no doubt be willing to pay him for it."[14]

In October of 1870 Dowd distributed his bound pamphlet to the industry, along with a "practical" testimonial and a "scientific" one. The first testimonial listed two dozen railway superintendents, who wrote that the educator's plan "seems *simple and practicable*, and worthy of the attention of Railroad Managers." Among the five who signed the second, scientific testimonial, were four astronomers—and all agreed that Dowd's plan "is based on *correct scientific principles, and is commended* to the attention of Railroad Managers."[15]

Dowd sent his materials to Cassatt, asking for the railway official's support at a forthcoming convention of trunk-line managers. Cassatt's reply was noncommittal. Soon after, in a note to Samuel Langley's mentor, William Thaw, Cassatt wrote that he wanted a "Pennsylvania institution" to gain the credit for introducing such an innovation. More than likely this influential railroader then blocked discussion of Dowd's plan at the convention, held a few days later. Early the following year Cassatt announced the Pennsylvania's decision to use the Allegheny Observatory's time services. Charles Dowd never suspected a link between the railroad convention's October 1870 rejection and the Pennsylvania's January announcement. Unlike Langley, the Saratoga Springs educator was an outsider.[16]

Having failed to gain a hearing from the trunk-line managers, Dowd began what became years of toil to sell his proposal to the country's regional railroad associations. In 1872, in response to one railway group's criticism, Dowd switched his plan's basis from Washington's meridian to one exactly five hours earlier than the Greenwich meridian. This alteration shifted the hour zones eastward by eight minutes, making their boundaries conform more closely to the country's rail lines. But doing it gained him nothing. A later proposal (1877), with railway zones based on the New York City meridian, also failed to win industry endorsement.

By 1873 Dowd was fading into obscurity. A major railroad association dismissed his plan with the remark that "the disadvantages the system seeks to avoid are not of such serious consequences as to call for any immediate action on the part of the railroad companies." This association recommended that "the question of uniform standard national time be deferred until it more clearly appears that the public interest calls for it," and further noted, quite correctly, that "the great body [of the public] travel only short distances, and to them the proposed uniformity is of little or no importance." Left unsaid, of course, was the fact that railroaders saw little need for greater time uniformity. With some forty years of railroading experience, most professionals were convinced that multiple times were good enough for the safe running of trains.[17]

Also left unexplored was the question of whether two times were actually needed everywhere—an absolute requirement in Dowd's eyes. By 1873 many cities and towns hard by railway lines were already using the railroads' "invented" times. For these citizens, using one time for all civil purposes was far simpler than Dowd's two-time plan. Why change back? Of course this simpler approach meant that often a town's official time did not square with its longitude. But who cared?

An answer came soon after 7 April 1874: geophysicist and sometime-astronomer Cleveland Abbe cared. On that date an aurora borealis lighted the night sky, and the event heralded Abbe's interest in national timekeeping.

Numerous people across the United States—from Eastport, Maine, well into Indian territory west of Minnesota—saw this spectacular and intense display of northern lights.[18] Responding to Abbe's official request, twenty Signal Service offices and eighty volunteers submitted reports. From these raw observations the scientist toiled to determine the aurora's height above the earth—an unknown quantity then—and attempted to correlate specific features with concurrent weather observations and magnetic measurements. Abbe wrote in his summary report: "The errors of the Observers' clocks and watches and *even of the standards of time used by them* [emphasis added], are generally not stated . . . so that the uncertainty of this vitally important matter will be found to throw obscurity upon some interesting features."[19]

The volunteers who responded were members of the corps of meteorological observers, a group established by the Smithsonian Institution in 1848 and recently transferred formally to the Signal Service. The written instructions to volunteer observers had not been revised for at least twenty years. Underlying them was the assumption that a locale's civil time was its local time, and thus civil time could be calculated from the longitude. But, as noted above, these relations were no longer always valid.[20]

Moreover, the volunteers' reports varied in quality. Few research scientists would have wasted their time trying to identify and correct the exact time biases in the eighty sets of observations. Abbe did not. But he took measures to prevent such reporting inconsistencies from happening again.

In considering the path that Abbe took, it is important to note how the Signal Service's own observers—soldiers assigned to locations around the country—did their work. These men had been trained to take their observations in specific sequences. The climatological observations were to be recorded at local times consistent with the Smithsonian Institution's extant instructions to the volunteers, and additional ones were to be made at noon and at the exact hour of sunset. Three other sets of observations, determined everywhere at specified Washington mean times, formed a *simultaneous* group. They were telegraphed by cipher as rapidly as possible to the Department of the Army's Signal Office in Washington for the development of the country's daily weather forecast. Abbe stressed the need for simultaneity in these latter observations, contrasting them with *synchronous* observations: those taken at the various stations across the country at the same (respective) local times. The Signal Service, whose clock telegraphed the Naval Observatory's Washington time to its stations, controlled the accuracy and consistency of these official observations by a rigid system of instructions and periodic inspections.[21] How surprised and upset Abbe must have been when he learned that some of the observer-sergeants were using railroad times for the supposed simultaneous observations![22]

Abbe's aurora study uncovered a previously unrecognized problem: Observers were obtaining their location's "mean time" from the railroad's clock—an accurate clock, but not necessarily displaying a time related to their longitude. Any set of observations was thus fundamentally in error, for it was neither synchronous nor simultaneous. Only with great difficulty could the raw observations be adjusted. Worse still, only these local observers, "who chance to be favorably located," could provide the sort of network needed for Signal Service studies of "atmospheric electricity, auroras, thunder storms, earthquakes, [and] meteors."[23] So the timing errors in the records would vary in an inconsistent manner.

Clearly, this issue of multiple times had to be resolved. Abbe's first step was to have new instructions prepared for the corps of volunteer observers. Next, he wrote to the American Metrological Society.[24] The society, founded late in 1873 by a group of educators and government scientists, debated various uniformity issues. In his May 1875 letter to its president, Abbe urged "action by the Society to secure the adoption of a uniform standard of time." Responding, the society established the Committee on Standard Time and made Abbe its chairman. That same year a draft report on the subject was prepared. It was returned to the committee.

Years passed with no Committee on Standard Time activity recorded in the society's proceedings. Then, in May 1879, the society approved for release its "Report on Standard Time," the key document in the process leading to uniform time for the United States.[25] Abbe began it with a discussion of accurate time. Then he addressed time uniformity, listing and charting the seventy-five standards used by the railroads, pointing out the advantages of adopting a much smaller number. He urged the immediate adoption of no more than five standards for North America—to be known as "Railroad and Telegraph Time." This set, indexed to the Greenwich meridian and differing by exact hours from each other, would be the first step toward the adoption of a single national standard of time for the United States. (Abbe's personal choice was a meridian through the Mississippi Valley, six hours earlier than Greenwich.)[26]

The Signal Service scientist further proposed that cities and towns discontinue the use of local time, adopting instead the time standard of their area's principal railroad. Implementation would come via the railroad and telegraph companies, which "exert such an influence on our every day life . . . that if they once take that step towards unification which they have been talking about for years, then every one in the whole community will follow."[27]

Adopting one time system for all civil purposes was not a new idea. What was new was Abbe's proposal that the national government be persuaded to make these new time standards legal for general use. To increase the society's influence, Abbe recommended that certain people be invited to join it to help effect the changes. He listed several astronomers, the presidents of two of the country's major telegraph companies, and "Mr. W. F. Allen, Sec'y Gen. Time Convention of R.R. officials."

The American Metrological Society worked at a leisurely pace: Two years after approving its report, it had not yet sent out the requested membership invitations. Abbe's own pace quickened, however. With society approval secured, he prepared the "Report on Standard Time" for publication; began a fruitful interaction with Canadian railway engineer and fellow uniform-time propagandist Sandford Fleming; laid plans for an international conference on uniform timekeeping; and proposed a standard time committee under the auspices of the American Association for the Advancement of Science. In October of 1881, he sent a copy of the report to a railway association.[28]

Many roads were leading to time uniformity now, for other scientists had entered the discourse as well. One branch of the route that Abbe was blazing led first to a public conflict with the Naval Observatory and then petered out soon after the International Meridian Conference. Nevertheless, Abbe's road, constructed to ensure the simultaneity of geophysical observations, ultimately led to the adoption of Standard Time in the United States.

8. SHAPING A NATIONAL TIME CIRCUIT

Cleveland Abbe eventually opted for a standard of time indexed to the Greenwich meridian. But in 1876 he was urging the Signal Service's volunteer observers to "adopt a uniform standard of time, for which Washington time is preferable, since it may always be obtained by telegraph from the Naval Observatory, and affords the most convenient standard for the United States."[1] By the year's end, a plan to have citizens receive Washington time directly—instead of their having to telegraph for it—was being pursued vigorously. This turn toward easier access may have come from Abbe's proselytizing in Washington, for steering the effort was Vice Admiral Charles H. Davis, the Naval Observatory's superintendent.[2]

Charles Henry Davis (1807–77) was a well-respected naval scientist whose training in mathematics came from Harvard's Benjamin Peirce, his brother-in-law. Davis's skills in geodesy were the result of a seven-year posting to the Coast Survey. In 1849 he was placed in charge of the U.S. Navy's Nautical Almanac Office, a new organization responsible for developing astronomical tables for mariners and other users. In 1863 Davis was a key player in the founding of the National Academy of Sciences. In 1876, his objective was an alliance with the Western Union Telegraph Company.

Western Union had used Washington time before.[3] Astronomer William Harkness arrived in Washington during the Civil War, and according to his account, recommended in 1864 that the Naval Observatory's time signals be transmitted to the city fire-alarm office. Then in 1865, a telegraph operator in the building housing the State Department began retransmitting the time to Western Union's Washington office, where signals were sent to various points in the city. Somewhat later the company sent the daily signal to mid-Atlantic coast cities, to the Baltimore & Ohio Railroad, and to local offices of the country's Southern railroads, whose operators telegraphed Washington time as far west as New Orleans and Texas.[4] The Naval Observatory often asserted the importance of these transmissions, although they were being sent merely as a convenience to Western Union's and the railways' telegraphers.

In 1873 Western Union began telegraphing the Naval Observatory's daily signal to New York City. There it was retransmitted to states in the west and to New England. This new activity worried Harvard College Observatory director Joseph Winlock, whose time service was scarcely two years old. Reportedly, Admiral Davis assured Winlock that the telegraph company's increased coverage "would not injure the interests of those private observatories [which were] supporting themselves" by selling local time.[5]

Despite Western Union's expansion and in sharp contrast to an earlier superintendent's grand claim that "there is scarcely a train whose movements are not regulated by the Observatory clocks," the Naval Observatory's time service was still regional in 1876. Business traffic always took precedence on telegraph companies' lines, so the observatory's time signals were not sent regularly to inland cities. This limitation on the distribution of Washington time impeded Abbe's efforts to promote uniform time.[6]

In Washington itself, staffs in the military departments were always on the lookout for ways to publicize their peacetime activities. For example, in the spring of 1876 Admiral Davis wrote his superiors urging the erection of a time ball on the roof of the Government Building at the Centennial Exhibition in Philadelphia. But his attempt to give the many thousands of visitors to the exhibition visual proof of the Naval Observatory's daily contribution to their welfare failed. In the fall the superintendent charted a different course, one designed to expand the observatory's activities.[7]

The superintendent's handiwork can be seen in the inaptly titled, "Report upon the astronomical instruments of the Loan Collection of Scientific Instruments at the South Kensington Museum, 1876," which Naval Observatory astronomer Edward S. Holden wrote after his return from England. Addressed to the secretary of the navy, nearly half of Holden's thirty-seven-page account dealt with timekeeping, a topic not included in the London exhibit. Evidently, Admiral Davis commandeered this report and used it to help bring his plans to the secretary's attention.[8]

The trip report contained the observatory's first printed account of "distributing public time for the use of ship-masters and others, and the steps now taken by various countries to furnish accurate time to navigators." Holden inserted details of the world's time balls; highlighted the controlled clocks and time signals supervised by the observatories at Liverpool, Greenwich, and Edinburgh; mentioned the electric clocks controlled by the Pulkova, Paris, and Neuchâtel observatories; and summarized six American observatory timekeeping systems. But he was apparently unaware of the observatory time services operating in New York, Cleveland, and Ann Arbor, for they are not mentioned.

From this information Holden concluded that England was far ahead of the United States in providing the public with accurate time. Noting the relative simplicity of distributing signals in Great Britain—due to the modest distances involved and government ownership of the telegraph system—Holden asserted that to have a common time in this country, there must be "the *certainty* that the time-signals which are now regularly sent from the Naval observatory [*sic*] shall reach each railway station daily, at least."

Holden proposed that branch telegraph lines be erected at the five Atlantic-coast navy yards. Once in place, officers at chronometer offices could receive the Naval Observatory's daily noon signal, thereby leading to more reliable ratings as well as giving each site the capability to fire a time gun or drop a time ball. The astronomer estimated that a one-year experiment at the New York and Norfolk yards would cost about eight hundred dollars. He recommended that binding contracts be made with the telegraph companies, for success hinged on being able to receive the observatory's daily signal with no interruptions.

Agency plans require approval at higher levels before being submitted to Congress. Admiral Davis transmitted Holden's report to Secretary of the Navy George M. Robeson early in November 1876. On 4 December, the superintendent formally reminded his immediate superior that Holden's trip report contained a plan, "the further carrying out of [which] . . . rests with the Department."[9]

The very next day Superintendent Davis wrote Dr. Norvin Green, vice president of the Western Union Telegraph Company. The ostensible reason for this letter was to support a Leavenworth, Kansas, jeweler who was responsible for standardizing the town's time. The jeweler was anxious to receive Washington time monthly and was willing to pay for the service. After Admiral Davis passed on the jeweler's inquiry, he seized the opportunity before him and proposed a new enterprise: selling Washington time throughout the United States.

Without even a mention of U.S. Navy needs, Admiral Davis emphasized the desirability of enhancing the country's use of standard (uniform) time. "Any method by which it can be done will receive the intelligent support of the Observatory," he promised. Properly cautious with regard to any chain-of-command issues, the superintendent ended his communication with this statement: "The first object of this letter is to inaugurate a plan of operations. When that plan is concerted, I will ask the authority of the Department to carry it into execution."[10]

The superintendent's letter to Western Union altered completely the environment in which the Naval Observatory time service operated. Not only was a new goal—time uniformity throughout the United States—expressed, but in addition the Naval Observatory was taking action to achieve this goal. In effect, the national government was offering a guarantee: to provide the telegraph company with an accurate, reliable time signal. Although that was certainly not considered a complex task, much less a new one, if any problems cropped up, Naval Observatory funds would need to be diverted from other activities in order to solve them.[11]

In poor health, Admiral Davis died in mid-February 1877 and so did not inaugurate what Western Union would later term "a novel service." However, both his planning and his timing were superb: this joint program for distributing accurate time lasted nearly a hundred years.[12]

Holden had been sent to New York to negotiate with Western Union officials. Nine days after Admiral Davis's death, the telegraph company accepted the Naval Observatory's proposal and requested assistance in bringing the new activity into being.[13] (Navy Department officials were not notified formally of this "no-cost" arrangement until mid-April.)

Word of the joint effort reached the small community of astronomers even before Western Union's public announcement. Samuel Langley wrote at once to Edward C. Pickering, the new director of Harvard College Observatory. After learning of the telegraph company's fees, Langley charged the Naval Observatory with unprofessional behavior, complaining bitterly to Pickering that it "had not waited for a popular demand but [had] sent out canvassers so to speak to get business in a spirit more mercantile than scientific."[14]

Private observatory directors certainly had reason to worry. Western Union's prices for a daily time signal from Washington had been set at $300 to Chicago, $150 to Boston, and $75 to Pittsburgh; with this annual fee prorated among a city's subscribers, any single organization need pay only a very modest amount for the time. At that moment Pickering was charging the city of Boston $500 per year for a daily signal—equivalent to two-thirds of the salary of the assistant who operated the timekeeping system. Langley was charging Pittsburgh $1,000 for the same service—a sum equal to the royalty he was receiving and representing a third of his annual income. Both cities were already questioning the observatory directors' prices.[15]

In his remarks to Pickering, Langley considered ways to counter the threat.

He concurred with the Harvard astronomer: to alert the secretary of the navy of the near-certain damage to astronomical science should the private observatories lose their timekeeping revenues. To demonstrate that large regions of the country were well-supplied with accurate time, Langley suggested that both observatories make arrangements to have Western Union's rival, the Atlantic & Pacific Telegraph Company, transmit their time signals on its network—unquestionably a most naive idea at this late date, when Western Union dominated the market.[16]

Ironically, Langley's letter denouncing Naval Observatory mercantilism arrived just days after Pickering had sent his own proposals to both the Atlantic & Pacific and Western Union. (Harvard College Observatory's letter to the Atlantic & Pacific arrived just as Western Union was about to acquire a controlling interest in its faltering rival; A&P officials forwarded the correspondence to Western Union's New York headquarters.)[17] Pickering was already facing a financial crisis. Winlock, his predecessor, had developed the observatory's time service by using some eight thousand dollars in research funds, and Pickering knew that donors had intended their gifts to support only scientific endeavors. Purportedly, the service was still operating at a loss. Expensive improvements in instrumentation also loomed.[18] Pickering later claimed that he considered terminating the Harvard College Observatory Time Service at this juncture. Actually, he had decided months before to increase revenues by a two-pronged approach: publicity and lobbying.[19]

While these two directors worried—Langley that he was about to lose much of his observatory's discretionary income, and Pickering that his expansion decision was starting to look like a serious blunder—Western Union announced its new services.[20]

In its publicity, the company highlighted the device that would demonstrate its new role as the country's purveyor of accurate time: a time ball was to be erected on the Broadway tower of its headquarters building in New York and dropped each day at noon by a telegraph signal from Washington. So reliable would this new visual time signal be that henceforth ocean navigators could use it to rate their chronometers. Equally important, all New Yorkers would have daily access to a precise standard. But several months passed before reality matched Western Union's words.

Dropping a time ball via a transmitting clock 240 miles away raised technical problems. Many tests were run before the company's engineers were convinced that the distant signal triggered the ball's daily drop reliably. The ball itself, designed by the Naval Observatory for the telegraph company, turned out to be too heavy; a new one had to be designed, constructed, and installed. Consequently, Western Union's public time signal did not become operational until 17 September 1877.[21]

Fig. 8.1. Western Union's New York City time ball. From *Scientific American* 39 (1878): 335, 337.

More cautious now about untried technologies, Western Union's managers scrutinized all time-related activities. A month after inaugurating the New York time ball—and to Pickering's great relief—company officials finally acknowledged receipt of the Harvard astronomer's March time proposal and promised a decision. In mid-November 1877, Holden wrote to tell him that the telegraph company's operating superintendent was convinced "that the middle states & Southern Seaports would be quite enough to keep us [Western Union and the Naval Observatory] busy." Holden added, "I hope that you, Langley & Washington can eventually cover all the Eastern and Southern States and I think the system will grow." The market for time was being split into three, albeit unequal, pieces.[22]

An additional remark is appropriate here. Two weeks after receiving Holden's letter, Pickering held his first meeting with the Harvard Corporation's overseers. The observatory director confessed the financial sins of his predecessor, announced his own expansion plans—now essentially risk-free—and showed off the publicity materials that Leonard Waldo, his new assistant for time services, had prepared. His reviewers must have been pleased. Starting in May 1878, Harvard College Observatory time, adjusted for longitude, was telegraphed to the U.S. Signal Service observer-sergeants charged with dropping Boston's new time ball.[23]

From the beginning, Western Union had its own ideas regarding the distribution of time. Company officials refused to drop the New York City time ball at Washington noon. Instead, they released it 12 minutes, 10.47 seconds earlier—at noon, New York time. Western Union wanted to sell time in the city, and to do so it needed a New York–indexed signal. The Naval Observatory obligingly agreed to telegraph two "noons" daily to Western Union's Washington office—not at all difficult technically, but inconsistent with its views on uniform time.[24]

In its April announcement of New York time services, Western Union described electrically operated synchronizers that could be attached to privately owned clocks, terming such devices "the Bain system" of control. The company offered to rent telegraph lines so that its correcting pulses could be received directly by these clocks. The market for Western Union's new enterprise included railways, clockmakers and other manufacturers, and owners of public clocks, such as city churches and retail stores.[25]

Like that of Western Union's New York time ball, the inauguration of the new service was delayed. On this occasion, the delay stretched for almost two years. One source of delay was a dramatic improvement in clock-control technology. When, in October 1877, the editor of Western Union's *Journal of the Telegraph* printed excerpts of an article by Naval Observatory astronomer Edward Holden describing the company's new time services, he added that "a new, simple and effectual method for synchronizing clocks has been brought out by a well known firm . . . in London."[26] Given the significance of this advance and the company's tardiness in learning about it, it certainly behooved Western Union to make haste slowly.[27]

Additional factors contributed to Western Union's slow pace. Not until the following June, fourteen months after the first announcement, did company officials secure permission to install a "time circuit" wire on the city's fire-alarm telegraph poles along Broadway, a line that ran from company headquarters on Dey Street to Thirty-fourth.[28] Deciding where to place responsi-

GOLD & STOCK TELEGRAPH COMPANY.

TELEGRAPHIC TIME SERVICE.

For the Distribution of Correct New York Time.

This Instrument beats every Two Seconds, with these exceptions:

I. *At the beginning of each* **Hour** *and* **Quarter Hour,** *the Instrument will* **strike** *like an* **Ordinary Clock,** *after which it will pause until the beginning of the next minute. The quarters are indicated by one, two, or three* **Double Beats,** *as required, after the repetition of the hour striking.*

II. *A pause of one beat (interval of four seconds) immediately before the beginning of each minute.*

III. *A pause of nine beats (interval of twenty seconds) immediately before the beginning of each period of five minutes.*

The Hours and Quarter Hours are struck with increased power.

JAMES HAMBLET, Manager Time Service,

Room 48, Western Union Building, New York.

Fig. 8.2. Announcement of Western Union's New York City time service. Courtesy of the Harvard University Archives.

bility for the new operation also took time; Western Union management eventually assigned all time services to the Gold & Stock Telegraph Company, a wholly owned subsidiary. Even acquiring someone to manage the effort became a drawn-out affair. As late as mid-September 1878, the company was still in negotiations with James Hamblet.[29]

In November of that year, Western Union announced its long-awaited local time service. But by now, despite their earlier statements, company officials had rejected clock synchronizers, selecting instead quite dated technology: sounders in subscribers' shops connected to a master timekeeper at company headquarters via dedicated telegraph wires.

Western Union's master clock was one of E. Howard & Company's Transit-of-Venus series of timekeepers, first designed for the government parties sent abroad in 1874 under the aegis of the Naval Observatory. This master clock was compared with—but not automatically corrected by—the noon signal from Washington. Its electrical attachments were like those in Howard clocks installed years before by Hamblet in Chicago and Philadelphia and at Allegheny Observatory and Harvard College Observatory. In this arrangement, the master timekeeper produces continuous pulses (every two seconds). When transmitted, some pulses are dropped, thereby creating at the subscribers' sounders a pattern of "ticks" by which the start of every hour and the beginning of specific minutes can be identified. Once familiar with this coding, customers were able to adjust their timekeepers precisely.[30]

Western Union's New York time service required dedicated wires, which represented a significant initial expense. Why the company decided against using the now-patented synchronizers developed by Lund is not known. Perhaps Hamblet, an expert on the technology even prior to his days at E. Howard & Co., predicted that synchronizers would evolve rapidly now that their long-standing reliability problem had been solved. In any event, Western Union opted for a proven American system.

Reinforcing the view that Western Union was proceeding very cautiously is the fact that it did not own its time-service enterprise. As Hamblet himself confided before the city service began operating, "I am starting this business with my own resources, and have got to show the company, that a profit can be made &c., &c." Although his title was that of manager, his actual position was that of a contractor to the communications giant. As technical supervisor of the Western Union Time Service, he became the telegraph company's gatekeeper, reviewing all inventors' and would-be developers' proposals.[31]

Before Hamblet's arrival, Western Union officials had targeted businesses that could be served by clocks that ran a trifle fast (so that their periodic synchronization would be certain). Now, with a time line delivering seconds pulses continuously, New York clockmakers and jewelers started to use the service as their primary source of accurate time. This unanticipated application exceeded the timekeeping needs of astronomers, for the star-gazers were satisfied with accuracy over short intervals—a single evening's worth of observations, say—while clockmakers and chronometer dealers demanded accuracy over days and weeks. Responding to this unforeseen requirement would have an enormous impact on the Naval Observatory.

The biggest commercial prize was the country's rapidly expanding rail network. How to penetrate that market was still problematic in 1877. New York–centered companies could receive New York time via Western Union's local circuit, of course, and roads elsewhere now had daily access to the Washington noontime signal. But only a smattering of the country's railways had ever bothered to install a sounder linked to an observatory. Jeweler-supplied time, transmitted via a road's own telegraph wires, was judged adequate for operations.[32]

Not until railway superintendents became aware of the imprecision inherent in the manual adjustment of station clocks was Western Union able to sell time services nationally. That critical information came from Plimmon Henry

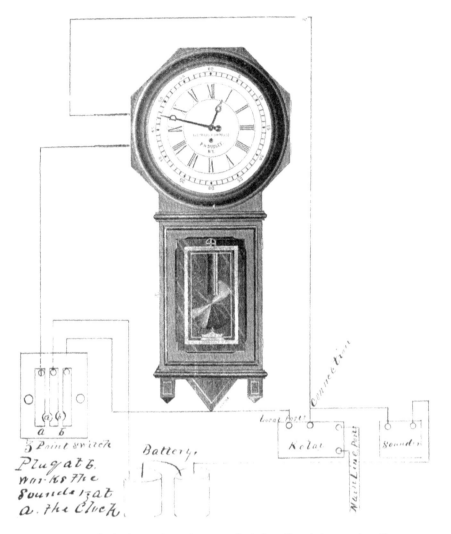

Fig. 8.3. P. H. Dudley's electrical-synchronizer clock for railroad time service. Courtesy University of Wisconsin-Madison Archives, # x252857.

Dudley (1843–1924), a gifted civil and metallurgical engineer whose judgments were respected by the railroad industry.

Industry standard practice was for a railroad company to transmit its operating time once every day or two from a central location, following which station agents adjusted all local clocks. Dudley showed that for a typical railroad, a station clock could be up to three minutes off the correct time, with the average among all station clocks offset from the road's correct time by half a minute. For operating officials, such variability in station times was troubling, even though it lay within the boundaries of satisfactory railroad timekeeping.

Dudley also demonstrated that the variability of station clock displays would essentially vanish when all timekeepers were reset simultaneously by electricity. Of even greater importance, by correcting all clocks automatically, operating-time displays were now under the control of the railroad's dispatcher, the person directly responsible for controlling the movements of trains.

An innovation in railway operating practices began in February 1879 when Dudley's synchronizer-equipped clocks were installed at the main-line stations of the Philadelphia, Wilmington & Baltimore Railroad. By September of the following year his clocks were also being used on the New York Central & Hudson River Railroad, with the synchronizing signals coming from Western Union's master clock.[33]

In the case of the New York Central, one of Western Union's sounders was placed in the chief dispatcher's office in New York City. At convenient hours, a railway telegraph operator began the synchronizing process, signaling for the shutdown of all message traffic via a premonitory warning message. All other operators ceased tapping and waited to switch their local clock circuit into the main telegraph line. The central operator signaled again, whereupon they switched their clock wires into the main circuit. Exactly on the hour, Western Union's pulse came. Instantly, all local synchronizers were actuated, and all clocks now displayed the same time.[34]

With the system in place on the Hudson River Division of the New York Central, its station clocks now varied "from one to [a maximum of] twenty seconds." Any accumulated error in the railway's time was "zeroed out" automatically, a decided safety enhancement in the running of trains. Timekeeping at railroad depots no longer depended on busy and often distracted agents.

Dudley laid the direction the railroad industry would take. His synchronizers were reliable, and his technically sound analysis of improvements resulting from centralizing the control of time convinced the country's railway superintendents. Other railroads began purchasing his clocks. Yet suddenly, they vanished from the market.

Apparently Dudley turned away from horology, an area of only peripheral interest to the railroad industry. Indeed, the record indicates that after 1882 the engineer applied his considerable skills to many of the other difficult problems associated with efficient railway operations. His departure left the market for high-quality synchronized clocks wide open.[35]

In 1882 two more time-service companies were founded, each with its own brand of patented electrical clocks. One company, closely coupled to Yale College's fledgling observatory time service, had only a brief corporate existence. The other, incorporated in New York, evolved rapidly; eventually, it became Western Union's partner in national time services. Both companies began as scientists were debating a variety of uniform time issues, including the accuracy of Naval Observatory time signals.

9. GAUGING TIME ACCURATELY

When in 1879 Cleveland Abbe presented his "Report on Standard Time" to the American Metrological Society, he began with accurate time, for that was his most important issue. Of the six American and two Canadian observatories he identified as time-givers, he described three—those at Washington, Allegheny, and Cambridge—as being able to provide "astronomically accurate time, to any customer on this continent regularly throughout the year."[1]

This judgment was Abbe's shorthand way of describing the country's timekeeping environment, for the "big three" were of course dependent upon the telegraph companies. Not even one was actually sending time regularly throughout the United States, however. Two of the observatory services were regional ones, and though its arrangement with Western Union was now two years old, the Naval Observatory still warned a traveling scientist that his "chances of getting accurate time at the Railroad Stations in the West are small."[2]

Abbe did not attempt to estimate the accuracy with which observatories and their distribution partners could deliver time. In the rush to complete a long-promised document designed to convince nonexperts of the importance of uniform time, he assumed that observatory time was accurate to about 0.2 second a day. This was a judgment of long standing among astronomers, based on the precision of stellar observations and the stability of the observatory's primary (sidereal) timekeeper, with a nod to the loss of in-house accuracy during periods of cloudy weather. This quality level certainly exceeded

Abbe's current needs as a Signal Service scientist: time accurate to within one second was more than sufficient for most geophysical studies.

The issue, however, was not how well time was being kept—or even the quality of time signals as sent—it was the accuracy of those signals *as received*. In May of 1879, only a few people knew that a problem even existed: the astronomers at the "big three" observatories and James Hamblet, time service manager for Western Union.

In September 1878, as Hamblet was installing the telegraph company's master clock, he began comparing its going to two sources of time: the Naval Observatory's noon signal and the signals that Harvard College Observatory was transmitting to Boston. Soon after, he added Allegheny Observatory's time signals to this set.[3] His purpose was to ensure the highest accuracy for the company's master clock—to be certain that the signals it was sending conformed to New York's true time.

From a timekeeping standpoint, Hamblet was doing something new: recording data from three independent sources of accurate time at one location.[4] The results were startling. Instead of observatory times differing from the master timekeeper by a few tenths of a second, he was logging one- and even two-second differences. Further, the differences were not systematic ones, but varied in a highly irregular, quite unpredictable manner. Which time base was correct?[5]

Hamblet was extremely worried by his results. His company's publicity emphasized these comparisons, asserting his ability to measure them correctly with a chronograph. Faced with such discordant values, all he could do was hedge, writing: "It must not be inferred that the [Western Union] clock in question is kept in exact accord with either or all of the observatory clocks, that being a mechanical impossibility." With regard to this master clock, Hamblet stated, "The range of [its] variation is kept within a few hundredths of a second."[6]

Once his immediate problem had been attended to, Hamblet sent all three observatories copies of his January 1879 comparisons and promised to keep sending them monthly reports. His approach was low key, and the Naval Observatory's response was scarcely more than a polite acknowledgement.[7] Over the next months nothing more happened, suggesting that no one there was concerned by the accumulating data. So he sent a blunter letter.

Annoyed, John Rodgers, admiral and superintendent of the Naval Observatory, wrote to Western Union's president and complained about the critical comments that his employee had sent. At the same time, Admiral Rodgers penned a brusque reply to Hamblet. He began by noting that the observatory's noon signals were being sent by hand. While conceding that small differences between observatory time signals and Western Union's clock were

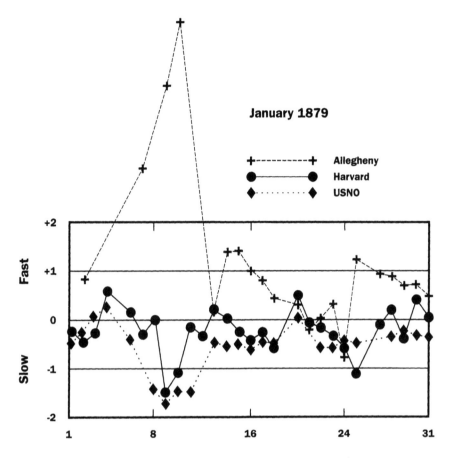

Fig. 9.1. Time signals from three astronomical observatories compared against Western Union's master clock in New York; differences are in seconds. Data in James Hamblet to Leonard Waldo, 1 February 1879, Harvard University Archives (UAV 630.20.10).

measurable, he also claimed that these were "not practical [that is, useful] quantities as regards single time signals." The superintendent added: "Our chronograph does not confirm your criticisms. The errors lie either in your means of measuring time, or in your telegraph lines, mode of working, interruptions by way stations, or otherwise." After a year of trying, Hamblet, Western Union's time-giver, had finally touched a nerve.[8]

Admiral Rodgers's strong defense of his agency's program did not alter the facts. The joint time-services arrangement that had been agreed to three years before placed the same burden on the Naval Observatory that had been placed on Western Union. The dismissal of Hamblet's analysis and his complaint to the company's top management notwithstanding, Superintendent Rodgers had to investigate these criticisms in detail—at the technical level.

So what had earlier been touted to Navy Department superiors as a "no-

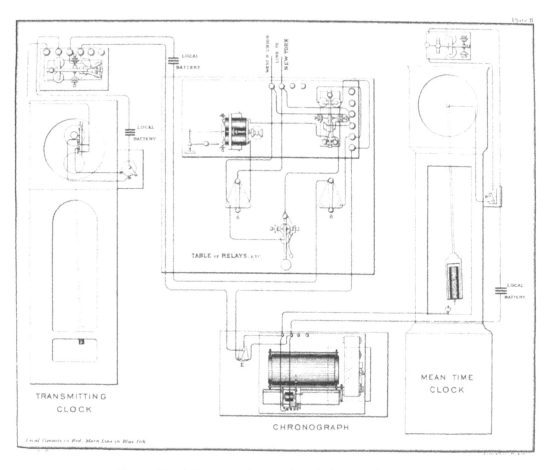

Plate II

Fig. 9.2. Naval Observatory's mean-time clock system for transmitting daily noon signals to Western Union's Washington office. The mean-time clock is compared against the observatory's sidereal-time standard (not shown) and its error determined; then the transmitting clock is compared with the mean-time clock and adjusted so that its noon pulse is error free. From *Notes on Navigation and the Determination of Meridian Distances for the Use of Naval Cadets at the U.S. Naval Academy* (Washington: Government Printing Office, 1882), plate II.

cost" program began to siphon off funds appropriated for other Naval Observatory tasks. In the fall of 1880, Admiral Rodgers announced the installation of automated equipment to reduce signal transmission errors, thereby "making the time signals almost absolutely correct." He also remarked that the observatory's mean-time, Transit of Venus–series transmitting clock "may vary slightly—say a second, or in extreme cases even two seconds."[9] By May of 1881 Hamblet could write that "Washington is doing much better than *formerly*," for the time ball's drop signal, as received in New York, was now agreeing more often with the telegraph company's master clock beats.[10]

Exactly when Abbe learned about Hamblet's comparisons is not known. He could have become aware of the monthly tabulations some time in 1879, for the community of American astronomers was a small one and gossip a fertile subdiscipline. More likely, however, Abbe became privy to the details of the "big three" discrepancies in early 1880, while he was in the midst of proofing the American Metrological Society's "Report on Standard Time."[11] Abbe's informant was Leonard Waldo (1853–1929), who was inaugurating a time service at a most unlikely place: Yale College's nonexistent Winchester Observatory. Though Waldo was little known in his field at this time, his energy and enthusiasm must have impressed Abbe. Certainly, Waldo's ambition was coupled to a flair for publicity, and thanks to this, the public soon learned about time-giving and -keeping.

The young astronomer had arrived in New Haven late in 1879 with superior credentials: his experience included the post of assistant astronomer with the 1874 Transit-of-Venus expedition to Tasmania, four years at the Harvard College Observatory, where he was assistant for time services, and a just-awarded Sc.D. from Harvard.[12] Waldo began by inaugurating a horological bureau for the testing of watches. Next he established Yale's time service, which distributed New York meridian time to New Haven's City Hall and to the New York, New Haven & Hartford Railroad. He followed that with a campaign to determine the longitude difference between the Horological Bureau's transit house and Harvard College Observatory.[13]

In August the AAAS held its annual meeting in Boston. Abbe's long-awaited report was finally in print and was having an influence on some of the country's astronomers. Ormand Stone, Abbe's successor at the Cincinnati Observatory, read a paper: "On Uniform Time." Those present concluded that the time was ripe for the creation of a committee on standard time, and one was formed. Chaired by Stone, its nine members—Waldo among them—were all astronomers directly involved in the distribution of observatory time.[14] No doubt they attended P. H. Dudley's talk, "Railway Time Service," at which the railway engineer described his synchronized clock systems. If so, his talk certainly alerted them to a new commercial opportunity in railroad timekeeping.[15]

At the end of the year, an article by Waldo appeared in the *North American Review*; it signaled his move to a more public venue.[16] In the article Waldo coupled Yale's new time service to his own business plans—albeit carefully and indirectly. Along with comments regarding accurate observatory time, he included numerous criticisms of the U.S. Naval Observatory, and particularly its partnership with the Western Union Telegraph Company.[17]

Any observatory has "a monopoly of the best article," the astronomer began. So it ought to be able to supplant everyone, even the "respectable jeweler who takes good care of a good clock" and who "has acquired the art of determining his time carefully." But if an observatory's time service cannot be "relied upon within a single second," then it should not be trying to propagate any standard of time. Waldo continued: "Unfortunately, the example set [by] the time services of the country, by that under the direction of the Naval Observatory at Washington, is not of the best; and . . . [so] the services organized under the control of the universities will occupy the first place for accuracy."

Waldo turned to the time ball operation in New York City, where "a [time-distribution] service is performed . . . though not with the assurance of accuracy we have a right to expect in such a Government work." Taking a leaf from Edward Holden's 1876 report, he urged that additional time balls be erected at coastal sites, underscoring the device's importance for marine navigators. As the exemplar of proper performance, he cited Boston's time ball—with its accuracy based on Harvard College Observatory's time and its connection with the U.S. Signal Service. Such quality time balls "have thoroughly ingratiated themselves in the public favor," he claimed.

Yale's time-giver also denigrated a specific type of electrically controlled clock, using the one at the Treasury Department maintained by the Naval Observatory as his example. He, too, he reported, had used its Jones system of control; however, he found it wanting in perfect reliability, as "errors of from two to ten seconds were sometimes found to exist in the controlled clock."[18]

Given these remarks, no one could have misunderstood Waldo's theme: the Naval Observatory's vaunted national time service was an inferior one. But no general reader would have been able to discern the astronomer's lack of objectivity in his zeal to deflect the impact of the Naval Observatory's expansion of its time service on universities. While Waldo highlighted shortcomings, his supporting data were out of date, insofar as the Naval Observatory's current program was concerned. Yet that hardly mattered; Waldo's fellow time purveyors must have been pleased with the article.

Waldo was now in the front ranks of private observatory time-sellers; this meteoric rise must have dazzled Abbe, for he had incorporated many of his younger colleague's ideas into his uniform-time plans. Already the Signal Service was considering the problems associated with telegraphing the Naval Observatory's Washington time to its far-flung stations; given Hamblet's results, the accuracy of the Signal Service's derived signals, as received, was suspect.

A careful scientist, surely Abbe concluded that time errors of unknown magnitude were affecting the Signal Service's own weather and geophysical programs. Unfortunately, he did not attack this technical problem directly. Rather, he subsumed it within a larger, more visible effort.

A few days after Waldo's article damning the Naval Observatory was reprinted in *Science*, Abbe attended the winter meeting of the American Metrological Society. He began the Committee on Standard Time's progress report by highlighting Waldo's efforts in advancing time-signal accuracy at both Harvard and Yale.[19] Next he proposed that the country's private observatories provide the signals necessary for dropping time balls. Doing so would increase the general public's access to accurate time, strengthen the efforts of the Signal Service's corps of volunteer observers, and tend to further uniform time in the United States.

Abbe then went on to urge the society to lobby various "Chambers of Commerce, Boards of Trade, Maritime Exchanges," and others "on the importance of establishing Time-balls." But surprisingly his focus was on seaports. Given Abbe's oft-stated arguments on the need for accurate time in the country's interior, one might conclude that his desire to establish uniform time finally overwhelmed his logical processes. Or, just as likely, it may be that he had concluded that Greenwich time for navigators was salable as an agency function, while Greenwich-based meridians for public time was not.

Abbe concluded by urging that any society communication on the subject "should suggest . . . that it is probable that the observatories, telegraph and railroad companies, and the *U.S. Government through the Signal Service* [emphasis added], will co-operate with any effort to provide Greenwich time to navigators and the public generally." To the society's officers he submitted drafts of what would become the Signal Service position regarding assistance to observatories.[20]

Abbe's December proposal, however, was just that: a proposal. In order to expand their public services the observatory directors needed resources. Assistance was promised officially on 1 March 1881, when General William B. Hazen, the army's new chief signal officer, signed two memoranda. The first one listed conditions under which the Signal Service would "co-operate with others in the maintenance of a public Standard Time Ball." One condition was that the drop signal be "given automatically by telegraphy from the [particular] . . . Observatory, which shall alone be responsible for the accuracy thereof." The chief signal officer agreed "to publish such portions of the annual reports of the observatories in charge of time balls as relate to the accuracy of the signals"—welcome news to cash-strapped directors. General Hazen's second memorandum listed fifty-five Signal Service stations where a time ball could be maintained; a dozen of these were located near private observatories.[21]

The Signal Service continued along its new path of improving public time distribution with a formal study of time balls. Among other questions directed to the country's time-service experts, Hazen solicited their judgments regarding signal accuracy.[22] Western Union's James Hamblet supplied current observatory comparisons, the data documenting the inherent difficulty of improving the quality of as-received time signals. Once again, none of the "big three" directors could tell his customers much more than that his observatory's "authoritative time" was not wholly accurate.

Indeed, given such an uncertain time base, the accuracy of a time ball's drop could be no better than that of these as-received signals. Yet several of the nine observatory directors who responded were quite optimistic in this regard. Edward Pickering, for example, asserted that a time ball would almost always "fall within one quarter of a second of the intended time; and after cloudy weather for five days the error should rarely exceed half a second." One suspects that the Harvard College Observatory director had not reviewed his assistant's current results, as documented by Hamblet's data.

Samuel Langley was less sanguine. He had never dropped a time ball, but drawing on his decade of time-service experience and his awareness of Hamblet's comparisons, he wrote: "The mean error of dropping such a time-ball here throughout the year would be less than a second"; but, with no stellar observations for a week, the mean error would "lie as often above as below one second, but would rarely reach two seconds. . . . These errors will seem large . . . [since observatory time can be determined precisely within a few hundredths of a second]; nevertheless, such errors as these do manifest themselves in spite of all precautions, when we are delivering time outside the observatory, whether we have had fresh observations or not."

Langley's sobering analysis (and a similar one from Waldo) should have alerted the Signal Service's cadre of scientists. But, unfortunately, Abbe was already sure that he knew the cause of the time-signal errors, as well as the solution. For among the questions posed in the Signal Service questionnaire had been: "Would you recommend that in the case of a ball dropped by its own controlled clock, the latter should be regulated by means of signals received daily from two or more observatories so situated that one of these should be enjoying good weather, and therefore certain of the accuracy of its time-signal?"

Given such biased guidance, almost all respondents embraced the notion that the large differences between observatory time signals was a cloudy-night problem—and that a more accurate value would result from ignoring a projection based on their own clock's rate data during poor seeing periods, using instead the (longitude-adjusted) time signal from an observatory "in the clear." These would-be time-givers were unable, or unwilling, to consider that a signal received via telegraphy might be corrupted as badly as their own outgoing one, albeit for different reasons.

At some stage, the Signal Service suggested that time signals be averaged. This "technical" proposal was actually a political decision, a sop to the private observatories.[23] Averaging meant that all could participate in providing a uniform standard of time for the United States; by implication, they would be given resources to ensure that participation. Those astronomers with little time-distribution experience, or with inadequate timekeeping equipment, would require both time and resources. Not merely for form did Abbe write that time-signal accuracy "will eventually be greatly increased" by the cooperative effort.[24]

Of course, Abbe wanted the most accurate time for scientific uses. But he also had friends, and their local-time distribution was seen as necessary to achieving his complementary goal of establishing uniform time throughout the United States. As the political process unfolded, the issue of time-signal accuracy became less prominent.

The Signal Service continued to march along its expanded program path. A primary clock, against which the observatories' time signals could be compared, had to be designated. Of course Western Union's master timekeeper was unsuitable, considering its New York City location; neither, according to Abbe, was it a "clock of the highest perfection."[25] On 1 November 1881 the Signal Service placed an order with the American Watch Company of Waltham, Massachusetts, for an astronomical clock able to run, and be rewound, in a sealed, constant-pressure environment. Next, the agency began the construction of a constant-temperature vault in a subbasement of the War Department building in Washington.[26] Then it attacked the national government's time-giver, the U.S. Naval Observatory.

PART IV. CONFLICT WITHOUT RESOLUTION (1879–1884)

10. CLASHING OVER TIME BILLS

On 13 December 1881, Representative J. Floyd King introduced H.R. 594, a bill "to provide for placing time-balls on custom-houses at ports of entry, and other cities, and for other purposes." His bill sparked a messy public confrontation between the Signal Service and the Naval Observatory. For when Congressman King linked the time balls' drop to the latter's noontime signal, in effect he designated Washington as the country's prime meridian.

Why would a seemingly obscure congressman from Vidalia, Louisiana, show any interest at all in timekeeping, much less in the Naval Observatory's signal? In 1879 King was serving his first term in Congress, so he had little or no influence. His main concern was to improve navigation of the Mississippi River. None of his committee assignments were germane to the Naval Observatory, and he was a Democrat in a Republican-dominated Congress and administration. Perhaps other factors influenced him: in the late 1840s his father had been chairman of the powerful House Committee on Naval Affairs.

All we actually know about Congressman King's link to timekeeping is that Rear-Admiral John Rodgers, the Naval Observatory's sixth superintendent, answered the legislator's request for information in September of that year, just after the 46th Congress's first session had ended. Furnishing "Washington Observatory time to the city of New Orleans every day" is not a problem, the superintendent wrote; any American city can make arrangements with Western Union to receive it.

Of course Admiral Rodgers provided additional information, reminding the congressman that in New York, Western Union used the Naval Observatory's signal to drop a time ball—an event that not only signaled the city's noon but also "enable[d] mariners to rate their chronometers." Then, in the normal way that agencies telegraph their desires to Members of Congress, the superintendent observed: "Foreign governments send their Observatory times to their seaports as an aid to commerce, but our government has never done this, and the Observatory has no funds for the purpose."

In December, Congressman King acknowledged Admiral Rodgers's reply and apologized for the long delay. The congressman then reported that he had conferred with authorities in New Orleans regarding the erection of a time ball, as well as the use of Washington time there; he anticipated a favorable response. "I have offered a bill that will cover the whole subject," King added, "and believe that it comes within the purview of your suggestions."[1] One may well conclude that the fledgling legislator was already skilled at reading between the lines.

On 21 January 1880, King introduced his first piece of time ball legislation, H.R. 3769. In it was language providing twenty-five thousand dollars to erect time balls and to pay the telegraph companies for their time-transmission services, these monies "to be expended under the direction of the Secretary of the Navy."

The bill was referred to the House's Subcommittee on Commerce, where it died; it had been introduced too late in the session.[2] But on the very first day set aside in the 47th Congress for introducing legislation, King placed his time ball proposal in the hopper again. Ten days later, the battle to select America's central meridian erupted.[3]

That a conflict existed between the Signal Service and the Naval Observatory is understandable enough; but that it was fought in public is remarkable. Throughout 1881, General William Hazen's first year as chief signal officer, relations between the two agencies had been correct—and no different from those of the decade before. In 1871 a controlled clock had been installed in the Signal Office (U.S. Signal Service headquarters) and wired to the dedicated telegraph line that ran from the State, War and Navy Building to the Naval Observatory; the time displayed by this clock became the Signal Service's official time.[4]

In June of 1881, Chief Signal Officer Hazen wrote asking for Superintendent Rodgers's views regarding the erection of time balls at various seaports, dropping them at the moment of noon along the seventy-fifth meridian from the Royal Observatory at Greenwich. General Hazen asked specifically "to know whether such work would be of any service to the Navy Department."[5]

Admiral Rodgers responded that "the plan of disseminating Greenwich time is not the best for general use by the public." He added that Congress might legislate Washington time for the railroads, a remark that should have been a clear signal to anyone familiar with the niceties of official correspondence. The superintendent closed his reply with the comment that "the philosophers can easily apply the differences of time," adding that "the learned . . . sometimes overestimate their functions"; this remark was aimed at Cleveland Abbe.[6] Like his predecessors, Admiral Rodgers favored Washington time as the basis for the nation's timekeeping, and held just as strongly to this position as Abbe did to his.

Having spent over six years considering the problem, Abbe probably saw himself as "owning" the issue of uniform time. Ownership meant, of course, that his opinions must be the correct ones. Moreover, he had accomplished far too much to slacken his pace and consider other views—even when they were the official policy of a sister agency. The American Metrological Society, the sponsor of what Abbe considered "his" committee, was in the midst of distributing circulars that included General Hazen's memoranda—setting the Royal Observatory at Greenwich's meridian as America's time basis, and outlining the Signal Service's support for the university observatories in dropping time balls. These circulars had already arrived in the New York office of the *Nation*.

After objecting first to the notion of a Greenwich meridian as "the perpetuation of the bad habit of sacrificing practical convenience to some supposed scientific universality," the *Nation*'s editor wrote:

> We hope that in carrying out any arrangement of this sort [time balls] the Society will take care that there is no such conflict between branches of the Government service as frequently occurs, to the detriment of scientific interests. Many of our readers know that an arrangement was made some time since between the Naval Observatory and the Western Union Telegraph Company by which time-balls should be dropped . . . and . . . a ball is now dropped regularly every day. . . . The question whether the Signal Office or the National Astronomical Observatory is the one to be entrusted with the work is one which we do not propose to enter; but we do earnestly trust that a clear understanding on the subject will be had before any steps are taken which may lead to a conflict between the two branches of the service.

Unfortunately, response to his concern came from outside the government.[7]

Leonard Waldo of Yale College Observatory defended the plans. Noting first that the choice of meridian would naturally "lie between New York and Washington"—a view that conflicted with Abbe's—he turned to the time ball proposal. Fear "of a possible conflict between the Signal Service and the Naval Observatory . . . must originate in a misconception as to the part the Signal Service offers to perform," he wrote. The latter organization would cooperate "only in those places where the [local] observatory desires to drop such a ball," since "it is hardly possible to suppose that any other time-ball than the one in New York would be gratuitously dropped by the Western Union Company." Waldo added: "The Signal Service is very commendable in offering to aid in the establishment of time-balls more generally throughout the country than they are at present."

The *Nation*'s editor was satisfied. Having "overlooked the fact that the Signal Service's proposal was restricted to transmitting time-signals to be furnished by observatories," he now saw "nothing in the way of friendly cooperation" between the two agencies. Clearly, the issue of uniform time in the United States was such a minor one that it concerned him no longer.

Over the next several months, as the Signal Service continued its planning, both General Hazen and his chief scientist ignored the Naval Observatory's opposition to many of their ideas. Also during this period, Waldo began negotiations with one of Western Union's competitors to have a new time ball erected in New York, this one to be dropped on Greenwich time via signals transmitted from New Haven. When in the fall of 1881 a tentative agreement was reached, the Yale astronomer conveyed the news of his achievement to Abbe, and departed soon after for England to negotiate commercial arrangements for distributing time.

In mid-October Admiral Rodgers responded to General Hazen's letter regarding the Signal Service's plan to establish, temporarily, "a central clock and chronograph to which the time shall be daily transmitted" from the country's observatories. "I shall be very happy to send you Time from the Naval Observatory," wrote the superintendent.[8]

Considering the fact that these multiple-observatory time signals could be sent to Washington at virtually no cost, the Signal Service's idea was a good one. Although the retransmitted average was unlikely to be accurate enough for all scientific uses, it was more than adequate for the general public. Indeed, early on, Admiral Rodgers had even congratulated General Hazen on his plan to send time throughout the country, terming it "a boon which the Observatory would offer if it had the means of paying for the messages."[9]

No evidence has been found to suggest that Admiral Rodgers contacted Congressman King during the summer or the fall of 1881. But only an incompetent manager-bureaucrat would fail to make a courtesy call on a second-term congressman—who was now so clearly a friend of the U.S. Navy—when that legislator arrived back in Washington for the start of the new Congress. In any event, King introduced H.R. 594. In 1880, Abbe may have decided that the congressman's first proposal was unlikely to pass. Regarding H.R. 594, his actions indicate a far different view.[10]

Two days before Christmas, Abbe, as the Signal Service's chief scientist, held a news conference. He began by announcing that a soon-to-be-erected time ball in New York City would be dropped hourly on Greenwich time, using signals transmitted from Yale College Observatory in New Haven.[11] The Greenwich meridian was the proper time basis for dropping the ball, Abbe pointed out, since the signals were to benefit the city's shipping interests.

Reiterating the Signal Service's cooperative stance toward observatories throughout the country, Abbe outlined the plan to have their directors transmit time signals to Washington. After averaging these values, a "standard time signal" would be telegraphed back, the Signal Office acting as a daily "clearing house." Only in this manner, the scientist claimed, would it be possible to have a time accurate enough to permit navigators to rate their chronometers. (And with essentially all of the participating observatories inland, the resulting uniform time—based on the seventy-fifth meridian from Greenwich—would become the standard for American railroads as well.)

Abbe claimed that the necessary arrangements with the observatories would be in place "around the first of February [1882]," a date soon enough to suggest no need for any time-service legislation. (Abbe never even alluded to Congressman King's bill.) Moreover, he argued, the Signal Service was the only government organization capable of coordinating time throughout the country.

Then, with a most remarkable display of hubris, Abbe justified his agency's decision to enter the field of timekeeping by comparing Boston's time ball with the one maintained by Western Union in New York City. Boston's noon signal was transmitted under the supervision of Harvard College Observatory's astronomer, and its performance was excellent, for "the greatest error ever made . . . did not probably exceed half a second." In contrast, the New York time ball's controlling signal came from the Naval Observatory, and "an examination" of the device's performance showed a "number of large errors in the falling of this ball, sometimes amounting to five and six seconds."[12]

Abbe's remarks were published in the New York papers on Christmas Eve; the *New York Times* followed up on New Year's Day with an article that supported the Signal Service's plans. Abbe was probably quite happy throughout the holiday season. One certainly hopes so, for the month of January was to be among the very worst of his entire career.[13]

On 5 January 1882, Naval Observatory superintendent John Rodgers responded to the Signal Service's criticism by sending a carefully worded letter to the secretary of the navy. After noting the several articles in the public press, Admiral Rodgers claimed—correctly—that the performance of the Naval Observatory's time services compared quite favorably with those of the national observatories at Greenwich and Paris. "Professor Abbe's statement in regard to the [quality of our signals for the] regulation of chronometers is apparently erroneous," he concluded. Regarding the purported time ball errors, Admiral Rodgers quoted Western Union's James Hamblet, who stated that the vast majority of these were due to "disturbances on the wires" and thus beyond Naval Observatory control.

After reminding the secretary of the observatory's support of the chief signal officer's earlier plans, the superintendent wrote:

> He [General Hazen] now proposes, according to Professor Abbe, to send out Greenwich time for shipping. . . . And [thus] . . . to usurp the functions which the Observatory has performed, spending . . . for rating ships' chronometers, money given to him for meteorological services. General Hazen's change of base necessitates change of mine, and I now earnestly protest against his meddling with the time service . . . given exclusively to the Observatory.

He concluded with a formally worded complaint:

> What the Chief Signal Officer proposes is at once indelicate, illegal, and against official comity, and therefore the Superintendent of the Observatory respectfully asks that the honorable Secretary of War be requested to inquire of the Chief Signal Officer by what authority of law he acts in the foregoing premises.[14]

Once Admiral Rodgers's letter reached the press, Abbe never had a chance. The scientist might have argued that the Greenwich time signals from Yale's observatory were for maritime, not naval, needs, and also that the Signal Service would not be rating chronometers directly. But the revelation of his rash, unfounded criticism—in public—of the Naval Observatory was far too damaging to make any defense possible. Ridicule and criticism of the Signal Service soon began, with the newspapers looking forward to what one reporter judged would be "a pretty fight."[15]

Secretary of the Navy William Hunt forwarded the Naval Observatory's complaint to Secretary of War Robert Lincoln, who endorsed it and directed it to the chief signal officer. General Hazen had only one option. In his response to Secretary Lincoln he wrote: "I respectfully submit that the superintendent's letter, based as it is upon my alleged intentions, published without my authority in an article which I never saw until after it was printed, is of such a character that I cannot reply to it. . . . I ought not to be asked to make an answer. I have neither issued orders nor committed any act interfering in the slightest degree with the Observatory work."[16] Upon receipt of the official replies, Admiral Rodgers responded: "Consider my words unsaid."

The battle for time was over—at least in public. Within the executive, the secretary of war enforced an uneasy peace. In his reply to a congressional request for comments on Congressman King's bill, Secretary Lincoln wrote: "There is a misapprehension that there was ever a conflict between the two bureaus," and he furnished copies of the correspondence to the committee's chairman to demonstrate the truth of this statement—a nice bureaucratic touch, indeed. Secretary Lincoln ended the department's required response to Congress with: "The question of placing time balls has not been considered by this Department," and returned H.R. 594 without comments. The Naval Observatory had swept the field.[17]

Abbe, who never forgot General Hazen's disavowal of his actions, suffered one more public indignity before the affair ended.[18] A New York editorial writer noted that "it is really difficult to conceive that a man like Prof. Abbe, in so extended and detailed a conversation as the one published, should either have totally misrepresented his superior officer or should have succeeded in making himself totally misunderstood." Nevertheless, the writer continued, "the practical result [of General Hazen's disclaimer] has been that the menaced quarrel has been averted."[19]

Since the public clamor resulting from Abbe's December press conference had not focused on the legislation before Congress, Samuel Langley did not catch wind of King's bill until mid-February. "I do not know whose measure it really is," he wrote his friend Edward Pickering, "but I conclude it must be the Naval Observatory's since Abbe tells me it has not his interest." Justifiably concerned about the bill's effect on his revenues, Langley stoked his anger with his view of the Naval Observatory's four decades of American timekeeping: "Had . . . [it] originated the system here as Greenwich did in England; had it done anything but profit by the labors of others in popularising it; it might urge the measure with more grace." Langley called for concerted action by the astronomy community and urged that astronomers either communicate with the House committee to inform it of the facts, or drop the matter. However, he ended, "the latter means definitely the end of the local services."[20]

Congressman King's bill contained several flaws, and the House Subcommittee on Commerce set about correcting them. As written, H.R. 594 covered all American cities with a population of fifteen thousand and larger. Its language directed that all such cities be equipped with a time ball apparatus, and that each one must receive the Naval Observatory's telegraphed time signals daily. Nearly 150 cities came under this provision, so capital costs would have far exceeded the appropriation of twenty-five thousand dollars called for in the bill.

After receiving the Naval Observatory's estimate of twenty-four "principal maritime ports," committee members modified the bill's language so that only those ports of entry having a customs house would receive a time ball and controlling-clock system. With this change, equipment costs became $15,840, with funds left over to pay the telegraph companies for their services.[21] In an attempt to limit the number of participants still further, language was inserted requiring municipal authorities to request the observatory's time signals formally. Finally, the committee added a section directing the Treasury

Department to detail a person at each customs house as a time ball attendant, and specified compensation of five dollars a month for the extra duty—$1440 annually for all.

The committee's work was over by March. Along with a unanimous report on the initial timekeeping proposal, it provided its substitute, H.R. 5009, "a bill to provide for transmitting the meridian time of the Naval Observatory at Washington to ports of entry and other cities, and for placing time-balls on custom-houses, for the protection of commerce, and for other purposes." Brought up on 9 March, the bill was referred to the Committee of the Whole House, and the report ordered to be printed. The first hurdles in the legislative process had been easily cleared.[22]

Even in its now-restricted form, the bill looked like a gold mine for timekeeping enterprises. The Mutual Union Telegraph Company—Western Union's competitor, and associated with Waldo's proposed New York time ball—assured Admiral Rodgers that it would "name a very low price" for transmitting signals. "We'll also undertake the construction and maintenance of lines to time-ball stations," the telegraph company's superintendent wrote. Hamblet, who had just negotiated a new contract with Western Union that gave him freedom to work for others, offered to supply clocks and time balls and to supervise the installations.[23]

Waldo, whose writings and alliance with the Signal Service had done so much to fuel the just-ended controversy, wrote to the Naval Observatory on Standard Time Company letterhead. As secretary of this newly incorporated enterprise, he offered to erect, at no cost, a complete system on the Naval Observatory grounds, including a controlling clock, six synchronized ones, and a "model time ball." The company's offer was forwarded to the department with the observatory's strong endorsement and a request for permission to accept it.[24]

Despite widespread interest by the business community in expanding the distribution of time, in Abbe's view, H.R. 5009 still contained one intolerable flaw: its specification of the Washington meridian. His plans included establishing an international basis for public timekeeping; he and others had already spent many months gaining signatures on petitions to Congress to have it authorize a conference on the matter. The past weeks' controversy had been a setback, of course. But in his official response General Hazen had insisted that none of the Signal Service's current actions were interfering with Naval Observatory timekeeping. Exercising more caution now that concerns regarding the Signal Service's functions had surfaced in the press, the chief signal officer and Abbe pressed on.[25]

Memorials began arriving in congressional offices, the first in late March. One, signed by "3200 prominent people," asked Congress to pass a joint resolution authorizing an international conference. The American Metrological Society's own submission arrived in April. On both occasions, Congressman Roswell Flower of New York requested that the memorials be referred to the Commerce Committee, and they were.[26]

Aghast, Abbe drafted a joint resolution; General Hazen presented it to Congressman Flower. Introduced on 10 May, the Joint Resolution (H.R. 209) called for an international conference, and it was referred to the Committee on Foreign Affairs. Similar legislation introduced in the Senate was referred to the Committee on Foreign Relations. During May and June additional petitions from learned societies arrived, including a resubmission from the American Metrological Society; all were referred to the "correct" congressional committees. As expected, both produced favorable reports.[27]

Parliamentary maneuvering began in earnest late in June. Congressman Belmont of New York failed badly—twenty-three to fifty-eight—in his first attempt to move forward with H.R. 209, which authorized the president to call an international conference. But two weeks later he succeeded, when the House voted favorably on the measure. Rushing toward adjournment, the Senate passed the House's version of the joint resolution on 28 July, and the president approved it on 3 August. The first session of the 47th Congress ended five days later.

Abbe had won the war. Admiral Rodgers's death on 5 May had deprived House sponsors of a strong champion for the passage of H.R. 5009, the competing time ball legislation. Brought up before the Committee of the Whole House in the very last days of the session, final consideration of the measure and its passage required near unanimity. Eleven members objected. So the bill died.[28]

Two additional attempts would be made to enact time ball legislation. Both failed; no funds for distributing time from the Naval Observatory were ever appropriated. Timekeeping companies—now keenly interested in expanded national and municipal time services—had to be content with the heightened public awareness that the conflict had generated.[29]

Abbe's victory was a bittersweet one. Concerned that the Signal Service was an agency out of control, Congress added a proviso to its appropriations bill. Funds could be expended for various activities, "*Provided,* That the work of no other Department, Bureau, or Commission . . . shall be duplicated by this Bureau." The agency was being warned that its activities were under scrutiny.[30]

Since the Signal Service's assistance to private observatories had not been a part of the Naval Observatory's encroachment protest, the agency continued to work with them. Time balls were erected in Cincinnati, St. Louis, St. Paul, and New Orleans, with time coming from the country's private observatories; like the one in Boston, many of them were armed and dropped by Signal Service observer-sergeants. But in November 1884 the Signal Service withdrew its manpower support. Time balls were also being erected by the U.S. Navy's Hydrographic Office, and the nonduplication proviso in the Signal Service's legislation was still in force. Without the agency's support, the country's inland time balls slowly withered away.[31]

The Signal Service's program for observatory time-service comparisons also ceased at this time, an innocent victim of the law. In 1883 the Signal Service had mounted its new precision timekeeper, and, after testing it for a year against Naval Observatory time signals, returned it to the maker for alterations and improvements. It added that "if possible, arrangements will be made which will secure to the principal stations of the service correct time for observations." Meanwhile, the temperature-controlled clock vault in the basement of the War Department was turned into a storage room.[32]

Abbe was understandably devastated by these unexpected events. Adding to his current woes had been the inauguration, in 1883, of formal comparisons of time signals from observatories at Toronto, Quebec, Saint John, and, eventually, McGill University in Montreal, a program designed to improve Canadian meteorological data. As senior scientist of America's weather services, Abbe had been scooped.[33]

Later, Abbe gave his agency unwarranted credit for discovering the unexpected differences among observatory time services, telling an audience of meteorologists that "the performances of this [Signal Service] clock were so perfect that the irregularities of the noonday time signals became apparent and furnished a final argument for our action relative to a proposed system of distribution of standard time."[34]

Moreover, Abbe neither forgot nor forgave the Naval Observatory's protest of his actions, always blaming the termination of the Signal Service's timekeeping programs on observatory machinations. While this view may be correct, by 1882 the Signal Service's troubles with Secretary of War Lincoln and with Congress had become so extensive that it seems simplistic to assign blame to a single source.

Twenty years later and still blinded by anger, Abbe summarized the events of the 1880s:

> In 1884, shortly before the clearing house system was to go into operation, some evil-minded person seems to have induced the secretary of the navy that the Signal Service, in its desire to secure accurate time, was exceeding its legitimate duties, and an order came from the secretary of war forbidding further actions, and even requiring the removal of the elegant clock from its ideal location in the basement of the War Department building. . . . [T]he U.S. Naval Observatory continued its struggle against the use of standard time.[35]

Of course, Secretary Lincoln had issued no such order.

11. INVENTING STANDARD RAILWAY TIME

By the calendar, the battle between the Signal Service and the Naval Observatory lasted twenty days, halted by General Hazen's disavowal of his subordinate's press conference. The observatory remained the agency for time signals. But other facets of timekeeping were left unresolved, so scientists and engineers flooded the public arena with their pronouncements on matters of time.

Most Americans must have been a bit bewildered by all this consensus-in-the-making. Many newspaper and magazine editors had already voiced skepticism, particularly with regard to basing an American time system on Greenwich time. Yet almost no one opposed the idea of a single time within some large region(s) of the United States. The problem was which time to choose.

From the beginning, Cleveland Abbe saw both transportation and telegraph companies as likely sponsors of uniform time. In the "Report on Standard Time" he proposed three such officials by name for election to the American Metrological Society, as a way to speed the adoption process. But the society had its own pace. Still Abbe pressed on.[1]

Not until October of 1881, as the General Time Convention was preparing for its scheduled fall meeting, did a railroad industry association even receive materials proposing the adoption of some form of uniform time. They included Abbe's report, American Metrological Society circulars, and the U.S. Signal Service's memoranda regarding the dropping of time balls throughout the country and cooperation with university observatories. The convention also received a letter from Ormand Stone, chairman of the AAAS Committee on Standard Time, who commended the issue to the members. All documents were turned over to the General Time Convention's secretary with instructions to report back at the next meeting in April. Fortunately for these general managers and superintendents, that official was William F. Allen.[2]

A railway engineer by training, Allen was in a unique position. He was the editor of the *Travelers' Official Guide*, a monthly compendium of railroad companies' timetables and listings of their company officials that included a short editorial section as well.[3] Allen also served as permanent secretary of the General Time Convention (1872–85), holding the same office in its minor companion, the Southern Railway Time Convention (1877–85). Both associations had been established "to settle questions of running time for through trains."[4] Under Allen's aegis, time uniformity became the first issue to be brought before either association that did not involve scheduling concerns.

Allen had already printed the American Metrological Society's circulars in the *Travelers' Official Guide*. But, busy with his editorial duties, he did not turn to the rest of the documents immediately. Reading Abbe's report for the first time, Allen finally learned of his two-and-a-half-year-old nomination to the American Metrological Society.[5] He attended its December 1881 meeting, held in the president's rooms at Columbia College.

This meeting, which took place just four days after Abbe's Washington press conference, must have been a revelation to someone like Allen, who was not familiar with the time standardization efforts of the past two years. Several international legal and scientific societies had just discussed time reform, and more planned to do so. American scientific and engineering societies were focused on the subject. The American Metrological Society itself was in the midst of a campaign to petition Congress for an international conference on uniform time. A government program to help private observatories distribute Greenwich time was slated to begin in six weeks. And a bill to erect time balls and distribute Washington time to the country's largest cities had been introduced in the House of Representatives just two weeks before.

Allen was seeing the steps that influential people in no way connected with the railroads were taking to secure a uniform time system. He must have felt that they would succeed, for in the *Travelers' Official Guide* for January 1882, he wrote the first article on timekeeping to appear in a railroad industry publication in a decade. Over the next two years, almost every one of its issues carried some mention of uniform time—a marked contrast to the industry's prior lack of interest.[6]

Given the high level of effort on the part of outsiders in an area critical to the railroads, one would have expected a lively discussion at the General Time Convention's April meeting. But a rate war raged among the country's trunk lines, and discord among railroaders was running high. Allen, who had solicited information on uniform time and might have already prepared his own plan, announced the meeting's indefinite postponement on 7 April, five days before it was set to begin.

By the end of the summer of 1882, Congress passed the international meridian legislation, which promised to affect the railway companies' operations. At that critical moment, Allen, paying scant attention to the scientific community's drive to reach consensus, seized control of the issue of standard time.

The rate war finally ended. In October 1882 the General Time Convention held a truncated meeting at which there was talk of disbanding in order to reduce the likelihood of conflict. Allen told those assembled of his conclusions regarding uniform time and discussed the outside events—the legislation just passed, and the scientists' ongoing efforts. Convention attendees responded by calling for a meeting to be held in April 1883 to discuss Standard Time.

Allen's April 1883 editorial in the *Travelers' Official Guide* and his subsequent report to the General Time Convention urged immediate action: "We should settle this question among ourselves, and not entrust it to the infinite wisdom of the . . . State legislatures." He reminded them of Connecticut's uniform-time law, and emphasized the state railroad commissioners' current warning regarding more coercive legislation.[7] Nor was he sanguine about Congress, writing later: "Congressional action . . . is to be deprecated, as . . . there is little likelihood of any law being adopted in Washington, effecting [*sic*] railways, that would be as universally acceptable to the railway companies."[8]

On 11 April 1883 the General Time Convention met in St. Louis specifically to consider Allen's "Report on the Adoption of Standard Time," which emphasized feasibility. At the meeting he showed two maps. The first one depicted in color the overlapping nature of the forty-nine railroad operating times currently in use, and the other showed the country's rail lines in terms of five times, each time a different color. The visual contrast was enormous; in Allen's plan the rail lines' touching and crossing points were reduced in number from close to three hundred to about forty.[9] Remarking frequently that his focus was on railroads and solutions in terms of the companies' needs, Allen discussed his plan for clustering rail lines. He grouped ten of the most widely used operating-time standards into an Eastern set and a Midwestern one, and located geographically the endpoints of the railway lines in each set. Then he calculated the mean or central meridian of each pair of extremes, and found that the two values were "almost exactly one hour apart" in time. He then proposed that these two sectional meridians be located precisely one hour apart. Finally, Allen defined three more meridians—two in the West and one passing through Canada's maritime provinces, each an exact hour away from the two major ones—to encompass the remaining 20 percent of North American railroads. Thus the concept of hour-differences for timekeeping purposes, favored by many American scientists, was affirmed and made real in a practical context.

Allen listed the cities and towns where a one-hour time change between sections would occur, locating them "where they change at present, and at the terminus of a road, or at least at the end of a division." By doing so, he assured his audience that the proposal would lead "to no practical difficulty whatever in the construction of time-tables."

Not surprisingly, the General Time Convention adopted Allen's report and proposal unanimously. As secretary, he was directed "to endeavor to secure the acquiescence of all [railroad] parties to the plan proposed," and to report at the next meeting.

Two issues transcended the railroad managers' routine professional interests. Breaking with the worldwide tradition of basing a country's timekeeping on a national meridian—one contained within its borders—Allen referenced his timekeeping system to the Royal Observatory at Greenwich's meridian. His selection was one of several possibilities being actively debated by scientists. Indeed, as several opponents of the Greenwich meridian were arguing, there was no reason to base a time system for railway operations in America upon the location of an English observatory's transit telescope.[10]

Allen claimed that those scientists who wanted Greenwich had not influenced his choice.[11] Rather, his clustering of the rail lines had revealed what he termed "a curious fact": The central meridian for the Eastern section coincided (within six seconds) with the seventy-fifth meridian west of Greenwich —exactly five hours away from the Royal Observatory.[12] Since Allen could have chosen almost any mid-Eastern city's meridian as the basis for his partitioning, one might consider this an extremely fortunate accident of world geography.[13] More telling, Allen's selection of Greenwich removed intercity rivalry as a possible source of conflict and won him the support of the scientific community, particularly among his fellow American Metrological Society members.

The second nonrailroad issue was cities' local times. Allen asserted that systems of hour-based standards led to time differences of no more than "about thirty minutes." He further argued that numerous such cases already existed "without detriment or inconvenience to any one." Like others before him, Allen suggested having one's watch equipped with a second pair of hands, "if local and standard time *must* both be kept." However, he predicted that should the railroads adopt the new system, "local time would be practically abolished."

Actually, the system was not quite as Allen described it. Its central cluster of railroads spanned nearly two hours, and the zones he was creating overlapped along the Pennsylvania and New York shores of Lake Erie. But such "slight" anomalies were perfectly acceptable to the railroads, and city problems were not the companies' concern. Still, the cities' problems had to be addressed. And Allen, now the industry's spokesman, had to convince at least some municipalities to adopt his proposed system. Having them do so would reduce initial confusion among rail passengers—in particular, the country's rapidly growing ranks of commuters.

First, however, he had to convince the railroads. Late in August, Allen sent 570 company managers a detailed circular and maps, asking each for a decision on the new time system. An enormous volume of mail and telegraph traffic ensued; Allen later described the process as a battle campaign. At the General Time Convention's 11 October meeting in Chicago, he announced that roads representing 79,041 miles of trackage favored the change to the new system. The convention set the conversion moment for noon on Sunday, 18 November 1883, a date consistent with the railways' usual shift to their winter schedules.[14]

Although astronomers had had little say in the matter, Allen and the New England railroads did secure the assistance of the Harvard College Observatory in time for the convention's Chicago meeting. Agreeing on the merits of dropping Boston's time ball on seventy-fifth meridian time—sixteen minutes earlier than the city's official time—Harvard astronomers and other scientists wrote newspaper articles and lobbied local governments in support of Allen's proposal. Despite some opposition, the City Council of Boston passed an ordinance changing its official time to seventy-fifth meridian time and scheduled the change for 18 November.

New York City's government was far easier to persuade, for there the shift from local to railroad time was only four minutes. In mid-October, Allen and James Hamblet met with the city's mayor, who promised his cooperation in securing an ordinance. New York's scientific community then led the necessary lobbying effort, which included newspaper interviews, public lectures, and quiet discussions with members of the city's Board of Aldermen.

Allen had earlier contacted the superintendent of the Naval Observatory regarding the railroads' plan to change to this new system of multiple times. He asked if the observatory would signal the dropping of the New York time ball at noon on the new time. While disagreeing with the choice of four meridians for the country—"[W]e think that a single standard would be better, from a scientific point of view"—Admiral R. W. Schufeldt was quite receptive to the change. In addition to dropping the time ball in New York on the new time, the superintendent promised that "we will . . . try to secure the immediate adoption of the same time, as the local [that is, civil] time for the whole section in which it will be used by the railroads."[15]

A few days after the railroads' decision, Allen wrote again. Superintendent Schufeldt forwarded the correspondence to the secretary of the navy, along with his own position that "it [is] desirable that the Standard Time of the 75th

Meridian . . . be adopted as the local time of Washington City." Such a shift meant a change in government working hours, so the superintendent asked the secretary to query the heads of the departments. And at that point a hitch arose.[16]

Only one incident had marred the adoption process thus far. In September a forgotten Charles F. Dowd called on Allen in New York. He informed him of his own efforts with the railroads and reminded him of correspondence sent to the *Travelers' Official Guide* years earlier and of the then-editor's favorable response. He implied that Allen's main ideas were actually his. According to Allen, who cannot be considered unbiased in this matter, Dowd then argued "quite strenuously that the proposition to abolish the use of 'mean local [*sic*] time' could not and should not be carried into effect, he consistently preferring his own position." So began yet another squabble over priority, one that has only recently been resolved.[17]

As the day for the time change approached, a few voiced skepticism. Others took a wait-and-see attitude. By 18 November, almost everything was ready. Allen had distributed Translation Tables—time-conversion lists—to speed the rescheduling process. Circumstances differed among the railways. Those roads whose new operating time would differ ten minutes or less from their current one planned no schedule changes and simply adjusted their clocks. Some roads, in a spirit of extreme caution, decided to have their trains stop completely during the "extra minutes" generated by retarding their clocks. A few of the observatories transmitting time signals issued their own detailed information.

Allen witnessed the switch to the new time at the Western Union Building in New York City. He described the event: "Standing on the roof of that building . . . I heard the bells of St. Paul's strike on the old time. Four minutes later, obedient to the electrical signal from the Naval Observatory . . . the time-ball made its rapid descent, the chimes of old Trinity rang twelve measured strokes, and local time was abandoned, probably forever."[18] At that moment, Standard Railway Time, the first uniform time system spanning the entire country, began.

Writers have described that November Sunday as the "day of two noons." So it was in the eastern parts of each new "time belt," as the zones were being called then. Sandford Fleming, who had been urging a worldwide time system, noted that "a noiseless revolution was effected throughout the United States and Canada."[19] Others remarked that no train accidents occurred on that day: the change went smoothly for the railroads.

In most major cities, the change went smoothly, too, for the new way of reckoning time was simple: It required only advancing or retarding clocks a geographically defined number of minutes. New Yorkers retarded their clocks

YALE COLLEGE OBSERVATORY.

RAILROAD STANDARD TIME CIRCULAR.

INSTRUCTIONS TO TELEGRAPH OPERATORS AND OTHERS CONCERNING THE TRANSMISSION OF THE NEW STANDARD TIME, WHICH IS PRECISELY FIVE HOURS SLOW OF GREENWICH, ENGLAND, AND IS DESIGNATED "EASTERN TIME."

On Saturday evening, November 17th, at nine o'clock, the observatory telegraph instruments now transmitting the State Standard, or New York City Hall Time, will be disconnected at the observatory, and no Time Signals will be sent from the observatory from Saturday evening at nine o'clock until Sunday morning, November 18th, 1883, at nine o'clock by the New Standard, or Eastern Time. The clock beats will be sent every two seconds, beginning at nine o'clock precisely, and with the same arrangement as heretofore, concerning the insertion of the 56th, 57th, 58th and 59th second of each minute, and the omission of the last 20 seconds of each five minutes.

To insure the correct setting of time pieces out of New Haven to the New Standard Time, the operator at the New Haven Railroad Depot will carry out the following arrangement:—

SUNDAY A. M., NOVEMBER 18th, 1883,

At $11^h\ 44^m\ 30^s$ The word "Time" will be telegraphed, and the lines must be closed.

At $11^h\ 45^m\ 0^s$ The Observatory Clock will be switched in.

At $11^h\ 57^m\ 0^s$ The Clock will be switched out, but the lines must be kept clear.

At $11^h\ 58^m\ 0^s$ The Clock will be switched in, and the usual twelve o'clock signal will be received.

At $12^h\ 5^m\ 0^s$ The Clock will be switched out by the New Haven operator, and the regular business of the lines resumed.

Beginning with Monday, November 19th, 1883, only that part of the programme beginning at $11^h\ 58^m\ 0^s$ will be carried out.

On and after Sunday, November 18th, at 9 A. M., all public time signals from the observatory will be as above in "Eastern Time," except that, until further notice, the observatory will send the present State Standard Time for one minute each day, beginning at about $6^h\ 56^m$ A. M. Eastern Time, and ending about $6^h\ 57^m$ A. M. Eastern Time, and sending every second from $7^h\ 0^m\ 0^s$ State Standard Time to $7^h\ 1^m\ 0^s$ State Standard Time, to comply with the State law on this subject. The present State Standard Time, or New York City Hall Time, is $3^m\ 58\frac{4}{10}^s$ fast of the New Standard or Eastern Time.

LEONARD WALDO,
Astronomer.

YALE COLLEGE OBSERVATORY, November 9th, 1883.

Tuttle, Morehouse & Taylor, Printers, 371 State Street, New Haven, Conn.

Fig. 11.1. Yale College Observatory announcement of its 18 November 1883 switch to Standard Railway Time. In William Frederick Allen Papers, Manuscripts and Archives Division, The New York Public Library, Astor, Lenox, and Tilden Foundations.

four minutes, while Philadelphians, New Orleaners, and Denverites did nothing at all.

A myth persists that changing Chicago's official time proved a difficult and lengthy process—caused by too little advance lobbying there, the city fathers' pique at being asked to use "New Orleans time" (requiring a nine-minute change), and so on. But official records document a prompt acceptance, with the mayor proposing, and the City Council immediately adopting, the changeover on Monday, 19 November.[20]

Some confusion did occur in the District of Columbia, however. Responding to the Navy Department's request, Attorney General Benjamin H. Brewster had prepared an Official Opinion. "A change of time in Washington . . . can not be effected by mere executive authority," he summarized. "It can only be done by appropriate legislation." So on Monday, 19 November, the government departments stayed on Washington (local) time, while many city businesses advanced their clocks to conform to seventy-fifth meridian time. All waited for Congress to convene and pass the necessary legislation.[21]

As a result of Washington's two-time system, newspapers had a field day. One even reported a "scoop": Because Attorney General Brewster had not accepted the proposed time change, he missed his Sunday train to Philadelphia by eight minutes and twelve seconds. Even earlier the *New York Times* had characterized him as the only one in the executive branch opposed to the adoption of the new time standards, while grudgingly conceding that congressional consent might be required for adoption because "Congress legislates for the District."[22]

Elsewhere, the mayor of Bangor vetoed the railroad's time for his city. Dubbed "Mayor Dogberry" by the new time system's proponents, he was also the subject of a well-known lampoon. Too long to be reprinted here, it ends with these lines: "Remember what Leonidas was at Thermopylae to Greece, what Bruce was to Scotland, what all great leaders have been to the liberties of the people in all ages, he [Mayor Cummings] is to the town clocks of Bangor!"[23]

Here and there a few clergy argued that the local time of their region was "God's Time," and that the new time was a falsehood—not based directly on the earth's rotation and the locale's meridian. And one tongue-in-cheek article in an Indianapolis newspaper is still cited as indirect evidence of the public's fundamental mistrust of the railway companies' power in the 1880s: "The sun," the article lamented, "is no longer to boss the job. People . . . must eat, sleep and work . . . by railroad time. . . . People will have to marry by railroad time. . . . Ministers will be required to preach by railroad time. . . . Banks will open and close by railroad time; notes will be paid or protested by railroad time."[24]

Despite the generally effortless transition, small pockets of opposition to

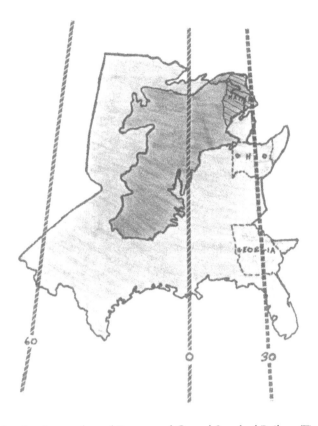

Fig. 11.2. Map showing overlapped Eastern and Central Standard Railway Time zones, both centered on the central (time-defining) meridian, with meridians drawn sixty minutes earlier and thirty minutes faster than the time-defining one. From Bartky, "Adoption of Standard Time."

the time change remained. Diffuse in nature, scattered about the country, and seldom reported in newspapers, such intense local resistance prevented total uniformity in American timekeeping, a situation that continues to this day.[25]

Explanations for the public's opposition have not been based on careful analysis. For example, ministers residing in Chicago or New York or Philadelphia did not denounce the new time, so it is difficult to accept resentment at the loss of traditional values as the prime motivation for resisting the changeover. Public mistrust of railroad power has also been suggested; yet, in those regions where the extant local time changed by only modest amounts, no denunciations of the railways have been found. Given the history of intense antirailroad feelings during those decades and the public's awareness of the change, one would certainly expect to find evidence of opposition nationwide. A more likely explanation may be found in the one feature common to all incidents of known opposition: it took place along the boundaries between time belts, and was strongest, and longest-lasting, in the eastern regions of the two most populous belts.[26]

Allen had said that his plan would produce relatively minor changes in which "the greatest discrepancy between local and standard time can be but about thirty minutes."[27] However, in the country's new Eastern zone, Standard Railway Time ranged from thirty-two minutes slow on the eastern edge to thirty-eight minutes fast on the western side, compared with the established local times. In the Central zone the new time was up to forty-five minutes slow and up to sixty-six minutes fast. These differences would scarcely have mattered except for one thing: people at the extremes of both these zones experienced a significant shift in their sun-lighted hours.

Consider the plight of Savannah. Allen's plan assigned it, like all of Georgia, to the Central zone. When Central Time was adopted, Savannah's sun set thirty-six minutes earlier than people had been used to. Citizens tried the new time briefly and then rejected it.[28] Residents of Maine cities like Bangor and Bath foresaw the change bringing earlier sunsets, and—four years later, when the entire state switched—experienced them every winter. Citizens in Ohio, too, balked at the change; similarly, residents of Detroit and Port Huron, Michigan, denounced the new system and stayed with local time.

Nevertheless, by April of 1884, Allen could report that seventy-eight of the one hundred principal American cities had adopted the new time standards. The public's attitude generally confirmed the prediction made as Standard Railway Time began: that "in a short time the new standards will be accepted everywhere . . . and then all intelligent persons will ask why the change was not made years ago."[29]

In December 1884 Allen reported that all railroads but two short ones around Pittsburgh had adopted Standard Railway Time. To achieve this level of acceptance, he had extended the Pacific zone east to Deming, New Mexico, and El Paso, Texas. Thereby he created the anomaly that three time zones touched in west Texas. By 1 January 1887, when even the Pittsburgh holdouts (and the city) adopted Eastern Time, Allen could consider that his railroad task was complete.[30]

One time system now reigned; after years of debate among scientists, Greenwich time and time zones had become America's norm. Even acknowledging the enormous economic and political power of the railway companies in that era, one still remains awestruck by the rapidity and smoothness of this change: in only twenty-seven months, cities and railroads in North America had dropped one time system and adopted another. William F. Allen had wrought well.

Yet, four Greenwich-related times had been chosen, and many scientists still clamored for only one. So we turn to the concept of a universal time.

"Yes," he continued . . . "the blowing up of the first meridian is bound to raise a howl of execration."

—Joseph Conrad, *The Secret Agent*, 1907

12. A FAILURE IN TIME

On 1 October 1884, promptly at noon, seventy-fifth meridian time, thirty-three diplomats and scientists were presented to the Hon. Frederick T. Freling-huysen, the secretary of state. These men represented twenty-one nations, and, along with a few late arrivals from four other countries, were in Washington "for the purpose of fixing upon a meridian proper to be employed as a common zero of longitude and standard of time-reckoning throughout the globe."

Included among the five-member American delegation to the International Meridian Conference were Cleveland Abbe, William Allen, and gifted amateur astronomer Lewis Rutherfurd of New York. Great Britain's official contingent totaled four. Spain and Russia each sent three representatives, while the other nations of the "civilized world" were content with one or two delegates each.[1]

For Abbe, the conference was a personal triumph. In 1880 correspondence he had suggested holding an international convention to consider the issues associated with uniform time. Two years later he had maneuvered to ensure that House and Senate committees friendly to his proposal would consider it. But success remained in doubt, and final passage of the resolution calling for such a conference came only in the very last days of the legislative session.

Convening the Meridian Conference was also the culmination of eight years of heroic effort by Sandford Fleming, the now-retired senior Canadian railway engineer and recently elected chancellor of Queen's University in Kingston, Ontario. Fleming's fascination with timekeeping was aroused in

1876 by an error in an Irish railway guide. A train's departure time was listed there as 5:35 P.M., which caused him to arrive at a deserted, rural station almost twelve hours late. Outraged by the resulting delay and subsequent embarrassment, Fleming cast about to find a solution for all time. He began studying how time was measured and reported, and later in the year published his first tract, "Terrestrial Time." Fleming later claimed that in 1878 the British Association for the Advancement of Science (BAAS) accepted his paper on the subject, but BAAS officials snubbed him when he tried to present the already-accepted paper at their annual meeting.[2]

Some time during this period Fleming coupled his ideas regarding unambiguous time reckoning throughout the world to a subject already being considered: the use of a single meridian—that of Greenwich or its antimeridian—in the preparation of every country's series of ocean navigation charts. By linking his time proposals to navigation and geography, Fleming had found the means to gain an audience receptive to his ideas. In February 1879 he presented two papers, "Time Reckoning" and "Longitude and Time-Reckoning," to fellow members of the Canadian Institute in Toronto.[3]

Fleming's proposals suffered from an unnecessarily ponderous framework of definitions. Despite that impediment, his key idea was important: the adoption of a single time—he called it "universal," "cosmopolitan," or "cosmic" time at various stages—to be used by scientists and the world's transportation companies, as well as for certain public activities. Fleming coupled this universal time to a single meridian that would also serve as the longitudinal basis for all the world's maps and charts. Along with this so-called zero of time, he identified twenty-four meridians spaced uniformly around the globe—each precisely one hour away from its neighbors. Suitably identified, these would be used to specify universal time for technical purposes. They also could be used by national governments as the basis for civil time in their respective regions.

With Fleming's emphasis on technical applications, it was natural to consider specialists' views first. A few months after his February 1879 presentations, the governor-general of Canada sent copies of both papers to Great Britain's Colonial Office, where they were distributed to British and European scientific societies and organizations along with a request for comments. Although some dismissed Fleming's universal time ideas as wildly utopian, all agreed that his prime meridian concept raised important issues. While the societies' reviews were under way, Fleming and Abbe inaugurated their own fruitful interaction. Canadian authorities included Abbe's "Report on Standard Time" in their second distribution of Fleming's materials.

During 1880, Fleming became a member of the American Metrological Society and joined his Canadian Institute forces with those of F. A. P. Barnard, who was president of the society and also president of Columbia College in New York. The unification of time had legal implications as well as technical and practical ones, so in 1881 Barnard introduced the subject to the Association for the Reform and Codification of the Law of Nations, an international group of legal experts.[4]

Barnard attended the association's ninth annual conference, which was held in Cologne, Germany. He began a lengthy discourse by detailing Fleming's ideas on timekeeping, and he presented a resolution supporting a world time system with its associated base meridian; he proposed that the meridian be Greenwich's antimeridian, 180 degrees away on the other side of the globe. Several members voiced opposition, one arguing that Barnard's entire discussion raised "matters of a purely scientific character, and beyond the scope of the objects of the Association." The educator's resolution was held over. Somewhat later, influential members of the association maneuvered to disregard an agreement that a committee be formed to study the timekeeping and prime meridian proposals and make recommendations. Barnard's complaints were duly recorded.

At the association's 1882 meeting in Liverpool, England, attendees noted that the U.S. Congress had just passed a joint resolution calling for an international conference and voted to "await the result of that [International Meridian] Congress [sic] before acting definitely on the subject."[5]

Never again would the Association for the Reform and Codification of the Law of Nations consider the subject of world time. No matter. For the time-uniformity advocates, success had been the process itself: an international group of legal experts had listened to a presentation on the subject, and though they had dismissed it, thirty-five pages of text were now a part of its *Report*. Henceforth, these advocates cited only the Cologne meeting, not the one in Liverpool.

Just as Barnard's attempt to move time reform into an international venue met with almost no success, Fleming (and several American scientists) achieved little at the Third International Geographical Congress, which met in 1881 in Venice. At this forum, in which specialists from eleven countries participated, Fleming presented his meridional and time uniformity proposals. Subsequently, the Geographical Congress voted to recommend that national governments appoint an international commission "for the purposes of considering an initial meridian" that could be used not only for longitudes but also for a worldwide system of uniform hours and dates. Although steps were taken to inform the Italian government of the vote, no follow-up government-to-government overtures occurred.[6]

During 1881 and 1882, Fleming marshaled support for an international conference on world timekeeping, particularly among members of the AAAS and the American Society of Civil Engineers. Officers of the American Metrological Society and the New York Geographical Society did the same among their respective constituents. Petitions were sent to Congress, Abbe maneuvered, and success came late in July of 1882 with the passage of the Joint Resolution authorizing the president to call for an international meridian conference. In October the secretary of state began querying countries, soliciting their views on convening a conference and the likelihood of their participation. Advocates of time reform must have been frustrated by the seeming delay; in their view, holding a conference immediately was a necessity.

The deliberate pace of the diplomatic process, however, actually helped the reformers' cause. In October 1883 the Seventh General Conference of the International Geodetic Association met in Rome. In addition to geodesists, astronomers concerned with the preparation of nautical and astronomical tables participated in the sessions. From the discussions came a forty-eight-page summary of concerns surrounding both the adoption of a single meridian for all purposes and the introduction of a universal time. Pros and cons were identified and supported by detailed analyses, discussions were summarized, and the votes on specific recommendations were recorded.

By an overwhelming majority, those meeting in Rome selected Greenwich as the world's initial meridian—their judgment resting on practicality, not science. As others had pointed out before, many nations were using this meridian on their technical charts and maps even now, and most ocean navigators already calculated their position in terms of Greenwich. Adopting this meridian as the basis for all countries' ocean and coastal charts and converting non-Greenwich-based astronomical and navigation tables would minimize the overall costs of such a change. Given this view among the specialists who were meeting at Rome, the selection, at some future conference, of a meridian other than Greenwich became highly unlikely.[7]

The issues associated with universal time were also addressed at the Rome Conference. Emphasizing its specialized nature, delegates linked it directly to longitudes and to the Greenwich meridian. They also adopted a resolution that the hours of a universal day be counted from zero to twenty-four—doing away with the use of A.M. and P.M..

Given the breadth of the discussions and the substantial support for reform at Rome, U.S. president Chester A. Arthur decided that it was finally appropriate to convene an international conference. The president's decision meant that at last those advocating time reform had a government forum. Even though the delegates to this conference would be merely advisors without authority to commit their governments, an official record would be created. In December the State Department issued invitations, announcing that the International Meridian Conference would begin on 1 October 1884.

The conference began on schedule. Welcomed in the name of the president of the United States, the delegates—eventually forty-one in number—held their sessions in the ornate and imposing Diplomatic Reception Hall of the State, War, and Navy Building. After the formal welcome, participants elected American rear admiral C. R. P. Rodgers conference president. Altogether the delegates held two ceremonial and six working sessions. Judging by the amount of time actually devoted to their announced purpose, attendees enjoyed a most pleasant October holiday in Washington.[8]

Most participating countries included technical specialists in their official delegations. With so many experts, many of whom breached diplomatic protocol by inviting comments from distinguished friends in the audience, discussions were wide ranging and filled with detail. Each country had only one vote. Many delegates had received official instructions on particular issues, so their positions were necessarily identical to the ones adopted by the specialists in Rome. Several attendees presented set pieces for the evolving record.[9]

Not surprisingly, conference delegates unanimously recommended adopting a single meridian for all nations. And with near unanimity they recommended that this prime meridian be the one passing through the center of the transit instrument at the Royal Observatory at Greenwich.

Turning to matters of timekeeping, delegates heard from Allen, who reminded them of the enormous change that had been made in the North American public's time, the result of the railroads' adoption of Standard Railway Time. He extolled the virtues of the system the industry had adopted ten months before, calling it a noncontroversial replacement for American locales' civil times. He displayed a worldwide set of hour-section time standards indexed to the Greenwich meridian. Here was proof that a public system of time could mesh well with a system designed to meet the needs of specialists.[10]

Allen's concern was the implications of the resolution adopted in Rome the year before. Delegates there had gone on record as supporting "the utility of adopting a universal time . . . for certain scientific wants, and for the internal service in the great administrations of ways of communication, such as those of railroads, lines of steamships, telegraphic and postal lines." But they were also on record as viewing a country's "local or national time" as one "which will necessarily continue to be employed in civil life."[11] The successful adoption of Standard Railway Time in North America brought into question the actual need to continue using local, or even national, time in civil affairs. It also raised the question of whether the scientists' view of public inflexibility on timekeeping matters was correct.

Responding to Allen's concern, delegates to the Meridian Conference recommended, by a vote of twenty-three ayes with two abstentions, the adop-

tion of "a universal day for all purposes for which it may be found convenient, and which shall not interfere with the use of local and other standard
time where desirable." Standard Railway Time was safe from change, and local time could be superseded.

Enlarging on the subject of timekeeping, conference delegates recommended that this universal day begin "for all the world at the moment of
mean midnight of the initial meridian, coinciding with the beginning of the
civil day and date of that meridian; and is to be counted from zero up to
twenty-four hours." This start of the universal day was inconsistent with the
resolution adopted the year before at Rome. Delegates there had linked the
proposed universal day to the astronomical day, which began at noon. Given
the fact that Meridian Conference delegates had to adhere to the instructions
received from their governments, this recommendation was not so well received, the vote being fifteen ayes, two nays, with seven abstentions.

Conference delegates tried to speed the unification process by approving,
without dissent, a recommendation expressing "the hope that as soon as may
be practicable the nautical and astronomical days will be arranged everywhere
to begin at midnight"—that is, to make these specialist passages of time consistent with the public's civil day. Among the world's astronomers, "as soon as
may be practicable" took forty-four years.[12]

The last session of the conference took place on 1 November. As the delegates extended their congratulations to Abbe and Fleming, they must have
felt that a treaty binding the world's nations was only a few years away. But
worldwide adoption of a prime meridian and a time coupled to it proceeded
at a snail's pace, a country-by-country process unaffected by the Washington
deliberations and recommendations. In this sense, the International Meridian
Conference of 1884 failed.

With the exception of Japan (in 1888), no country's decision to adopt a
meridian indexed to Greenwich can be traced back directly to the sessions in
Washington. Nonetheless, almost all English and American writers view the
International Meridian Conference as a watershed event. Indeed, the centennial of its convening was celebrated by an international symposium held at
Greenwich and by the issuance of British stamps to commemorate one hundred years of the Greenwich meridian as the world's "zero," or prime meridian.[13] I believe that this misplaced emphasis on the 1884 deliberations stems
from a misunderstanding of the U.S. legislative process.

Holding the conference did have one impact: it created a record, so that the
American government could take steps to promote a treaty among nations.
Submitting the official minutes, the executive branch urged both houses of
Congress to consider a treaty. Still, only Senate lawmakers passed a resolution
designed to initiate that process. Three years later, a "Message from the Presi-

dent" was sent to Congress. Nothing happened. So in 1888 the executive branch's responsibilities in the matter ended.[14]

Responding to the lack of progress, Fleming marshaled his Canadian and American forces, and in 1891 and 1892 bills were introduced in both houses of Congress to legalize standard time. If not yet the world, then the United States (and perhaps Canada). Again, nothing happened.

The world, it seemed, was not ready.[15] Adopting a single meridian for geodesy and geography, ocean navigation, and astronomical tables was a long-considered issue, and the experts had now reached consensus. But promulgating a universal time for scientific and public purposes was still a radical proposition.

Certainly, U.S. lawmakers felt that way. Even those senators who in 1882 reviewed the petitions and supported an international conference reported that "the committee believe that the acceptance of . . . [a universal standard of time is] extremely doubtful. [With] . . . so many chimerical schemes proposed, and no thoroughly practical one suggested . . . the committee cannot urge this as a reason for supporting the recommendation of a convention."[16]

Indeed, Fleming and other time reformers were ignoring reality. Most European countries were content to use national time for civil purposes. Great Britain itself had adopted Greenwich-meridian time for public purposes only in 1880. And when Parliament passed the necessary legislation, two times were designated: Greenwich time for England, Scotland, and Wales; and Dublin time for all of Ireland. If the British Isles could not agree on one time scheme, how could the world?[17]

Changing to a new way of reckoning time necessarily affected the public. Newspaper and magazine editors had voiced strong skepticism regarding many of the advocates' choices; yet the reformers marched on. Even Astronomer Royal George Airy, opposed in 1879 to all of Fleming's universal time ideas, argued in 1881 that "you must begin with considering . . . what the mass of people want, and must think of means of supplying that want." Why alter people's habits by mandating twenty-four hour clock dials?[18]

Insofar as the American public was concerned, the adoption of Standard Railway Time in late 1883 left the United States with scarcely anything more to do—and those minor issues could be handled by specialists in their own forums. So the country's national legislators moved on, not mandating Greenwich-based timekeeping until World War I. But there was one loose end: Standard Railway Time was not an official system. That legal problems would arise from its use had been forecast by Attorney General Benjamin Brewster in his 1883 opinion. Responding, a few states enacted statutes that defined their civil time in terms of one of the four Greenwich-indexed meridians—the path that Abbe, Fleming, and others had blazed at the national and international levels. But the majority of states ignored the issue, and now and then a contract dispute centering on which time was in force came before some local court.[19]

Despite the failure of the International Meridian Conference, by 1888 the future of world timekeeping could be seen—albeit dimly. Achieving time uniformity would be a lengthy process, particularly where public timekeeping was concerned. New Zealand was the first country to base its civil time on a Greenwich-linked meridian, adopting the measure in 1868.[20] Twelve years later Greenwich time became the legal standard for much of the United Kingdom, but that was just another case of a European country adopting a national time based on a meridian within its borders. Greenwich's importance in public life expanded in 1883 when North American railroads adopted a system of operating times indexed to that meridian. But except for the District of Columbia, that system of timekeeping was not America's legal time.

Nevertheless, the specialists' acceptance of Greenwich as the prime meridian meant that any widespread system of public timekeeping would be linked to it. In the 1870s and throughout the 1880s, scientists and time-reform advocates had tried and failed to influence governments. Now, those involved with the world's railroads would lead the way from national and local times to Greenwich-indexed ones.[21]

PART V. EMERGING AMERICAN TECHNOLOGIES (1880–1889)

13. NEW COMPANIES, OLD BUSINESS

Electrical timekeeping was transformed in 1876 by John A. Lund, partner in the famous London watch and chronometer firm of Barraud & Lunds. Lund called his time-control device a "synchronizer"—perhaps the first use of the word in that sense. When installed on a clock running up to two minutes fast or slow, his device would forcibly bring the minute hand to the "12-position" a few minutes before the hour and hold it there. Exactly on the hour the current to the device ceased, thereby releasing the hand.

Lund's synchronizer was a significant improvement over earlier British time-correcting mechanisms, whose importance the inventor acknowledged, even though they had not remained in service for very long. It could be attached to existing clocks, so it was very attractive to would-be buyers. By wiring a group of such clocks together in a local (district) circuit, all of them could be synchronized simultaneously. In addition, transmitting the correcting current once an hour meant that even clocks of moderate quality could be set properly, making high-precision timekeepers unnecessary. Furthermore, Lund's system of a central clock sending signals along telegraph lines made it possible for distant groups of clocks—four hundred miles away, in one of his examples—to be kept in step. And, of course, the occasional use of a central network of wires—a city telegraph network, for example—for timekeeping purposes was extremely attractive. It meant that only modest sections of wires—say, those strung throughout an office building or factory—needed to be dedicated to the clocks. And even those wires could be used for other signaling purposes.[1]

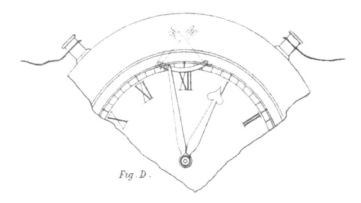

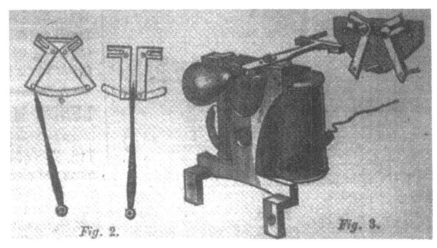

Fig. 13.1. John A. Lund's successful clock synchronizer. The partial clock dial showing the synchronizer's location is from *Journal of the Society of Telegraph Engineers and Electricians* 10 (1881): plate 1; the details of the magnet and control arms are from the *Horological Journal* 20 (April 1878): 106.

Around 1880 clock-synchronizer technology began to alter the conduct of American time-giving.[2] By decade's end, periodically resetting clock hands via electricity was the preferred method—not just for railroads but also for large stores, businesses, and a host of other public activities. When coupled to the fundamental change in public timekeeping that came to pass in November 1883 when railroad time "belts" went into effect, the advancing use of synchronizers halted forever the expansion of observatory time services. Ironically, introducing this technology into nonrailroad markets was the work of two astronomers: Leonard Waldo and Edward S. Holden.

Leonard Waldo's interest in synchronizers was aroused in 1877 when a notice in *The Mail* of London caught his eye. After reading the description of Lund's development, Waldo, newly appointed assistant in charge of Harvard

College Observatory's time service, wrote to Barraud & Lunds for additional information. He may have already been considering some sort of time service based on these British devices, and perhaps even approached prospective investors during his student days in Cambridge. He eventually found backers in New Haven.[3]

Waldo inaugurated Winchester Observatory's time service early in 1880; by midyear the expanding Connecticut Telephone Company had wired its downtown New Haven office directly to the time-distributing relay in Yale's North Sheffield Hall. Those who led Connecticut Telephone at this time were risk-takers fascinated by new technologies. Among them were national political figure Marshall Jewell, Yale-trained lawyer Morris F. Tyler, H. P. Frost, and Charles L. Mitchell; all are remembered today as extremely successful businessmen.[4] In April 1881, Tyler, in a talk before the newly formed National Telephone Exchange Association, described the telephone company's "auxiliary systems." Among them was one that would transmit continuous time signals to local subscribers for twenty dollars a year. Connecticut Telephone was also transmitting a noon time-signal to its central exchanges in half a dozen towns. By "arrangements with Prof. Waldo," it used this same noon signal to drop a time ball mounted on the roof of its Hartford office.[5]

In his speech Tyler was describing one of the world's first time-by-telephone services. Of course, the technologies that Connecticut Telephone employed were not new; they had been perfected and used for decades by astronomers and telegraph companies.[6] But Tyler went on to describe a new service based on the Winchester Observatory's development of "a special clock with a lever attachment, which sets itself at twelve o'clock noon every day without interference."

This description of a synchronizer is the earliest evidence linking Yale's observatory directly to clock construction. Tyler's remarks may have been premature, however. Robert Willson, who designed and built the device, was not appointed to the Winchester Observatory's Horological Bureau until late June 1881. Willson, a Harvard-trained physicist, was working simultaneously on several devices, including a mechanism to control a clock's pendulum electrically, a time ball, a clock for transmitting signals, and clock synchronizers. Yet premature publicity was not really a problem, for Waldo had already telegraphed his business associates' intentions in an article in the December 1880 *North American Review*.[7]

By the end of 1881 the outlines of a time-selling company had become clear. For a fee it would maintain the Winchester Observatory's signal transmission wires to New Haven's city hall, to the New York, New Haven & Hartford Railroad, and to the Western Union Telegraph Company. By paying an annual fee to Yale College, the company would have exclusive access to the observatory's time signals and would be permitted to solicit subscribers for its specific time services.

Moreover, the nascent company's investors had decided to sell synchronized clocks and timekeeping systems as well—a decision probably based on Waldo's (and Willson's) examination of the state of the art. Accordingly, on 20 January 1882, John A. and Arthur J. Lund of London agreed to assign to "all . . . persons who may now or hereafter constitute The Standard Time Company of New Haven, Connecticut" an exclusive license to manufacture and sell, in North and South America, the devices embodied in their two American patents. Further, both parties agreed that all device improvements would be communicated to the other party, and patents applied for.[8]

Connecticut's Standard Time Company (STC) became a reality on 9 March 1882 with the approval of House Joint Resolution No. 95. Capitalized at twenty thousand dollars but able to start business operations with only half that amount paid in, company officers included Waldo as secretary and Willson as engineer. Both men also served on STC's board of directors, where they were the only ones not also associated with the Connecticut Telephone Company.[9]

The time services company immediately issued three circulars describing its planned operations and the Lunds' patents. In one, the company offered exclusive licenses for its time distribution systems outside Connecticut, and the rental of clock-control and distribution systems within the state. In another, it announced the availability of continuous time ticks traceable directly to Yale Observatory's mean-time transmitting clock. Prospective customers were jewelers and watchmakers as far away as New London and Springfield, Massachusetts—the region served by Connecticut Telephone.

In its May circular, STC documented the clock system's nonexperimental nature by listing hundreds of Barraud & Lunds' subscribers and more than a dozen English testimonials. Company officers implied early commercial success in the United States by noting negotiations for installing 161 of its synchronized clocks in New York's elevated railroad stations. They also highlighted their manufacturing contracts with the E. Howard Watch and Clock Company of Boston, asserting that "the well known reputation of this house is a sufficient guarantee of the excellence of the supplies furnished to the local licensees of the [Standard] Time Company." In addition, the company placed notices in newspapers and magazines, and an article aimed at astronomers appeared in the *Sidereal Messenger*.[10]

Eager to sell products and franchises, Waldo courted the directors of private observatories. Topping his list of prospects was Edward Holden, director of the University of Wisconsin's Washburn Observatory, whom Waldo visited in April 1882. But the astronomer turned out to be a "tough sell."

A year earlier Holden had taken a leave of absence from the Naval Observatory in order to complete construction of the University of Wisconsin's observatory. From the very first, Holden expressed interest in inaugurating a public time service.[11] Aware of P. H. Dudley's system of controlled clocks, he obtained one on trial and began direct negotiations with the railway engineer for favorable prices. Holden purchased several of his clocks, which were installed at various firms and in offices of the state government in Madison.

During their negotiations, Dudley agreed to pay Holden "a fair commission on all R.R. orders you can get," if the astronomer would act as his sales agent. Clearly, Holden hoped to penetrate the Midwestern railroad market, for when he inaugurated the university's public time service in February 1882, he sent out Chicago time, even to the observatory's customers in Madison.[12]

From the first, Dudley had emphasized precise railroad timekeeping. His sales prospectus now included a "city time service" section, listing installations already operating in Philadelphia and Washington and noting that both banks of clocks were controlled by the Naval Observatory's daily time signals. (Later he included Washburn Observatory's time service in his promotional materials.) He also placed advertisements in the *Sidereal Messenger*, viewing time-givers in the astronomy community as prime customers for his clocks. And he was intimately familiar with the time ball legislation before Congress, and knew how much federal money would soon be available to fund locally run public time services.[13,14]

Waldo knew this, too. Already he had sized up Holden as an excellent sales prospect. Not only was Holden proposing to university authorities the purchase of an observatory clock system for ringing classroom bells; he had statewide expansion ideas as well. For Waldo, there was just one problem: the Washburn Observatory was using Dudley's clocks.

The young astronomer-entrepreneur went to work. Within days Holden sent a letter to the Standard Time Company outlining his plans; he ended it with: "This information is supplementary to a conversation with Mr. Waldo, & to serve as a basis for considering the abandonment of the system of Dudley Clocks and for introducing your control." Later in the month STC appointed Holden "its sole agent for the State of Wisconsin." But the appointment came with a condition: the observatory must abandon the Dudley clock system it had installed in Madison. Although Holden did not accept this condition, he did agree to adopt STC's products in the rest of the state, starting with Milwaukee.[15]

Having bested what was at that moment the only competition, Waldo arranged to have STC president Tyler meet with Holden and the other important Midwestern observatory directors: William Payne of Carleton College Observatory and Henry Pritchett, director of the observatory at Washington University in St. Louis. However, no clock sales resulted.[16]

Also in 1882 Waldo negotiated with Lewis Boss, director of the Dudley Observatory. Boss rejected the young astronomer's blandishments and installed three controlled clocks in Albany city offices, their regulation based on the decades-old Jones system. Inaugurating this municipal-supported service on 1 August, Boss alluded to startup delays caused by "negotiations with outside parties, who have contemplated an extensive system of regulated clocks for this city."[17]

At first, STC's directors must have viewed their meager success among astronomers as a minor setback. New York City was the primary business market for controlled timekeepers, and prior to the company's formation, Waldo had forged a time-signal distribution alliance with the Mutual Union Telegraph Company. Even though part of the Yale Observatory's expansion effort required support from the Signal Service—and Abbe's public blunder early in 1882 meant that no assistance could be expected—the Standard Time Company had on its sales books more than 160 synchronized clocks. Moreover, by mid-1883 Waldo was having some success in Chicago, another major public market for time services.[18]

Overall, the company seemed healthy, even though its sales force was minuscule. Its synchronized clocks were based on the patents of a very highly respected English firm, and the synchronizer itself was extremely reliable. Its American supplier of clocks was also a highly esteemed company. Fledgling franchisers were assured of expert assistance. The firm's technical experts had already generated patents and were hard at work on new, patentable devices that would keep the company at the state of the art. Nevertheless, on 19 September 1883—just weeks before the North American railroads' adoption of Standard Railway Time—STC collapsed. Its directors granted exclusive, ten-year licenses for its intellectual property, the firm vanished from New Haven's city directory, and Waldo became a consultant to the lessee, the Time Telegraph Company of New York. The Standard Time Company of New Haven had lasted scarcely eighteen months.[19]

Left in the lurch was the Yale Observatory, for the Standard Time Company neglected to mention its change in status to the observatory's professor-managers. Instead, in a letter to the college's president and fellows, Secretary Waldo stated STC's intention to terminate its fee-for-time contract with the observatory, promising only that its directors would "exert their influence to have . . . a contract with the company now organizing offered to you." Until now Waldo had been the astronomy community's finest public-time salesman. But the Standard Time Company's demise destroyed his hopes for financial success.[20]

Two complementary explanations for this rapid collapse are offered here. First, the major investors in the Standard Time Company had more important things to do with their venture capital. The Southern New England Telephone Company (SNET), Connecticut Telephone's successor, had been capitalized at half a million dollars, with authority to increase to five million, and that company was growing by leaps and bounds. Though Tyler, now SNET's president, had extolled the income-producing aspect of transmitting time, other members of the National Telephone Exchange Association responded with caution: "Telephone exchanges have . . . created new wants and are bringing about a revolution in the manner of transacting business. The consequent unsettled conditions of affairs evidently offers many opportunities for exchanges to engage in many branches of outside businesses. . . . [T]he establishment of . . . an exchange is an undertaking sufficiently ambitious and important to engage the best efforts of the members . . . and that the chances of its success will be enhanced by avoiding all doubtful auxiliaries."[21]

Competition in public time services was also growing by leaps and bounds. To remain viable, the Standard Time Company needed additional funds—money for hiring a full-time sales force, for example. So Connecticut's telephone pioneers sold the company's assets, terminated STC's contract with Yale, and moved on.

A second reason for the directors' decision to end operations was the ferment in time-distribution technology. SNET's system for transmitting time signals had been designed decades earlier for telegraph companies. Although STC's new patents for clock control and signal distribution were improvements on the Lunds' seven-year-old devices, they, too, were grounded in telegraphy. By the middle of 1883 other time-services enterprises, specifically the Time Telegraph Company of New York, had become the acknowledged leaders in this particular distribution mode. So learning that a device designed specifically for telephone exchanges was about to be tested, the Connecticut risk-takers turned to it as a more appropriate technology.

Already by early 1883, John Oram, a Dallas jeweler, had applied for patents on a battery-powered device that would subsequently be called the "Oram Time Machine," the "Oram Time Regulator," and the "Oram Time Indicator." His instrument took advantage of the dedicated wire connection between a subscriber's telephone instrument and the company's central exchange. Once installed, Oram's system signaled the time of day continuously. A subscriber could hear the time—a code of pulses—merely by holding his local receiver to his ear. For example, the code for 8:35 was eight pulses, a pause, three pulses, another pause, and then five pulses; at the exact minute, the subscriber would hear a buzz. The pulses themselves were generated via a clockwork-driven assembly that resembled a music box. Oram's device did not interfere with the telephone's primary use, for its pulses ceased automatically when a call was rung.[22]

Fig. 13.2. John Oram's Time Indicator for telephone subscribers. From *Electrical Review* 4 (24 May 1884): 1.

The New England Telephone Company, the first to test Oram's now-patented device, introduced its new service on 18 November 1883, the same day that the country's railroads switched to Standard Railway Time. In its publicity, the telephone company claimed that its minute buzz was accurate to a tenth of a second, for its central clock, which controlled the Time Indicator, was being regulated by the Harvard College Observatory's transmitted signals.[23]

In New Haven, trials began early in 1884. SNET's decision to use the Oram time system was a foregone conclusion, for the company's general manager had just been made president of the National Time Regulating Company. This Connecticut enterprise, founded in March and capitalized at sixty-five thousand dollars, had purchased control of the inventor's patents.[24]

SNET offered its time service to subscribers for a dollar a year. Unfortunately for the company, those not renting this additional service could also hear the time ticks, for cross-talk on the lines plagued telephony from its very start. So a "confuser" was invented, which added "a series of faint, meaningless ticks" to the ones being picked up via induction.[25]

With a well-tested, viable product, the National Time Regulating Company was ready to lease its time-distribution equipment to telephone exchanges throughout the country. Company managers inserted publicity items and ran advertisements in the country's burgeoning electrical journals. In

September 1884, Oram installed a complete system at Philadelphia's International Electrical Exhibition, at which the fruits of the ongoing worldwide revolution in applied electricity were displayed. Of the four types of time-distribution systems shown, his was the only one designed specifically for telephone exchanges.[26]

Over the next twenty-odd months, the National Time Regulating Company enjoyed some success in New England and large Midwestern cities.[27] However, the parent Bell system did not urge its franchise holders to make time a fundamental part of the telephone business. More important, equipping an exchange with Oram's instrument was costly: up to five thousand dollars. Telephone companies that had charged nothing during the equipment's test phase now found it difficult to convince subscribers to pay even a small charge for the service. And, though time information was seen as a way of maintaining customer satisfaction, its value in attracting new subscribers was not clear. Customers could simply ring up and ask, "Hello, Central. What's the time?" Consequently, Oram's system was of little interest to the rapidly growing industry.[28]

A system of coded time ticks could also be distributed via telegraph wires, and Oram Time Indicator entrepreneurs turned to this communications link. Many American cities and towns had dedicated wires that had been strung by district telegraph and messenger services—firms with whom the telephone exchanges competed. But others had already invaded this niche market for time signals, and one enterprise—the Time Telegraph Company of New York—was well along the path to dominance. Ultimately, in spite of its innovative technology, the National Time Regulating Company vanished. Oram returned to Texas.[29]

Sensing a growing market for time signals, other investors turned to pneumatic systems. One such system had been exhibited in Vienna in 1875; another was installed in Paris. By the 1880s the Paris system, with its miles of pipes, was operating successfully, competing directly with the French government's electrical distribution of time.[30]

However, pneumatic timekeeping technology, as developed in Europe, was limited to very large cities. Moreover, the time displayed on clock dials located at the end of the air-distribution piping differed by many seconds from that shown on dials close by the central air source. While such differences in time were not critical for the general public, those firms in America who sold competing (electrical) technologies emphasized this shortcoming. As a result, few pneumatic timekeeping systems were ever installed in America, most, if not all, being in individual buildings. And though patented pneumatic devices abounded, the major time-service companies protected themselves by patenting their own devices.[31]

Here and there a local market for time services would sprout, as some electrical inventor came to the fore. One of many examples was Louis Spellier of Philadelphia, who set great store by his gold medal from the Franklin Institute. His first patent was granted in 1882. Spellier patented improvements and struggled to interest investors, his devices eventually becoming the foundation for the Spellier Electric Time Company, incorporated in 1888 and capitalized at two hundred thousand dollars. Despite receiving favorable comments from both the Pennsylvania Railroad and the Naval Observatory, the company remained a local one. Competition from the Philadelphia Time Telegraph Company, which controlled yet another inventor's patents, probably scared off many would-be investors.

Seeing this ferment in public and business timekeeping, local astronomers began to see in their minds the fees they might garner by expanding their time-distribution businesses. Indeed, sales of franchises were ensuring growth in demand for authoritative time signals. But the country's observatory directors did not understand that a corollary to the expansion of franchises was an accelerating drive for national time signals. By 1890, they would understand only too well.

14. TWO INSTRUMENT-MAKERS

In the closing decades of the nineteenth century, two inventions dominated timekeeping systems for distributing time signals throughout the United States. One, a self-winding clock patented in 1884 by businessman Chester Pond, transformed electrically based timekeeping. The other was a clock synchronizer patented thirteen months earlier by Naval Observatory employee William Gardner; it became the basis for the government's System of Observatory Time. Both instrument-makers were skilled electricians, not clockmakers; and both profited from their horological creations.

Chester Henry Pond (1844–1912) formally entered the timekeeping business on 25 October 1882, when he and two associates filed incorporation papers in New York for the Time Telegraph Company. After authorizing the issuance of one million dollars worth of stock—an amount dwarfing all other time-service enterprises (and equivalent to more than sixteen million dollars today)—Pond and his fellow trustees created a company logo and prepared a twenty-page prospectus.[1]

To the city's financial community, Pond was a known quantity. Already holding seventeen patents, Pond was a director and secretary of the Gamewell Fire-Alarm Telegraph Company, the preeminent fire-alarm equipment manufacturer. He was also vice president and general manager of his own manufacturing company, which produced his patented fire-alarm devices for Gamewell. As this technically skilled and successful inventor-entrepreneur noted in a letter penned not three months before the Time Telegraph Company's incorporation: "My business never has been better or general prospects more encouraging."[2]

Time Telegraph was founded to exploit certain timekeeping patents: in particular, improvements in secondary clocks. One of these patents was for the components in an electrical circuit that dramatically reduced spark erosion of the switch contacts, the primary cause of failure in American electric-clock systems.[3]

The company offered "uniform standard time" distributed via subsidiary clock dials installed in groups of stores within a city block or in large office buildings, their hands advanced by a high-quality transmitting clock placed within the same bounded area. Subscribers paid twenty-five or fifty cents a month for each dial. The Time Telegraph Company offered exclusive rights for city districts, counties, and entire states.

Although the company's basic technology was a traditional one, its prospectus, which gave examples of more than 25 percent return on invested capital, was persuasive. Before the year ended, the Time Telegraph Company doubled the number of its investor-trustees, adding still more in 1883.[4] Acquiring growth capital was never a problem for the enterprise.

From the beginning, Time Telegraph protected its technology position by acquiring patent rights via license agreements and purchases. In 1883 it acquired rights to James Hamblet's just-granted time-distribution patents. Hamblet, who continued as manager of Western Union's time service, became both a trustee of Time Telegraph and its general superintendent. That arrangement not only gave the time-services company virtual control over which new technologies Western Union selected; it also provided easy access to the company's wires—a necessary test-bed for experiments in long-distance control of clocks. During this same period Time Telegraph acquired the assets of the Standard Time Company of New Haven, but chose not to use the firm's English technology. By the fall of 1883, Time Telegraph's trade journal advertisements were claiming that "this company now own and control forty-eight U.S. patents, covering every feature and branch of Electric Time Service." That number, forty-eight, meant that Time Telegraph controlled almost half of all the American electromechanical-clock patents issued up to that moment. The company reinforced its technical domination with threats of litigation in cases of patent infringement—a tactic also used by Gamewell and common throughout the electrical and communications industries.[5]

In October the company's New York directors forged a close association with the owners of the Rhode Island Telephone and Electric Company, an electrical supply house and construction firm. Henry Howard, Rhode Island's former governor, was the company's president; he and his business associates became the "Licensees of the Time Telegraph Company of New York for the

New England States." Along with a former governor of Massachusetts and Senator Joseph Hawley of Connecticut, Howard became one of Time Telegraph's nine trustees.[6] Through the Rhode Island firm, Time Telegraph dominated the New England market, with "time plants" in Boston, Providence, Hartford, and elsewhere. In addition, Time Telegraph reported that it had established similar plants in New York, Louisville, Nashville, Atlanta, and New Orleans.

During 1884, Pond became Time Telegraph's president. With the company controlling much of America's state of the art in electrical timekeeping, its display at the International Electrical Exhibition in Philadelphia was almost an embarrassment of riches. Along with a variety of patented slave clocks (dials) and synchronized clock systems, the company showed off devices and circuit improvements designed by its electrician, as well as an electric table clock invented by Pond. Of great importance to the company was its installation of an electrically impulsed master clock that controlled sixty subsidiary dials mounted throughout the exhibition hall.[7]

Those who organized the International Electrical Exhibition of 1884 had taken several pages from the history of the International Exhibition of 1876, also held in Philadelphia. A distinguished group of judges had reviewed that splendid display of technology. Now, exhibition organizers selected its Board of Examiners from a roster of American scientists and engineers involved in precise measurements and electrical matters.[8] However, in 1884 the devices assigned to Section XXII, "Electric Signaling and Registering Apparatus," were so broad in scope that the examiners created four subcommittees to judge them. The Time Telegraph Company's exhibit was reviewed by the subcommittee on "Time-pieces," chaired by Leonard Waldo of Yale Observatory.[9]

As astronomers intent on their own interests—high accuracy and consistency in timekeeping—the examiners' final report focused entirely on Time Telegraph's products. Operating within the context that "it is still an open question whether an electric clock can be regarded as an instrument of precision," they documented the excellent short-term performance of the company's master clock. Its synchronizer for keeping pendulum regulators at the same time was also studied: "practically perfect" was their verdict. Finally, all sixty slave dials driven by the master clock were sealed and their displays examined carefully several days later. With only three exceptions, two of which had been anticipated, "the coincidence of beats was exact in every case."[10] This unusual mix of a master and synchronized pendulum clocks, coupled to a large group of slave dials, was thus capable of providing uniform time to better than a second. The astronomers had provided the Time Telegraph Company with an invaluable testimonial.

The company's display at the Philadelphia exhibition represented technologies already protected by patents and license agreements. During this same period, another of Pond's inventions was under study by a different group of examiners—those at the U.S. Patent Office. The inventor was applying for a patent on his most famous invention: a "self-winding" clock. Although the concept of using power from a local battery to wind a clock's mainspring had been patented before, Pond's invention included placing a small electric motor within the clockwork. Periodically—once an hour, say—the motor's armature wound the clock's mainspring a constant amount. The Time Telegraph Company's president was granted a patent in only fifty-six days, one of the shortest application-to-grant periods for any horological patent.[11]

The importance of Pond's device cannot be overestimated. For the first time a timekeeper was available that could run for over a year without anyone having to wind it. When built into a high-quality pendulum regulator, his winding mechanism led to a clock that displayed true time, within a few seconds, for months. Moreover, if such clocks were equipped with an electric synchronizer, then all displays throughout an area would be consistent—to within a second. Although termed "a useful novelty" in a later trade magazine, the clocks' clear advantages to large businesses, owners of office buildings, and the like actually gave its manufacturers a significant market advantage.[12]

Time Telegraph's directors exploited Pond's innovation in a fascinating way. Early on they terminated the corporation and created the Telegraphic Time Company in its stead. Then through a series of incorporations, stock issuances, franchise repurchases, cross-licensing agreements, and the like, these entrepreneurs split the time-distribution market. One group continued to sell local and regional franchises for the less costly and already successful master clocks and associated slave dials; another group emphasized the sale and rental of self-winding clocks—with and without synchronizers. A third concerned itself with the manufacture, sale, and rental of self-winding clocks for railway and transportation purposes. Not surprisingly, several investors served as trustee-directors of more than one enterprise. In July 1885 the new Telegraphic Time Company showed off its extensive product line, which included a self-winding regulator.[13]

Like its predecessor corporation, the Telegraphic Time Company acquired rights to the patents for clock synchronizers and other devices, while its engineers developed and patented additional subsystems. Like all time-services companies, it offered regional licenses throughout the United States. Both Pond and Hamblet served as directors of the enterprise.

The inventor-businessman continued to develop and patent electrical clock components. Then in 1886 he and two associates incorporated the Self-Winding Clock Company, Pond having retained or repurchased some of his own patent rights. The aims of this New York enterprise were "the purchasing of

Fig. 14.1. Telegraphic Time Company pendulum clock incorporating Chester Pond's self-winding clockwork. The illustration is perhaps the first to depict this important American development. In *Electrical World* 6 (18 July, 1885): 25; and *Electrical Review* 6 (18 July 1885): 1.

necessary lands, materials, and machinery, erecting necessary buildings, and manufacturing, selling and leasing clocks and electrical appliances." Its capital was fifty thousand dollars divided among five hundred shares of stock—a very modest enterprise when compared with others Pond had established. Nonetheless, as Pond was fond of saying to his wife, he "had plans."[14] After publicizing its New York and Chicago locations in a full-page advertisement and offering a catalog of its products, Self-Winding announced its decision to move operations across the East River to a manufacturing site in Astoria.

Then, in a startling and mysterious turn of events, Pond suddenly resigned his key positions in both the Gamewell Fire-Alarm Telegraph Company and the Self-Winding Clock Company, sold his home in Brooklyn, and moved to Chicago. During his brief sojourn there the inventor tested clock synchronizers and managed the short-lived Central Standard Time Company, which transmitted both clock-synchronizing pulses and time signals. Although Pond's business career in horology ended in mid-1889 when he moved to the delta area of Mississippi—where he purchased thousands of acres of land and founded a prohibitionist community—others continued to manufacture and distribute his self-winding timekeepers.[15] By 1889 the transformation of American time distribution had begun.

Chester Pond's 1884 patent for a battery-powered clock-winding mechanism probably would not have transformed American time distribution so rapidly if John A. Lund had not already developed a reliable clock synchronizer. Lund's device reached this country in 1877, and in 1879 an American railroad installed P. H. Dudley's synchronizer clocks in its stations, thereby giving the dispatcher control over all its operating-time displays. Opportunity knocked. Beginning in 1882, within a few years several electrical inventors had been granted synchronizer patents. All of these devices had to be tested. According to one account, trials of a group of self-winding clocks equipped with synchronizers took place in 1885 in Chicago, on a Western Union telegraph circuit. Problems were reported, suggesting that reliable synchronization had not yet been achieved—particularly at the level of reliability demanded by the railroads. But the inventors persisted, for a lucrative market loomed.[16]

The nation's railroads were also in the throes of change. The clamor for national regulation of their business practices caused them to recast their locally generated operating rules, transforming them into national guidelines. Among these national guidelines were rules associated with railroad timekeeping. (Codifying industry "best practice" also provided the railroads a measure of protection against negligence lawsuits.) Two significant results of this process of revision and codification were the establishment of criteria for railroad watches, and a decision to obtain daily time signals from the country's astronomical observatories. Proposed in 1886, both rules were formally adopted the following spring.[17]

Rule No. 12 of the General Time Convention stated that "Observatory Standard Time is the only recognized standard and will be transmitted from the . . . . Observatory to the designated [railroad] offices." Moreover, the committee drafting this requirement considered it "of great importance that the time be obtained from some observatory of recognized standing." Given the climate of public concern and criticism in the mid-1880s, the railroad industry would have viewed the government's national observatory as the first among equals. Also, members of this voluntary association were comfortable relying on the Naval Observatory, having worked with its officials since 1883.

At one 1887 meeting the Naval Observatory was invited to describe its time distribution operations. The next year the association's members adopted the following resolution: "That, in the interest of obtaining and transmitting accurate time for railroad purposes, it is important that close relations should exist between the General Time Convention and the U.S. Naval Observatory, and to this end the Secretary of the Convention be directed to request the Navy Department to detail an officer to attend the regular meeting[s] of this Convention." This became one of the convention's standing resolutions. A bond had been forged.[18]

Of course the Naval Observatory entered the relationship with many technical advantages: ties to those managing Western Union's time-distribution operations, years of daily experience in transmitting time at the highest levels of quality, and an articulated national focus. But despite the assertions in its annual reports, the observatory was more regional than national in importance—insofar as railroad timekeeping was concerned. No real change in the Naval Observatory's status had occurred since Cleveland Abbe's report on American observatory timekeeping in 1879. Over those past eight years the country's "big three" of railroad timekeeping had actually become the "big five," with the private observatories at Carleton College, Washington University in St. Louis, and the University of Wisconsin's Washburn Observatory as important in the Midwest as the Allegheny and the Naval Observatories were in the East.[19] Nevertheless, those in charge of the government's observatory had an ace to play: their gifted instrument-maker, William Franklin Gardner (1843–98).

By all accounts, Gardner was one of those rare and talented individuals able to make scientific equipment function properly—a boon in any laboratory. Hired by the Naval Observatory in 1864, within two years this master mechanic was promoted to instrument-maker. His accomplishments, often cited in observatory annual reports, included improving the running of the mean-time clock, modifying the driving clock on the 26-inch refracting telescope, and a host of other alterations to astronomical equipment. And when the Naval Observatory displayed its equipment at regional and international exhibitions, Gardner was there, making sure that everything performed properly.

From the 1860s on, Gardner was involved with the maintenance of, and modifications to, the telegraph devices and wires for the Naval Observatory's time transmissions. In 1870 he supervised the placement of electrically controlled clocks in various government buildings. To install and maintain these clocks, the observatory used the Washington electrical supply house of Royce & Marean. (Morell Marean, also the assistant manager of Western Union's local office, became one of Gardner's earliest financial backers.)[20]

During Fiscal Year 1880, one of the Naval Observatory's Transit-of-Venus clocks was modified to send seconds signals automatically. This provided a decided improvement in the consistency of time transmitted to New York. Gardner's efforts were acknowledged.[21] The next year the observatory's superintendent noted that "several horological establishments in Washington [are] connected directly with the Observatory . . . [and] regulate their clocks every day." The Naval Observatory had become the source of true time for local businesses, via a private service probably maintained by Royce & Marean.[22]

Gardner's first patent carries a 22 August 1883 application date. The text describes the clock-synchronization system that would later bear his name, its level of detail implying that there must have been a lengthy testing period. Indeed, the instrument-maker may have decided to patent his devices in 1882, when the bill for the distribution of Washington time not only ignited controversy but also raised the public's interest in uniform timekeeping.[23]

Like Lund's earlier patented device, Gardner's synchronizer could be attached to any weight- or spring-driven clock. But unlike Lund's synchronizer—which was limited to minute-hand errors of plus or minus two minutes—Gardner's arrangement could "control the hour, minute, and second hands to an unlimited extent," a significant improvement resembling P. H. Dudley's 1879 synchronizer.[24] However, Gardner's first patent describes not simply a clock synchronizer but an entire time-distribution system: transmitting clocks that have been adjusted to an observatory's time as given by the primary (sidereal) standard, a drum-chronograph comparison of timekeepers, coded time signals, daily zeroing of the accumulated error, and so on. His subsequent patents were improvements and additions to this system.

Throughout these years of development, Gardner had at his disposal arguably the best test-bed in the country: the daily series of Naval Observatory signals distributed on Western Union's and other companies' wires. He also had access to the railroad timekeeping market. In October 1883 in Chicago, for example, those railroad managers who assembled to approve the adoption of Standard Railway Time listened to an associate of Gardner's urge their companies to adopt this particular system of controlled clocks. During that same meeting the young officer in charge of the Naval Observatory's time service noted that three Gardner-equipped clocks had been in use in Washington "for some time."[25]

Favorable government activity continued. In March 1884 the secretary of the navy proposed to the heads of several government organizations and departments that clocks with Gardner's synchronizers be installed in their offices. Such clocks were duly placed in the White House, the Senate wing of the Capitol, the National Museum, the Smithsonian Institution, and elsewhere, with a new clock line transmitting a daily correcting signal from the Naval Observatory's site. Within a year the number of public clocks controlled via the Gardner system quadrupled, growing from twenty to eighty-four.[26]

Eager to equip these clocks with Gardner's latest device, the Naval Observatory's superintendent interceded with the Patent Office, asking that the inventor's patent application "be granted at an early day, so that the Government may use it where necessary." Upon being informed that such a request required the signature of the secretary of the navy and a statement "that the invention is deemed of peculiar importance to some branch of the public service," the superintendent persisted. Gardner was granted a patent in 188 days, a bit faster than usual.[27]

Such anecdotes suggest rather a tangle between Gardner's legitimate business activities and those related to his government work.[28] Of course such tangles were not uncommon in technical areas dependent on the rare fabrication skills of machinists and instrument-makers.

Publicity for Gardner's system increased as the Naval Observatory expanded its time services in other areas. By 1885 eight time balls situated on the country's coasts—from Woods Hole, Massachusetts, and New York City in the Northeast, to Savannah and New Orleans in the South—were being dropped via the same signal that synchronized Washington's clocks.[29] Apparently it became policy for the observatory's naval officers to mention Gardner and his system at various railroad and scientific meetings.[30] So when the railroads decided to designate the Naval Observatory as "first among equals" for time signals, the government-assured quality of Gardner's time-synchronizing and -distribution system was a prominent feature in the observatory's package of benefits.

Gardner tried to capitalize on the unquestioned value of his system. One attempt was the appearance in Pittsburgh of the Standard Electrical Time System Company, most likely established as a franchise enterprise. Early in 1885 its president wrote to William Thaw, Allegheny Observatory trustee and Samuel Langley's major benefactor, urging the adoption of Gardner's system for the distribution of time throughout the Pennsylvania Railroad's vast network of lines. But Thaw ignored the proposal, and the company faded away.[31]

Gardner's 1886–87 correspondence with Edward Holden casts light on such franchises. Holden, now president of the University of California, was awaiting the completion of Lick Observatory and his appointment as its director; he was also in the process of establishing a time service at Lick. Responding to Holden's request, Gardner offered fifty clocks via Royce & Marean; in passing, he mentioned his business efforts in California. He was prepared to sell the "right[s] for San Francisco for $5000." Soon after, negotiations for the sale of the Gardner system for the Pacific Coast reached a critical stage, and the instrument-maker withdrew his fifty-clock sales offer. Although he preferred "to have the system controlled by those who have the management of observatory time," Gardner wrote: "In justice to myself I feel that I cannot give away a system, that has been in successful use for three years." Asserting that his system was more reliable than those being franchised by the Telegraphic Time Company, he ended a later letter to Holden in this way: "Now I believe that if you were to organize a Co. in San Francisco (where there is plenty of capital and energy) that your observatory would soon be sending time to thousands of clocks along [*sic*] the Western cities [served by] railways; & from which there would be some benefit to you & myself." But neither Gardner's California negotiations nor his business proposition to Holden bore fruit. By March 1887 he was offering Lick's director-to-be as many controlled clocks as he wanted.[32]

Early in 1888, Gardner made another foray into the business world. Forming a partnership with three Dobbs Ferry, New York, investors, the inventor assigned his patent rights to the Gardner Time System Company, a proposed corporate enterprise. The investors' business plan was similar to the many successful ones already executed by Pond (and others): clock sales, the operation of complete time systems, and the sale of franchises and company stock. Gardner's partners agreed to pay him twenty-five thousand dollars for his patent rights from the company's gross receipts; he would also receive 40 percent of the company's profits.

Gardner borrowed one thousand dollars in capital from one of the investors in order to establish at once a separate company to "sell time and clock systems to the railroads of the United States." The agreement between the two noted that Gardner "is himself without capital necessary to undertake" this specialized company's formation. In June both agreements were terminated, and the inventor regained exclusive control of his patents.[33]

Gardner may also have negotiated with the Seth Thomas Clock Company, the firm that manufactured his clocks. Some years earlier Seth Thomas had supported Waldo's unsuccessful quest for a patent on a time-transmitting system. Now it was starting to sell its "clock of precision," which the clock company described as a major advance in timekeepers. Much of the company's 1888–89 publicity—which included Gardner's laudatory comments at the

Fig. 14.2. U.S. Naval Observatory System of Observatory Time exhibit at Universal Exposition, Paris, 1889. In *Jewelers' Circular and Horological Review* 20 (October 1889): 35.

Universal Exposition at Paris—revolved about their new timekeeper's use with the Gardner system of timekeeping.[34]

That the inventor interacted with Seth Thomas in this manner is speculation. What is known about his business career is that on 15 February 1889, the inventor signed an agreement with the Self-Winding Clock Company of New York. Unfortunately, the scope of this agreement and the magnitude of any associated payments to the inventor are unknown. However, both must have been considerable, for a year later the company's trade material stated flatly that "the Self-Winding Clock Co. govern the patents relating to the Gardner system of standard time."[35] This legal document, which apparently assigned some rights, was followed by one dated 16 September 1890, by which Gardner agreed to "sell, assign and transfer unto the said Self-Winding Clock Company the *whole* right, title and interest" to his three patents (emphasis added). After five years of trying, Gardner's attempts to form a business enterprise were over.[36]

For the Self-Winding Clock Company, its 1889 agreement with Gardner was an important milestone. By then its trustees had approved a major increase in capitalization: from fifty thousand to five hundred thousand dollars. Self-Winding's trustees must have used their greatly increased investment to acquire control of Gardner's patents—and thereby control his future business dealings.[37]

However, via its predecessor firms, the Self-Winding Clock Company already had access to many of the country's synchronizer patents. Gardner's synchronizer, an add-on device, was not actually needed for clockworks that had integrated synchronizers and winding motors.[38] Nevertheless, the company's acquisition of his patents ensured that no one else could compete effectively in supplying these timekeepers to the railroad market. Equally important, the acquisition gave the company an invaluable marketing advantage, for "The Gardner System of Observatory Time" and the Naval Observatory's system of time signals from Washington were virtually synonymous in the public's mind.

The Self-Winding Clock Company planned to dominate other markets as well. Although acquiring Gardner's patents had been important, the crucial milestone in its overall strategy was the establishment of a close relationship with the largest distributor of time signals in the United States: the Western Union Telegraph Company. That milestone was reached late in 1888.

Nevertheless, the Naval Observatory instrument-maker's 1889 agreement with the clock company was seized upon as the country's directors of private observatories fought to protect their own enterprises. So begins a brief but sorry episode in the history of American astronomy.

PART VI. FINISHED AND UNFINISHED
BUSINESS (1888–1903)

> Your attention is hereby respectfully called to the injury inflicted upon various
> astronomical observatories in the United States.[1]
>
> —To the secretary of the navy, 15 April 1890

15. THE TIME PEDDLERS

In 1888 the number of observatories distributing time was at a plateau. Two Naval Observatory sites—one in Washington, the other in California—were sending their respective noon signals to large areas of the country via the nation's telegraph lines, while at least twelve of sixteen other observatories, all of them associated with educational institutions, were selling time to cities and railroads in their locales. By 1892 four of the private endeavors no longer distributed public time signals, and the remaining ones were all worrying about their long-term survival.

Dearborn Observatory was the first to cease operations. During 1888 the city of Chicago terminated its two-thousand-dollar annual contract and began purchasing time from the Western Union Telegraph Company—at a cost of three hundred dollars per year. Despite the enormous saving, the city's selection of another supplier could not have come any sooner. Although Western Union was regularly transmitting both Allegheny Observatory and Washington times to Chicago, and although its subsidiary, the Western Electric Manufacturing Company, terminated its contractual distribution of Dearborn Observatory time to city jewelers and railroads in 1885, the trustees who had negotiated Dearborn Observatory's contract in 1881 were influential Chicagoans. Not until Dearborn shut down its time service temporarily and began moving from the now-bankrupt University of Chicago campus to Northwestern University in Evanston did the telegraph company have any chance to win Chicago's business—no matter what it charged.[2]

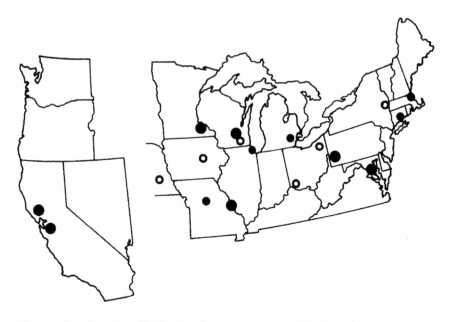

Fig. 15.1. American time-distribution observatories in 1888. The large dots represent "major" observatories, the small dots indicate "minor" ones, with the terms suggesting the number of miles of railroad trackage being served. The open circles represent minor observatories providing time signals only to their locales.

Western Union's gain and Dearborn Observatory's loss passed almost unnoticed. But in that same year the telegraph company moved aggressively in the Twin Cities area. In July a concerned William Payne wrote to George Comstock, the acting director of Washburn Observatory, telling him about Western Union's attempt to take away Carleton College's business with the railroads.[3] Payne proposed that he, Comstock, George Hough in Chicago, and Henry Pritchett in St. Louis all write to Western Union's management "asking that the Company desist from infringing on the rights of our Observatories by scattering Wash. time in our districts & underbidding local Observatories." He also suggested that one of them journey to New York to present the joint letter in person to Western Union president Norvin Green.[4]

Hard on the heels of Payne's alert came another from Asaph Hall, the Naval Observatory astronomer decrying his own institution's role in helping Western Union underbid the Midwestern observatories' time services. Since Hall mistakenly saw William Gardner and the observatory's assistant superintendent, Commander A. D. Brown, as being at "the bottom of this move," he recommended that Comstock, Payne, and Pritchett "unite in a letter to our Supt. Capt. R. L. Pythian . . . [who] is an excellent man, and you can rely on fair dealing from him."[5,6]

Immediately upon receiving these warnings Comstock checked on the situation, and he was relieved to learn that the telegraph company's "contemplated raid" was unknown to his railroad customers. As a result the Madison astronomer probably did not respond to Payne and continued with the planned expansion of his own time service. But in Chicago he was rebuffed by the Illinois Central Railroad, which was "rather inclined . . . to take our time from the Western Union Telegraph Co.," evidence that head-to-head competition with the telegraph giant would not be a good business strategy.[7]

Both Comstock and Payne worked to retain their local railroad customers, the latter even threatening legal action if Western Union used Carleton College's time signals without permission. Apparently both succeeded in staving off Western Union's immediate pressures, but neither foresaw that changes in technology were altering the market for time signals.[8]

Already by 1888 the Self-Winding Clock Company of New York had placed its clocks in the offices of several Eastern railroad companies. During 1888 it began field trials of its newest product: a self-winding clock equipped with a synchronizer. In 1889 one such timekeeper was sent to Pritchett, who mounted it in the extreme-vibration environment of an East St. Louis railway depot. Signals transmitted via telegraph wire from Washington University's observatory across the Mississippi River in St. Louis actuated its synchronizer. This particular field trial was quite important to the clock manufacturer, since a favorable report would certainly increase sales. Not only was Professor Pritchett in charge of what was arguably the country's largest railroad-time service; he was also the superintendent of time service and watch inspection for the Wabash Railroad.[9]

Pritchett lauded the Self-Winding Clock Company's product, for its greatest daily error had been only two and a half seconds. He ended his testimonial by remarking that "this clock has attracted considerable attention, and a number of roads are asking about the possibility of obtaining others."[10]

At this juncture, Pritchett may have known that Self-Winding and Western Union had a joint marketing plan, for other astronomers had already received brochures describing the venture.[11] But he seems to have concluded that his observatory's time signals would continue to be used throughout the Mississippi Valley.

In June a clock-industry trade paper noted that "Jas. Hamblet, manager of the time service of the Western Union Telegraph Co., has been appointed by that company to introduce the self-winding clock which it has contracted to put out on rental for the Self-Winding Clock Co. The Western Union Co. is ready to put them up wherever its wires run." A little later the trade press reported that Western Union was sending its agents "about the country soliciting orders" for clocks, offering to install them "on a rental basis of $1 monthly" per clock. According to the account, the company was guaranteeing that the timekeepers would not vary more than "one-sixtieth of a second from standard time"; this unprecedented claim of accuracy was sheer promotional puffery generated by the sales force.[12,13]

In August Pritchett learned the full import of the joint venture. "The . . . [Western Union] Company has recently served notice upon all railroads using our time that they would be expected to take time signals hereafter from Washington through the Western Union Company," he informed the Naval Observatory's superintendent in a hastily prepared letter. Facing financial ruin, for the telegraph giant was enforcing its service contracts with the railroads as a means of preventing his St. Louis observatory from running its own wires, the astronomer appealed to the superintendent: "I am led to believe that a request from you to the Western Union to limit the use of the Naval Observatory's signals . . . would have great weight not only with them but also with the railroads themselves," he wrote, envisioning the prohibition of the transmission of the Naval Observatory's noon signal to those "parts of the country already furnished with time signals." Pritchett supported his appeal with examples of Western Union's arrogance, among them that "the company assumes to control the government time service, for which it alone contracts."

Pritchett warned the superintendent that Western Union's use of these noon signals to synchronize regionally placed master timekeepers—that would in turn control banks of additional clocks—would bring discredit on the Naval Observatory. "As these signals will purport to give the [Naval] Observatory time the Observatory is certain to come in for the credit of the errors which are sure to come about in any such system."

Pritchett's statement diplomatically implied that a government time service carried a more impressive cachet than that of his own observatory's time. However, since that particular concern was in fact shared by all observatory time services, Pritchett's warning was simply fluff: "protecting" the Naval Observatory's good name really protected the astronomer's time service.

Capt. Robert Pythian, superintendent of the Naval Observatory, responded after he reviewed Western Union and Self-Winding Clock Company circulars. Indeed, the printed matter stated that the clocks would be "corrected daily by signals from the Naval Observatory." But with nothing more in these documents to support Pritchett's assertions, Captain Pythian requested additional information. (Some of Pritchett's statements had described Western Union's plans incorrectly.) He also asked the astronomer if he wanted his early-August letter to "be treated as official." If so, he would contact Western Union "for an explanation of the unwarranted use of the name of the [Naval] Observatory."

Pritchett's tardy response ended the matter. Western Union's statements had been made by "local canvassers" and by "the local [telegraph] superintendent in personal letters to railroad managers." Actually, the astronomer wrote, all his concerns vanished a few days after he posted his appeal. He, along with representatives of his customer railroads, met with Western Union officials, and the meeting's outcome gave him "reasons to expect that the matter will be arranged in a satisfactory manner." Pritchett closed his second letter to Captain Pythian by asking that both be treated as "personal communications."[14]

Although three Midwestern observatory directors had prevented Western Union from capturing their railroad business, the situation remained unstable. Self-Winding and Western Union continued to rent clocks whose time displays were linked to the Naval Observatory's noon signal. With nonrailroad business establishments also renting these timekeepers, the market for time was expanding—but that expansion was not helping local observatory services.

While Pritchett worried, Payne brooded.[15] Frustrated and apparently now suspicious regarding the lack of strong support from his Midwestern colleagues, he once again urged a united front. "I do not feel like giving up in this matter without a determined effort to withstand the wrong which has already been committed," he vowed, ending this note to Comstock with a threat: "Unless there is a united effort soon, I shall take the matter into my own hands . . . and if my plans interfere with you in any way I shall be sorry."[16]

Payne had already taken the matter into his own hands. His letter expressing resentment and anger toward the Naval Observatory's leadership was on the superintendent's desk. The Carleton College astronomer began by quoting from what he claimed was a just-received letter from an observatory director.[17] This unidentified astronomer reported to him that the Naval Observatory's superintendent felt "bound to furnish time signals to . . . the W. U. Tel. Co." Even if the telegraph company used them "to crush out private observatories it is something which the Supt. can not help and for which he is not responsible," the unidentified astronomer continued. Payne asked Captain Pythian "if you are rightly interpreted . . . for in using it I wish to represent your position exactly and truly."

Then Payne demanded confirmation of, or modification to, a statement attributed to Western Union: that "the Company was under contract relations with the National Observatory by which they were entitled to use the Washington Time throughout the United States for such purposes as they wished in their commercial business." Saying that the account had the ring of truth—else why would a Western Union vice president repeat it?—Payne expressed surprise at Captain Pythian's action to "make such a contract," and he voiced concern at the superintendent's inability "to perceive its effect on local Observatories," as well as his lack of sensitivity "in the interest of Astronomy and scientific institutions throughout the country." Hinting that he intended to bring the matter to the attention of his fellow astronomers and their friends, Payne closed his letter with a request for an early reply.[18]

Payne's concerns were already in print. In fact, his letter to the superintendent came only a day or two before the Naval Observatory received its copy of the December *Sidereal Messenger*. Headlining his remarks with "Observatory Local Patronage Threatened," Payne told his readers that Western Union officials "interviewed the Superintendent of the U.S. Naval Observatory, and sought and obtained of that officer . . . certain definite contract relations by which the telegraph company should have the privilege and right to use the [Naval Observatory's] time . . . for its own uses"—hedging his account with "as they say." The editor continued: "What arrangements the Superintendent of the United States Naval Observatory entered into [during the negotiations] with the telegraph company . . . it is not apparently very easy to find out."[19]

Stung by this two-pronged attack, Captain Pythian fired off telegrams—one to Payne and the other to Western Union's general manager and senior vice president, Gen. Thomas Eckert. To Payne he wrote: "In answering you[r letter], I assume that Pritchett is the director referred to. Am I correct?" To Eckert he outlined the assertions now in print, asking him to state if company officials had ever met with him (Pythian) to discuss the commercial aspects of telegraphing time, or if any of the company's officers had been authorized to say that they had done so.

Western Union telegraphed its denial on 13 December, following it up with a more detailed letter. Payne did not reply. In a cover letter attached to his formal response to Payne's query, Captain Pythian was blunt. Terming the editor's assertions "falsehoods," he called their publication without waiting for corroboration "an outrage you have perpetrated upon me, either by your journal or your informants." When Payne's acknowledgment suggested a drawn-out process, Captain Pythian insisted that his response not only be printed in its entirety but also in a single issue of the *Sidereal Messenger*. Payne complied, publishing the public parts of the interchange in February.

A private exchange between the protagonists also took place. Payne voiced his contempt of Captain Pythian for releasing and demanding the publication of Pritchett's private letters. Enlarging on his discussion of the superintendent's motives, Payne urged him to make public any relations between the "Naval Observatory, the Bedford Clock Company [sic], and others that may be mentioned." Payne was linking William Gardner's eleven-month-old agreement with the Self-Winding Clock Company to the Naval Observatory. Captain Pythian's response to Payne's demand for information shed no light on this new subject; the superintendent insisted that he had "no interest of any kind in any time transmitting company."[20,21]

While he was conducting his "fact-finding" mission at the Naval Observatory, Payne appealed to the country's observatory directors. One of his first letters went to his old Midwestern colleague, Edward Holden, now director of the University of California's Lick Observatory. Payne's late-December letter came at rather an inopportune moment. Holden was negotiating with Western Union, still seeking some *quid pro quo* arrangement by which the observatory's heavy cost burden—maintaining a seventeen-mile line of wire to San José—could be transferred to the telegraph company. Holden's ideas centered on the observatory's white elephant: its time service. He offered Western Union free signals in exchange for immediate repairs to the line—estimated to cost $1,000—provided they would then take over the line's maintenance, which was costing Lick $250 every year. (Holden, who inaugurated the time service in 1887, had already spent thousands of dollars equipping the enterprise; his major customer was the Southern Pacific Railroad, which paid one dollar a year for Lick Observatory's transmissions of 120th-meridian time.)[22]

Payne wanted a united front to deny the telegraph company free access to observatory signals; Holden was trying to give the signals away. But in an unthinking display of inconsistency, the Lick astronomer agreed with Payne's proposal—that both a circular outlining the telegraph company's encroachment be prepared and that a convention of observatory time-sellers be called. Holden underscored his views by writing: "I am heartily desirous of preserving the independence and support of all the various Observatories. But I think & always have thought that our time-services need unification & very likely the Convention can bring this about."[23]

Queried later on the issue of free time signals, Holden summarized: "My idea was [in 1876] & is that the U.S.N.O. should return to the U.S. what practical service it can. It should determine longitudes, drop time-balls in the principal ports, & do *some kind* of a time-service *free*. Just what kind etc. I don't think I ever said & I have not fixed opinions about it now." Again the astronomer was being inconsistent, for any increase in Lick Observatory's time service billings—and thereby a return on its investment—was blocked by the Naval Observatory's transmission of free signals from its Mare Island facility. Holden, made nervous by Payne's remarks in the *Sidereal Messenger*, now cautioned: "I prefer not to be named in this unfortunate dispute."[24]

Two leading figures within the astronomy community, Samuel Langley and Edward Pickering, finally intervened. As one of his biographers noted, Langley's dual roles as secretary of the Smithsonian Institution and director of the Allegheny Observatory made it difficult for him to take any public role in the controversy. Moreover, Langley still had a call—currently twelve hundred dollars annually—on the gross receipts of the Allegheny Observatory's time service, which he now administered from afar. Even though Langley asserted that he had not used these observatory revenues in the past two years—except for "actual expenses of travel, etc."—he was sensitive to the income that Allegheny Observatory derived from selling time. Early on Langley directed his observatory assistant to answer Payne's call for a convention—cautiously.[25]

Staying in the background, the Smithsonian secretary spoke with Captain Pythian. He also penned three letters to his old friend Pickering, a comrade in earlier time-distribution wars. "The real cause of the difficulty seems to lie not with the present superintendent of the U.S. Naval Observatory," Langley wrote, "but rather with the Telegraph and Time companies." Langley also voiced dismay that the Pythian-Payne correspondence and Payne's statements in the December *Sidereal Messenger* had "tended to a personal controversy that is not likely to lead to any satisfactory result."[26]

Langley proposed that someone not directly involved in the dispute prepare materials that could be presented to the secretary of the navy, and he urged Pickering to assume that role. In Langley's view, these materials should articulate the damage being done by the Naval Observatory's free signals. Noting that the secretary of the navy "will probably in the ordinary routine refer the letter to the Superintendent . . . for an opinion," Langley suggested that "an inquiry be first addressed to Captain Pythian as to whether . . . he would feel able to make such recommendations to the Secretary as would elicit an action preventing the telegraph and time companies from using the Naval Observatory signals in a way prejudicial to the interests of the other observatories."

Captain Pythian received two letters from Pickering.[27] The Harvard astronomer was quite direct regarding the enormous problem that Payne had caused, for he wrote: "Professor Langley . . . tells me that you have had reason to object to the tone of recent utterances by Professor Payne . . . [director of] one of the institutions which apprehend the loss of valuable patronage." He expressed the hope "that you will not allow any misunderstanding on the part of Professor Payne to interfere with your purpose of protecting him and others . . . from a danger the apprehension of which may have made him unduly sensitive." Voicing concern that progress in American astronomy would be hurt by the conflict, Pickering continued: "It would be most unfortunate . . . [if] any personal difficulties should interfere with harmonious actions among the directors of our observatories."

In his second letter, Pickering acknowledged that he was aware of the superintendent's sympathies toward the plight of those smaller observatories threatened by Western Union's expansion of its time services. He noted Captain Pythian's disposition toward restricting, if at all possible, the use of the Naval Observatory's signals as a way to counter the threat. He asked him to comment on the plan being proposed by directors of the private observatories to appeal to the secretary of the navy as a way of retaining their "local patronage"—the direct and indirect benefits that accrued from providing a local time service. Finally, Pickering invited the superintendent to meet informally with "a convention of astronomers called for the purpose of discussing the present difficulties of the smaller observatories and the appropriate remedies for them."[28]

The Harvard astronomer's letters reached Captain Pythian in the nick of time, for the February issue of the *Sidereal Messenger* was already at the printer and contained the initial correspondence between Payne and the superintendent—with additional remarks by Payne.[29] He enlarged upon his attack on the superintendent with questions implying collusion and possible payoffs: "When was the salary of the instrument maker [William Gardner] raised? How many times, and by whose recommendation? When did this instrument maker sell the rights of his patent electrically controlled clocks to the Bedford [*sic*] Clock Company of New York City? How much money was paid for the general control of this patent? Were there deferred payments? If so, were those deferred payments made dependent in any way on the use of the Washington time?"[30]

The editor continued his offensive into the next month, alerting *Sidereal Messenger* readers to an article in the jewelry industry trade press entitled "U.S. Government System of Observatory Time," which had been written by the U.S. Navy lieutenant in charge of the Naval Observatory's time service.[31] Payne termed many of the statements it contained "very erroneous and misleading" and asked pointedly: "By what authority are they published?" He raised the specter of a "tri-partite arrangement between the U.S. Naval Observatory, the Self-Winding Clock Company, and the Western Union Telegraph Company," citing its effect on both the Dearborn Observatory and Allegheny Observatory.[32]

The major article in the March *Sidereal Messenger*, however, was one by Pritchett. The St. Louis astronomer added fuel to the fire by lecturing Superintendent Pythian on the proper interpretation of the 1877 "tacit agreement" between the Naval Observatory and Western Union. Criticizing Western Union for attempting to take away his observatory's customers, Pritchett focused on the Naval Observatory's distribution of free signals, which were giving Western Union its competitive edge. Nonetheless, Pritchett pleaded for cooperation among the affected parties, asserting that "the interests of the private Observatories and the Western Union Telegraph Company in the matter of public service ought to be identical." He also predicted that "American Astronomy will be the loser if this matter is allowed to disturb that *entente cordiale* which has always existed . . . between the Government Observatory on the one hand and the private Observatories on the other."[33,34]

Pritchett's mixture of policy lectures, criticism, and calls for compromise did nothing to smooth ruffled feathers; Captain Pythian now lumped him with Payne. A few days later the superintendent responded to Pickering's January letters. He declined the Harvard astronomer's invitation to the proposed convention, saying that his "presence would not be conducive to harmony."

Captain Pythian then turned to the proposed memorial to the secretary of the navy. If asked, he would first provide a full history of the Naval Observatory–Western Union relationship. Should the department request additional comments in order to understand the astronomers' appeal, he would recommend that the Naval Observatory's time signal "be restricted so as not to interfere with the rights of other observatories." But, he added, "I have grave doubts as to the practicability of confining business of the Western Union Company to prescribed limits." Further, any "petition to the Department to forbid the use of the Naval Observatory signal for commercial purposes" must be a last resort, for the Navy Department was unlikely to grant such a request.[35]

Langley's private letter to Pickering was even less encouraging. "It looks as if the observatories . . . have to proceed to overt measures to protect themselves," the Smithsonian's secretary wrote. Observatory directors must "decide whether the appeal shall next be made to the Secretary of the Navy, and also whether . . . the final one shall be made to Congress. I regret to say that it now looks as though the best prospect of redress, was [*sic*] probably to be found in the latter."

Langley continued his pessimistic assessment by warning Pickering that "the people (Gardner & others at the Naval Observatory) interested in pushing the Gardner clocks, wish *the Naval Observatory time as an advertising card*, [and] will not improbably succeed in their aim, of getting railroads to substitute them for the continuous signal system of *Cambridge* and *Allegheny* unless some action is taken."

Events in March were happening much more rapidly than Langley realized. A week later found him informing Pickering that "the Pennsylvania R.R. Co. . . . has been approached with reference to exchanging their present service of continuous signals for that of the Gardiner [*sic*] Clocks with the so-called Government time"—although, he went on, "I do not personally care very much about it, and as Secretary of the Smithsonian, I cannot interfere." Langley was still director of the Allegheny Observatory.[36] Worried about protecting its finances, he wrote privately to the observatory's major trustee and to a vice president of the Pennsylvania Railroad.[37,38] Langley then journeyed to Boston, where he conferred with Pickering. A day or so later the astronomy community began its public campaign for redress with a rolling barrage in the national press.

Under such story banners as "Time is Money" and "The Time Pedler," in which Western Union's time-service practices were described, reporters quoted the country's astronomy directors at length.[39] Payne ridiculed the telegraph company's time, claiming a huge error of two minutes and naming the director of Allegheny Observatory as his source.[40] He also repeated his story of collusion and possible payoffs at the government's observatory.

Langley, identified only as the secretary of the Smithsonian Institution, supported the current situation of regional time services by noting the "importance of uniform accurate time in preventing disaster and in permitting [additional] traffic" in railroad operations. And Pickering proposed an astronomers' convention, to be followed by an appeal to Congress, whereby "it is hoped that this greedy giant [Western Union] will be forced to loosen its grip on American science."

One reporter questioned Captain Pythian. After first noting that Western Union had been using the Naval Observatory's time signal for over two decades, the superintendent remarked, "The charge of collusion with the Western Union made by Prof. Payne is the rottenest kind of rot. . . . I hope there will be an investigation. It will please me, and the sooner it comes the better, as it will illustrate what a talented liar Prof. Payne is." Payne termed this statement as one "unfit to copy, much less to answer. The language of the slum and the brothel which the superintendent relishes so well shall have no place in THE MESSENGER."[41]

One enterprising reporter found a crack in the astronomers' unanimity. Lewis Boss, director of the Dudley Observatory, asserted that no reason existed why "one observatory, owned by the government, should not furnish the time for the whole country," for "persons . . . needing accurate time . . . are not willing to pay" the actual cost. Consequently, "If a certain portion of the means of an observatory, in addition to what is derived from customers, must be devoted to maintaining such time service, then such maintenance is a direct injury to the scientific work of the institution. The function of an astronomical observatory is scientific observation and investigation, and it should be distracted from the pursuit of this object as little as possible."[42]

This remarkable statement by an observatory director raised a fundamental issue for astronomy. Which observatories were predominantly research institutions, and so perhaps deserved public concern? Which ones were simply business enterprises that found themselves at a major competitive disadvantage? But the time for reflection and thought had passed.[43]

The astronomers rushed to complete their memorial, and by mid-April copies were in print.[44] Representatives of one Canadian(!) and twenty-three American observatories attached their names; Langley's assistant at Allegheny Observatory signed in the director's stead. "Solidarity" had been the watchword among these college and university observatory directors. Fewer than half of their institutions were operating public time services, and, of the directors themselves, perhaps half were engaged in any kind of research.[45]

Pickering sent the petition to George Hoar, the respected and influential senator from Massachusetts. The Harvard College Observatory director asked him to lay it before the secretary of the navy, to whom it was addressed. Naively wise to the ways of Washington, he requested that Senator Hoar include "the personal support of such Senators and Representatives as take interest in the promotion of science." (Apparently Pickering believed that the senator or his staff routinely performed such lobbying tasks for constituents.) To reinforce his plea, Pickering added: "I should regard the loss of the support which . . . [directors] now receive from their time services as a severe blow to the progress of astronomy in this country."[46]

The astronomers' memorial focused on one fundamental issue: for years the astronomers had provided time signals as a "constant public service." Now, suddenly, Western Union's time service was causing the observatories to lose both the income and the goodwill of their local patrons. As a remedy, the signers proposed that the Naval Observatory cease transmitting its time signals to Western Union for subsequent use in commerce. They asked the secretary to direct this action if he found that current practice was "injurious to the interests of American astronomy." Their document contained not a single word about a tripartite arrangement involving any time-service company, Western Union, and the Naval Observatory.[47]

The memorial arrived in the office of Secretary of the Navy B. F. Tracy accompanied by a cover note from a somewhat bewildered Senator Hoar. As Langley had surmised in January, it was forwarded through the usual channels to the Naval Observatory, Captain Pythian receiving the package in mid-May. In his official response, the captain outlined the twenty-five-year history of Western Union–Naval Observatory cooperation, summing up their thirteen-year time-service agreement as one "perhaps informal" in nature—under which the observatory furnished a time signal daily, with Western Union distributing it to "time-ball and other public stations without charge," while charging a fee to distribute the time to corporations and individuals. He insisted that Western Union "does not enjoy, nor does it claim, the sole right to the Observatory signal"; his specific example was the use of the signal by the now-defunct Baltimore & Ohio Telegraph Company. "Any other [telegraph] company could have obtained the same privilege by going to the same expense of making the connections [at the observatory]," he argued.[48]

Captain Pythian turned to the conflict between Western Union and the regional observatories, terming it a recent one caused by the telegraph company's aggressive business practices: prohibiting "the use of its wires to competitors," urging "patrons of private Observatories to receive over its wires time from this Observatory," and promising subscribers "a [time] service at lower rates." Clearly, Western Union was gaining a major share of the time-service market. Further, wrote the superintendent, the astronomers' memorial "does not overestimate the injury to the progress of astronomy" that would follow if the company were to lure away all the observatories' customers.

The superintendent saw little hope, however, of assisting the private observatories. Western Union's success, he insisted, derived, not from its receipt and use of the Naval Observatory's time signal, but "from other considerations": (1) its ownership and "control of the patents for the best known devices for transmitting signals, and for automatically winding and correcting time-pieces"; (2) its "vast system of wires, its large corps of skilled operators and its complete outfit of electrical apparatus . . . [which allows it to] operate a time system without any appreciable additional cost"; and (3) the fact that "many of the Observatories . . . depend . . . upon the Western Union wires" for the transmission of their own time signals, an arrangement that could be denied "at any moment."[49]

Obviously, the Naval Observatory could cut off its time signal to the telegraph company, an "action that would seriously embarrass . . . the Western Union Company." But that action would probably cause the telegraph company to restrict the other observatories' transmissions immediately, thereby destroying their local businesses. "The struggle would then begin for the reestablishment of what is known beyond question to be a lucrative business. Can any one doubt who would be the victor in such a contest?" Captain Pythian asked.

The superintendent concluded his twenty-page report with a recommendation that "the Department lend . . . its kindly offices to effect a compromise" by creating a commission "composed of the Secretary of the Smithsonian, one or two distinguished Directors of Observatories, and a representative of the Navy Department . . . to meet and confer with . . . officials of the Western Union Telegraph Company." By taking this stand, Capt. Robert Pythian, superintendent of the U.S. Naval Observatory, had totally supported his fellow observatory directors.[50] Sent to the secretary's office, the official report and its several appendices were transmitted to Senator Hoar. The senator returned them, along with his observation that "the suggestion of Commandant [*sic*] Pythian that a Commission be appointed . . . is wise and desirable."[51]

The astronomy community waited. Late in July Pritchett wrote Congressman Henry Cabot Lodge of Massachusetts, asking for word on the matter, which is "of [such] great importance to the Observatories of the country I trust the [Navy] Department will give it immediate attention." The congressman transmitted the astronomer's inquiry to the secretary of the navy, and his office answered by referring to the Naval Observatory's mid-May report and the department's response to Senator Hoar.[52]

More time passed. In November, Payne, identifying himself as a journal publisher, wrote asking if a "reply had been made to . . . [the 15 April memorial] and [subsequently] sent to some one of its signers." The secretary of the navy responded: "No further action has been taken by the Department than to call for a report from the Superintendent of the Observatory, which report with the memorial was submitted to Senator Hoar, who on behalf of the Astronomers was interested in the subject, for his information and by him returned to the Department."

Although Payne was probably not in that small group of observatory directors aware of Captain Pythian's position with regard to their situation, and most certainly was not one of those very select few who may have seen the superintendent's official response, even he understood the Navy Department's meaning: it was ignoring the astronomers' petition.

Very likely, Secretary Tracy's decision was influenced by someone on his immediate staff, for a contemporary analysis of the situation exists. Although the text places strong emphasis on Superintendent Pythian's report, the unknown author's remarks show no sympathy at all for the astronomy community.

"That the people and the government [of the United States] are benefitted by the present arrangement with the Western Union Co. is apparent," the analyst begins. The government's time signal is being sent free to navy yards and branch hydrographic offices and used by "the navy and the vessels of the mercantile marine"; stopping the signal would result "in embarrassment and [immediate] loss to the navy." The navy would then be forced to pay the costs of sending time to those places.

"That [Western Union] . . . benefits . . . is nothing against the arrangement," the analyst continues. Moreover:

> The request of the petitioners is not based on any broad principles, but in the growth of the idea that the money now made by the Western Union might be squeezed out of the Company by various local observatories. That the petitioners have not regarded the subject in a practical business way is apparent to any one understanding the subject.

The secretary's staff assistant ends with a warning:

> Should the navy in this matter disregard its own and the public interest [by cutting off the government time signal], local observatories would not benefit thereby as the petitioners suppose . . . [for Western Union] would still control the paying time service. That the power of the local observatories to prevent . . . [Western

Union from doing so] would be futile, is clearly shown in the letter of the Super-intendent of the Naval Observatory.

Langley, Pickering, Pritchett, Payne—and Pythian—had all failed.[53] A dozen institutions now stood defenseless against Western Union's onslaught. What actually happened to these "minor observatories," as they were being called by Commodore George Dewey, the Navy's spokesman?

16. A SEVERE BLOW TO THE PROGRESS OF SCIENCE

Looking back, it is easy to smile at the dire warnings of disaster in astronomy. All assertions by the community in that regard proved false.

It is not so easy to smile about how a threat to research in astronomical science was invoked to protect what were in fact no more than faltering business ventures. During the time-service controversy, no leader voiced concern over the very meager scientific contributions from the plethora of small, poorly funded institutions that sold time locally. In the nineteenth century, true progress in astronomy required large telescopes coupled to photographic and spectroscopic instruments and maintained by well-endowed institutions. And although positional astronomy was important at a few research sites—with public time services its spinoff—it was scarcely a cutting-edge activity. Yet all observatory endeavors were lumped together as the community pursued a futile course of action—all in the name of science.

In 1890, astronomers portrayed the Western Union Telegraph Company as a predator. Their one major example was the Dearborn Observatory's loss of its contract with the city of Chicago. Indeed, the observatory was the prototypical impoverished institution—until its 1888 move to Northwestern University in Evanston. That move led to a new observatory building, an endowment, and a salaried professorship for George Hough including a house on campus—transforming completely both his professional and financial circumstances. No longer was he the director of a "minor observatory."

Hough remained convinced that private observatories had a public responsibility. In late 1889 he began distributing time again, controlling self-winding clocks of his own design placed in the president's and treasurer's offices. If not Chicago, then Northwestern's campus![1] When this time service finally ended in 1909, at Hough's death, the astronomer had been providing time signals to customers for half a century. Clearly, once such activities start, they are almost impossible to terminate—especially when such a "public service" is self-defined by its supplier.

By 1890 Edward Holden had been involved in public timekeeping for a dozen years. Now he was director of the Lick Observatory, the largest research observatory in the country. Thus far, publicity had been about the only return on the thousands of dollars invested in the time service's equipment. But Holden had founded the time service, and James Keeler, who had constructed the system, succeeded him as observatory director. So Lick Observatory Time Service signals to the Southern Pacific continued well into the twentieth century.[2]

Edward Pickering found the U.S. Naval Observatory a convenient whipping boy when he terminated Harvard College Observatory's time service. In his 1891 announcement he asserted that his facility stood ready to continue transmitting its accurate, continuous signals even if that meant a financial loss. But with Naval Observatory time signals being offered at such low rates, Pickering could claim that "the Harvard College Observatory is therefore relieved of this duty." He took a parting slap at both the government observatory and its telegraph company partner, warning his Boston public that they might lose quality and reliability by using this cheaper service.[3]

Pickering scarcely mentioned the other factors that influenced his decision. First, Harvard College Observatory's time-distribution equipment was wearing out and needed to be replaced. As early as 1885 Pickering had informed the Board of Visitors of the unsatisfactory quality of time transmissions, writing that "the time service . . . has not fully succeeded at all times during the year in maintaining the high standards of former years." By 1891 the situation was critical.[4]

Western Union was indeed offering low rates for time signals. But it was not the difference between its twenty-five or thirty dollars a year and the observatory's charge to city jewelers of fifty dollars for its continuous signals that forced Pickering's decision. Rather, it was the certain loss of thousands of dollars he faced from the Boston-centered railroads' cancellation of their annual subscriptions. After one road decided to switch to Western Union, Pickering's assistant warned, "Are any of them likely to continue this practice [of using our service] after learning that they can get similar signals at perhaps one tenth the cost from a company having the largest facilities for maintaining telegraphic connections?"[5]

Also in the foreground of Pickering's decision was Boston's horrific Thanksgiving Day Fire of 1889. At 6:48 in the morning, a time-service company's transmission line crossed with a high-voltage streetlight wire; the energy surge along the former's unfused, low-voltage line initiated a building fire. Spiraling out of control, the resulting conflagration caused millions of dollars in property losses and interrupted business. Worst of all, four firemen were killed. Fire insurance underwriters strengthened the region's electrical safety codes, while city authorities adopted stiffer regulations and instituted tighter inspections.[6]

Harvard's president, Charles W. Eliot, also aware of the near-certain loss of the railroads' subscriptions, voiced his concerns about safety. Pickering agreed that they were significant. Thus in his printed announcement terminating the Harvard time service, the astronomer touched on the danger of "fire, bodily injury, or even loss of life" from electrical lines, remarking that "the financial officers of the University regard such risks as more than offsetting the receipts for the time-signals."[7] Competition from Western Union gave Pickering a way to exit gracefully, and he took it. Harvard College Observatory time signals ceased after 31 March 1892.[8]

With a nonresearch program over and its protection no longer a burning issue, Pickering must have felt some relief. Indeed, in a postscript to the observatory's abandonment of "the Harvard system of time signals," he noted that "the action of the U.S. Naval Observatory in furnishing the [New England] public with time signals of moderate accuracy has not proved to be as injurious to this Observatory as to many others." He also reminded his audience of the dangers of electricity. In this one short paragraph, the director of the Harvard College Observatory thus demonstrated institutional pride, anger at interlopers, the avoidance of a possible disaster, and the prospect for a rosy future.[9]

Allegheny Observatory's threatened loss of revenues had been the accelerator in the simmering mixture of anxieties that exploded in the spring of 1890. Its research status was already precarious, and many believed that the observatory's primary telescope—sited in the smoky Pittsburgh atmosphere—was obsolete. Samuel Langley had already directed his assistant away from astronomy to aerodynamic experiments.[10]

Langley's enormous prestige in Pittsburgh and his lobbying of Pennsylvania Railroad officials delayed Western Union's onslaught. But his special relationship with the community ended in April 1891, and James Keeler became director. Late in 1893, Keeler announced that Allegheny Observatory "suffered a heavy financial blow . . . by the withdrawal of some of the railroad lines from their contract with the Observatory for its time signals." He added: "Nearly all the observatories in the country have suffered, in the same way, by the action of the U.S. Naval Observatory in supplying its time signals free to the Western Union Co., which sells them to the railroads." This assertion by Keeler has long lent credence to his peers' history of events.[11]

As Keeler's biographer has detailed, Allegheny's loss was not an unmitigated disaster. Observatory time was still sold to the Pennsylvania, but at reduced rates. In the first years, subscriptions from the university's president and several trustees made up most of the loss, and receipts from the time service eventually stabilized at seventeen hundred dollars—half of the gross during Langley's reign, of course—but no more than a loss of six hundred dollars in operating income when Langley's "cut" is taken into account. Observatory records show that this amount was still sufficient to hire an assistant and pay for a bit more than just new library books. And perhaps the revenue loss was a blessing in disguise, for the community eventually realized that counting on such sums was simply no way to run a research facility.[12]

Other research-oriented observatories continued to sell time, with varying degrees of financial success.[13] In the maneuvering for local revenues, Brown University's observatory director showed that Western Union was not invincible. In 1893, under the banner "Providence Has Local Standard Time [sic]," the trade press announced: "Hereafter the Ladd Observatory time will be the standard for Providence. . . . The Western Union Telegraph Co. have heretofore had charge of this [distribution] system, but it will in future be controlled by the local [fire and burglar alarm] company."[14]

The observatory at Carleton College in Minnesota continued selling time, and its railroad revenues apparently remained unaffected by Western Union. But when the facility was rechristened "the Goodsell Observatory," William Payne enlarged his list of complaints to include the sly jokes being made about the new name. Carleton College's telegraph-based time service finally ended in 1931.[15]

These examples demonstrate the enormous influence that the private observatories, allied with their benefactors, had in their respective regions. Payne's clarion call, "Observatory Local Patronage Threatened," was certainly wildly overplayed. The number of local time services still operating after 1890 shows that the threatened elimination of university time services posed by Western Union's use of the Naval Observatory's noon signal did not happen. Of even greater significance, the corollary notion—that the Naval Observatory–Western Union endeavor provided the entire United States with uniform time—is revealed to be only a myth.

The time wars continued into 1891, with a final skirmish led by Henry Pritchett.[16] In a front-page story, the *New York Times* reported that the country's astronomers "are now uniting their forces for a struggle from which they confidently expect to emerge victorious," urging Congress to "transfer the control of the United States Naval Observatory . . . to the hands of a purely scientific and astronomical board." After naming the six observatories involved, the writer remarked that "with the telegraph company receiving the signals gratuitously from the Naval Observatory . . . [it] can . . . greatly underbid the observatory service[s]." As a result, "the universities say that . . . the only [financial] means that they have of advancing the cause of science will be withdrawn."[17]

There was no ambiguity in intent here, for the article was printed on the same day that one astronomer's public demand for a civilian director reached the Naval Observatory. Further, Allegheny Observatory's new director noted privately that protecting the income derived from time services would require that "the Naval Observatory passes into civilian hands."[18,19]

The continuing attacks on the Naval Observatory had a chilling effect on its time-service activities. Early in 1892, its officers reiterated their responsibility to diffuse results "as widely as practicable," since all taxpayers had "the right to participate in the benefits arising from its work." But after citing its long history of distributing time signals "to all . . . who may apply and provide the necessary apparatus," it now saw its role limited "to the furnishing of an accurate daily signal at noon." Moreover, the country's center for timekeeping stated that it had *nothing to do with the distribution of these signals* [emphasis added] and is alike indifferent to the extension or contraction of the [private] time-services." So the Naval Observatory turned inward and shut its gates on received-signal accuracy concerns.[20]

While this response was certainly a victory for those interested in advancing astronomy, it was a severe blow to other disciplines. As Cleveland Abbe had argued many years earlier, studies in meteorology and the burgeoning fields of seismology, terrestrial magnetism, and atmospheric electricity required a well-characterized, national system of accurate time. James Hamblet's data, now covering a period of a decade and a half, showed the great need for efforts aimed at synchronizing the country's several time-signal systems. Moreover, significant improvements were possible, as Canadian astronomers were already demonstrating.[21]

The issue of nonuniformity of time signals was not, however, a pressing problem in astronomy. Simon Newcomb's 1884 review of astronomical instruments provides one example. Noting that "astronomical work of the first class" required the highest quality timekeeper attainable, this leading astronomer observed that only a very few clocks existed that preserved "their rate with remarkable uniformity through considerable periods." Outlining a test process to demonstrate a timekeeper's quality, Newcomb added: "A clock which stands this test well may be presumed beyond doubt to keep its rate during *short intervals* [emphasis added], which is generally the important

SELF WINDING TOWER CLOCK
OPERATED FROM BOOTH OF
SELF WINDING CLOCK COMPANY
SECTION N, MANUFACTURERS' BUILDING,
WORLD'S COLUMBIAN EXPOSITION,
CHICAGO, ILL.

Fig. 16.1. Self-Winding Clock Company's tower inside the Manufacturers and Liberal Arts building at the 1893 World's Columbian Exposition. From Self-Winding Clock Company, "What is Standard Time?" U.S. Naval Observatory Library, William F. Gardner collection.

point." And Truman Safford reiterated the issue: "A good clock [for astronomy] . . . [is] required to run uniformly for a few hours," while watchmakers' clocks, "more severely tested in their commercial use . . . are expected to keep good time over several days."[22]

In his 1893 review of geophysical needs and timekeeping, Abbe voiced the hope that "some one more fortunately situated will, in the future, accomplish that which the Signal Service had planned" with regard to comparing time signals from different sources. But Abbe knew that this hope was a forlorn one; time, and astronomy, had passed the problem by. The Naval Observatory, now the undisputed center for national time, did not return to the issue until well into the next century.[23]

Even though it truncated its time-service investigations, the Naval Observatory continued distributing signals to controlled clocks in government offices and, of course, to various communications companies and firms. In 1903 the observatory's superintendent wrote that "18 Government time-balls and some 40,000 public and private clocks [are] corrected daily by naval time signals."[24]

The coverage reported was actually the result of the partnership between Western Union and the Self-Winding Clock Company. Dominant from the start, both companies spent lavishly on publicity to increase their market share. For example, at the World's Columbian Exposition of 1893 in Chicago, Self-Winding erected an enormous clock tower in the center of the Manufactures and Liberal Arts building, while at its own pavilion it exhibited self-winding, synchronous clocks. According to one report, at that moment Self-Winding had in service "over fifteen thousand clocks"—which implies $180,000 in gross annual receipts. As those capitalists who initiated the joint effort in 1888 had suspected, the market was enormous: by 1932 Western Union had placed about 120,000 clocks on its circuits, and the two organizations combined were grossing $1.5 million annually.[25]

Among the publicity handouts at the Exposition was William Allen's copyrighted color map of the railroad time zones. Another was *What is Standard Time?* containing a slightly edited version of Lt. Hiero Taylor's 1890 article on the Naval Observatory's time service, as well as short sections on self-winding, synchronized clocks, on Western Union's time service, and on Self-Winding's products. Only the briefest mention was made of William Gardner's patents, "which are now owned by the Self Winding Clock Co." As Langley foresaw, commercial interests emphasized the link to the Naval Observatory's daily signal.[26,27]

Elsewhere on the exposition grounds the Naval Observatory displayed its now-mature timekeeping and distribution technologies—technologies that had satisfied the public's and the U.S. Navy's needs for a generation, and which would satisfy much of the public for several generations more. Also at

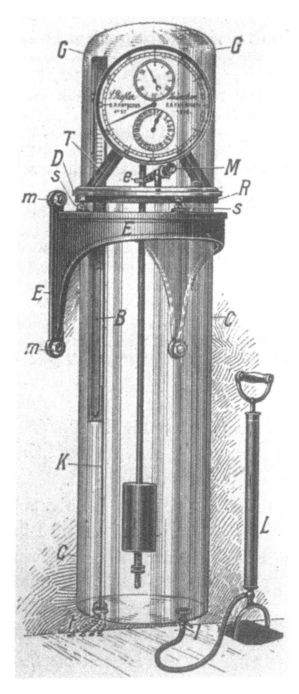

Fig. 16.2. Riefler's precision astronomical clock in its barometrically controlled case. From Sigmund Riefler, *Präzisions-Pendeluhren und Zeitdienstanlagen für Sternwarten* (Munich: Theodor Ackermann, 1907), 40.

the Columbian Exposition was a display of the next revolution in precision timekeeping: Munich instrument-maker Sigmund Riefler's clocks. Almost at once, these timekeepers would become the standard for the world's astronomical observatories and physics departments. Their long-term stability—the desideratum articulated by Newcomb—would also lessen the uncertainty inherent in the distribution of public time.

Waiting in the wings was the wireless transmission of time signals. Already voiced by respected British scientist William Crookes, in his seminal article "Some Possibilities of Electricity," was the prediction that the newly discovered Hertzian waves would be used "for transmitting or receiving intelligence" over great distances. A decade later the U.S. Navy would begin transmitting experimental time signals, inaugurating scheduled transmissions to the fleet in 1904 or 1905. By 1913 the use of radio signals as a time source for the country's jewelers and radio amateurs was in full swing.[28]

Now, at the start of the new century, public time services via wires was a most significant business enterprise, with several companies supplying time that was "good enough." The noontime signal from Washington reigned supreme, but it was not unchallenged. Still, for most citizens, there was only one uniform and precise national time: the "free" time signal of the U.S. Naval Observatory.

EPILOGUE

The phenomenon of astronomers selling time to the public and private sectors of society may be unique to the United States—or at least to North America. What little has been written about timekeeping in European countries indicates government control, whereby state-supported observatories provided time signals to railroads and state-run telegraph systems. Some might argue that America's sheer size made it inevitable that the nation would adopt decentralized timekeeping. Yet timekeeping in Russia was centrally controlled. Even with the countrywide adoption of time zones, trains on the Trans-Siberian Railroad run on Moscow's civil time. Thus sheer size alone cannot be the explanation. Until more national timekeeping studies are available, little more can be said on the subject.

However, size does influence time choices. All Western European countries were small enough that a national time based on the capital's meridian or, in the case of Sweden, based on the meridian that most nearly bisected the country's populated area, was good enough to be made official. True, the United Kingdom of Great Britain and Ireland maintained two times (until 1916), but crossing the Irish Sea was equivalent to crossing a national border.

However, the United States is so broad that no one official time, whether calculated from the meridian of the nation's capital, or even from some more central meridian—such as ninety degrees west of Greenwich, which was Cleveland Abbe's choice—would have been practical to use for civil purposes. Imagine if the time in the nation's capital had been declared the legal standard for the entire United States. Would San Franciscans ever be content to break for lunch when their clocks read 8:58 A.M., or to arise at dawn on March 21 (when day and night are equal in duration) with their clocks reading 2:58 A.M.?

Charles Dowd was right to conclude in 1870 that the United States requires more than one time for civil purposes. His plan for four (uniform) times—each exactly one hour different from its neighbor—was simple for anyone to comprehend. But Dowd was dead wrong when he argued that two times—local time and railroad time—would be needed everywhere. Moreover, he made his proposals at least a decade too early, when the railroads had other, more important issues to consider.

Even as late as the 1890s, if the federal government had legislated a civil-time system for the United States it would almost certainly not have been indexed to the Greenwich Observatory meridian. With one obscure exception—the colony of New Zealand—there was simply no precedent, legal or otherwise, for basing a country's civil timekeeping system on a meridian outside its own national boundaries.

Abbe certainly took a risk when he began lobbying for national legislation. The Naval Observatory was determined that the country's time be based on the meridian of the capital, as were the astronomical calculations in its *American Ephemeris and Nautical Almanac.* Without any doubt, Congress would have placed great weight on the Naval Observatory's position.

Abbe also took a risk by engaging the nation's railroads in the issue. He certainly was unaware of the detailed geographical distribution of America's railways. Neither could he have predicted how William Allen would use these data when the railroader began to create an Eastern railway time belt. Indeed, Allen could conceivably have calculated a mid-meridian thirty minutes and six seconds from the seventy-fifth meridian west of Greenwich, rather than the six seconds it turned out to be. Then, to keep the neighboring belt's mid-meridian an exact hour away from that of the Eastern belt, he would have ended with a set of what are today called "half-hour" zones: each an hour wide but halfway between the Greenwich-indexed hour meridians. (Currently, nine such zones exist.)

Before 1880, a small group of specialists wanted a timekeeping system based on the Greenwich meridian. They lobbied both publicly and privately, generated reams of propaganda, and simply ignored all the signs of general indifference or even outright opposition to their plans. Following the International Meridian Conference of 1884, they downplayed the fact that, insofar as the American public was concerned, the key event had taken place a year before.

Modern public timekeeping in the United States began on 18 November 1883. Could it have happened any earlier? Allen might certainly have launched his railroad industry campaign somewhat sooner—say, in 1879–80—but only if the American Metrological Society had been more prompt in sending out its membership invitations. Yet the society cannot really be blamed for its delay. Abbe, who proclaimed that scientists and citizens alike needed a uniform time system, took four years to prepare his report, although it consisted entirely of known material massaged into an advocacy document.

The U.S. might conceivably have adopted some sort of uniform system as early as 1869–70—if Samuel Langley had supported Dowd's plan, adding to it his own sales pitch: one time for all civil purposes. But human nature and Pennsylvania Railroad mogul A. J. Cassatt's desire that an institution within his state gain the credit for inaugurating such a change worked in harmony to sidetrack Dowd's proposal. And the lack of any urgent need kept it there.

Reliable synchronizers, invented in England and introduced into the United States in the late 1870s, certainly created new markets for telegraphed time signals. But even that innovative technology did not compel any move toward uniform time. Although using just a few times could certainly simplify operations, Gardner's 1883 patent describes how each timekeeper in a bank of clocks could be set to a different local time on receipt of a single pulse transmitted at noon from Washington.

In 1883, so little of the general public's world depended on having a uniform system of time that its adoption passed almost unnoticed. Today, of course, even proposing such a change would be front page news. The system's fundamental technical problem—the necessary one-hour shift at zone boundaries—would lead to a near-endless series of debates. Add the explosive issue of parental worry over added danger to schoolchildren created by shifts in daylight hours if the zones were changed, and not even Solomon could render a universally acceptable judgment.

Fortunately, our present system of civil time came into being over a century ago. The system is virtually transparent to those who use it, and our communications networks give us so many ways to check our timepieces that we seldom concern ourselves with accuracy or consistency. Granted the process that created the system was in no sense democratic. But the globally shared basis for Standard Time—a set of twenty-four equally spaced meridians tied to the daily course of the sun—is linked in a common-sense way to the cycle of daylight and darkness by which we live. I cannot imagine any more practical system of keeping public time.

REFERENCE MATTER

Appendix: American Observatory Public Time Services

	Observatory	Start (d/m/yr)	Region	Comments	End	Notes
1.	U.S. Naval	by 15/4/1845	Washington	time ball	16/12/1936	
		by 3/1856	Raleigh, N.C.	via telegraph		
		8/1865	Washington	city fire bells		
		1870s	South/East/			
			Midwest	via Western Union	ca. 1973	
		1926	Entire U.S.	via Western Union	ca. 1973	
2.	Harvard College	21/1/1851	Boston	to Wm. Bond & Son		
	(also as Cambridge	by 27/5/1851	New England	telegraph stations		
	Observatory)	8/1858	Boston	fire bells,		
				via Wm. Bond & Son	1862	
		1872		city and area RRs	1892	
		8/5/1878	Boston	time ball	1892	
		1902	Cambridge,			
			Boston		short-lived	
3.	Rutherfurd's	by 11/1852	Erie RR	also N.Y. &		
	(New York, N.Y.)			New Haven RR	ca. 1880s	
		1/1857	New York City	fire bells	short-lived	
4.	Dudley	ca. 22/1/1860	Albany	time balls, RRs		
	(Albany, N.Y.)	ca. 5/1860	New York City	time ball	short-lived	
		1882	Albany	city clocks	ca. 1890	
5.	Hudson,	after 1862	Cleveland		after 1870	
	Western Reserve					
	(Hudson, Ohio)					
6.	Detroit Observatory,	ca. 1863	Detroit	city, RRs, jewelers,		
	University of Michigan			jewelers' time balls	after 1885	
	(Ann Arbor, Mich.)					
7.	Washington University	1860s		local jewelers	intermittent	
	in St. Louis	6/1878	St. Louis	city clocks		
		ca. 1883		time balls, RRs	ca. 1901	
8.	Cincinnati	1868	Cincinnati	local jewelers	short-lived	
		after 1875,				
		before 1878		city clock, time balls	after 11/1912	a

Observatory	Start (d/m/yr)	Region	Comments	End	Notes
9. Dearborn, University of Chicago	1869 1874–75	Chicago	city, jewelers city, RRs, jewelers	8/10/1871 1888 & 1889	
10. Allegheny (Allegheny, Penn.)	1869 1/2/1871 1873	Pittsburgh Pennsylvania RR	local jewelers city clock	 3/2/1947	
11. Carleton College (Northfield, Minn.)	23/10/1878	St. Paul	RRs, time balls	12/1931	
12. Winchester, Yale University	4/1880	New Haven	city, RRs, telephone company, time balls	1918	
13. Morrison (Glasgow, Mo.)	1880	Kansas City, Mo.	RR depot, time balls	short-lived	b
14. Washburn, University of Wisconsin	2/1882	Madison	city, RRs	after 1892	
15. Boswell, Doane College (Crete, Nebr.)	1883	local	time ball	ca. 1900	c
16. Beloit College (Beloit, Wis.)	10/1884 11/1912	local	 time via radio	 after 1914	d
17. Mare Island Navy Yard (Vallejo, Calif.)	12/1884	San Francisco, Pacific Coast	RRs, telephone company, time balls	 1926	e
18. Buchtel College [University of Akron] (Akron, Ohio)	1886	local	time ball	after 1891	f
19. Lick, University of California (Mt. Hamilton, Calif.)	1/1/1887	Oakland, San Francisco	Southern Pacific RR, telephone company	after 1902, before 1923	
20. Iowa College [Grinnell College] (Grinnell, Iowa)	by 1889	local		after 1891	g
21. Underwood, Lawrence University (Appleton, Wis.)	ca. 1892	local		1894?	h
22. Ladd, Brown University (Providence, R.I.)	27/9/1893	Providence	local alarm company	after 1900	

NOTES

a. Ormand Stone, "Cincinnati," *Sidereal Messenger* 1 (1882): 33.

b. Henry S. Pritchett, "The Kansas City Electric Time Ball," *Kansas City Review of Science and Industry* 4 (1881): 720–23.

   Carr W. Pritchett, "Time Service of Morrison Observatory," *Publications of the Morrison Observatory* 1 (1885): 61–64.

c. [G. D. Swezey(?)], *Boswell Observatory* (Crete, Nebr., [1883]).

   "Boswell Observatory," *Sidereal Messenger* 3 (1884): 190–91.

   [Janet L. Jeffries], *Doane College's Boswell Observatory* (Crete, Nebr., 1996).

   Janet L. Jeffries (college archivist), personal communications, 1998.

d. Beloit College, "Local [News]," *Round Table*, 24 October 1884.

   Ibid., "Beloit Sends Time by Wireless," *Round Table*, 14 February 1913.

   E.A. Fath, "A Wireless Time Service," *Popular Astronomy* 21 (1913): 226–27.

   Frederick Burwell (college archivist), personal communications, 1998.

e. "Standard Time at Mare Island," *San Francisco Chronicle*, 8 November 1896, 12.

f. "The New Observatory," *Akron Daily Beacon*, 6 March 1886, 2.

   Buchtel College, *Catalogue: 1886–87 and 1887–88* (Akron, Ohio: Werner Printing, 1887), 12–13.

   "A time wire . . . broke . . . horse instantly killed," *JC&HR* 22 (24 June 1891): 17.

   John V. Miller, Jr. (University of Akron archives), personal communication, 1995.

g. "Observatory at Iowa College, Grinnell," *Sidereal Messenger* 8 (1889): 234.

   [Grant Gale], *College Time* (Grinnell, Iowa: Physics Department Press, [1935]).

   Leslie Czechowski (college archivist), personal communications, 1998.

h. William C. Winlock, "Progress of Astronomy for 1891 and 1892," *Annual Report . . . Smithsonian Institution, 1892* (Washington: Government Printing Office, 1893), 735.

   Carol J. Butts (university archivist), personal communication, 1998.

Notes

Introduction

1. Waldo, "Railroad and Public Time," 607.

2. The introduction of mechanical clocks in the fourteenth century began to alter the public's awareness of time. For two highly readable accounts of these changes through the eighteenth century, see Dohrn-van Rossum, *History of the Hour*; and Landes, *Revolution in Time*.

3. A 1997 map, "Standard Time Zones of the World," shows nine "half-hour" zones and two "three-quarter-hour" ones.

4. This remark was made by Sandford Fleming, a Scotland-born Canadian railway engineer and university chancellor. By contrast, in 1752, when England's Parliament voted to replace the country's old Julian calendar with the more consistent Gregorian reckoning, mobs took to the streets, howling, "Give us back our eleven days!"

5. Bartky and Harrison, "Standard and Daylight-saving Time."

Chapter 1

1. Additional tables provided other means to determine time; some—the rising and setting of the sun—were not precise enough for regulating a clock, while others—the moon's southing, for example—were more difficult to use. See Kelly, *Practical Astronomy during the Seventeenth Century*, 150–52; in the work's introduction, Silvio Bedini suggests that by the early eighteenth century, the sun-fast, sun-slow table was a common almanac feature.

2. Symonds, *Thomas Tompion*, 75. Here, "watch" means a timekeeper without a bell-striking mechanism.

3. In Shannon Spittler, "A Math Problem within an Antique Clock Label," *Pi Mu Epsilon Journal* 9 (fall 1991): 294–96; a portion repeated in *NAWCC Bulletin* 35 (February 1993): 86.

4. Whitfield J. Bell, Jr., "Astronomical Observatories of the American Philosophical Society, 1769–1843," *Proc. American Philosophical Society* 108 (1964): 8.

5. Some dates for the change are Geneva, 1780; England (London), 1792; Berlin, 1810; Paris, 1816—see Bigourdan, "Le jour," B.8–9.

6. Mean time is often considered as representing the motion of an imaginary sun

moving across the sky, the fictitious sun's meridional crossing and "noon by the clock" occurring at the same moment, day after day. For their professional work, astronomers are concerned with *sidereal time*—derived from the near-uniform rotation of the earth about its axis—as reflected in the apparent motion of the stars across the sky. Stellar objects arrive at the local meridian at the same moment of sidereal time, day after day; a well-regulated sidereal clock remains in synchrony with the motion of the stars. As the occasion demands, astronomers convert sidereal time into mean solar time, or to apparent time.

7. In addition to the several meanings of "true," this technical usage can also lead to confusion: "true time" is not the same as the "time of the true sun," which is apparent time. See Bigourdan, "Le jour," B.8, who traces the usage to, among others, Isaac Newton.

8. The idea of using a portable clock to determine longitudes was suggested in the early 1500s. Until the 1850s, the earth's rotation around its axis was viewed as uniform.

9. Harrison's triumph has been told and retold; one recent version, with astronomers cast rather as villains, is Sobel's *Longitude*. The most detailed, readable source is Andrewes, *Quest for Longitude*. The performance of H.4 quoted here is from Randall, "The Timekeeper that Won the Longitude Prize," 247. For a comprehensive summary of technologies to the present, see Williams, From *Sails to Satellites*.

10. Longitudes by lunars and the growing availability of chronometers meant that the work of ridding sailing charts of false islands and reefs could begin; see Henry Stommel, *Lost Islands* (Vancouver: University of British Columbia Press, 1984).

11. A navigator has to know both the chronometer's error and its rate, which are determined on land. Once his ship is at sea he can only pray that its going has not deviated much from its last recorded rate—or else have several marine chronometers on board so that a badly running one can be spotted.

12. Warner, "Astronomy in Antebellum America," 56; Whitney, *Ship's Chronometer*, 360–61, 380. West Coast observatories for rating chronometers included one established in 1848 or 1849 in San Francisco by dealer Joseph McGregor; see W. Barclay Stephens materials, NAWCC Library, Columbia, Pennsylvania. A competing observatory is shown in an 1852 newspaper advertisement; see David Myrick, *San Francisco's Telegraph Hill* (Berkeley: Howell-North Books, 1972), 30–31. Some American watch and clock manufacturers constructed timekeeping observatories equal in quality to those of universities and the federal government; see Waldo, "Longitude of Waltham," 175–82; Waltham Observatory, "Standard Time," 300–301; and Neidigh, *Elgin Observatory*.

13. The great economic significance of this advance is discussed in Albion, *Square-Riggers*, 247, 258–61, 265; and Albion, *Rise of New York Port*, 51, 320–25.

14. For the innovations in chronometers constructed by Edward J. Dent (1770–1853) during this period, see the summaries in Randall, *Time Museum Catalogue of Chronometers*, 37–40. For an appreciation of George W. Blunt (1802–78) and the family firm, see Bedini, *Thinkers and Tinkers*, ch. 16.

15. Dent, "Longitude between Greenwich and New York," 143–44; and BAAS, *Ninth Meeting*, 27–28. Also Blunt, "Experiments on Longitude"; a copy of this news-

paper clipping is in the Harvard College Observatory's scrapbook, a sure indication that William C. Bond was familiar with these trials. Blunt also informed the American government; see U.S. Department of the Navy, Board of Navy Commissioners, *A Report on the Subject of Chronometers*, 13–14, 26th Congr. 1st sess., 1840, H. [Ex.] Doc. 249.

16. Coastal charts based on chronometer values would have been adequate for the maritime industry, for ships' captains always posted lookouts when approaching land. Indeed, no sane one would ever have sailed solely by his navigator's reckoning. Hassler himself proposed both a trigonometric and a chronometer-based survey of the American coastline.

17. In a trigonometric survey astronomy plays a small but important role, for the latitude and longitude of only a few points have to be determined by stellar observations. The many stages of the overall charting effort are well delineated in an 1847 description—U.S. Coast Survey, *Annual Report for 1847*, 4.

18. Coast Survey, *Laws . . . with The Plan of Reorganization*, 9–14. Also, Cajori, *The Chequered Career of Ferdinand Rudolph Hassler*; and Charles H. Davis, "The Coast Survey of the United States," *American Almanac . . . for the year 1849* (Boston, 1848), 66.

19. Coast Survey, *Laws . . . with The Plan of Reorganization*, para. (1b), 9.

20. See, among others, the longitude articles written or sponsored by Sears C. Walker in the *Transactions of the American Philosophical Society*. With an articulated Coast Survey need, it was easier to support the science of astronomy, paying astronomers to collect observations and reduce the data.

21. Struve's data reduction schema became the Coast Survey's model for its own programs; his expedition is summarized in William Chauvenet, *A Manual of Spherical and Practical Astronomy*, vol. 1 (Philadelphia: J. B. Lippincott & Co., 1863), 323–24. In 1844 Struve conducted a chronometer-transport expedition with Astronomer Royal George Airy of England to determine the longitude between the observatories at Altona and Greenwich.

22. Hassler's importance with regard to the directions taken and the resulting quality of Coast Survey work cannot be overstated. For a useful summary see Dracup, "Geodetic Surveying. I."

23. Bache (1806–67) was a central figure in the advancement of American science; see Reingold, "Alexander Dallas Bache"; and Slotten, *Patronage*, the latter offering an extended view of Bache's management skills.

24. The Liverpool Observatory, established to rate chronometers for commercial maritime interests, was located in the docks area; astronomer John Hartnup was its director. See Robert W. Smith, "The Hartnup Balance," *Antiquarian Horology* 14 (1983): 39; and Bartky and Dick, "The First Time Balls," 163, n31.

25. Dent noted a similar east-west discrepancy in his 1839 New York–London trials. The Bonds' reports were published as appendices in Coast Survey *Annual Reports*.

26. U.S. Coast Survey, *Annual Report for 1856*, 12. The agency's own value, based in part on the chronometric results, was not superseded until 1867; see Gould, "Longitude between America and Europe." Members of the committee appointed by the overseers of Harvard University to examine the observatory's work were more positive

than Bache in their judgment of the effort. In their 1854–55 report they quote with pride Bond's own statement of the utility of his chronometric value; *HCO Annals* 1 (1856): clxxxiv. Reinforcing the view presented here—that Bond's overseers tended to be cheerleaders—is Chauvenet's comment (*Spherical and Practical Astronomy*, 324–25) that Struve's 1843 chronometer expedition resulted in a longitude difference five times more precise than the value obtained from the Bonds' last transatlantic trials.

27. Noteworthy here is the letter from Liverpool Observatory director Hartnup to Bond urging him to get out his chronometer results before the Atlantic cable value superseded them. Hartnup adds that the cable is expected to be laid during the next season—Hartnup to Bond, 19 December 1856, HCO-HUA, UAV 630.2.

28. House Committee on Naval Affairs, *American Prime Meridian*, 44–45.

Chapter 2

1. For some of the early American writers, see James A. Ward, "On Time: Railroads and the Tempo of American Life," *Railroad History* 151 (autumn 1984): 87–95. The quotation is from Heine's *Lutezia*, as given by Schivelbush, *Railway Journey*, 44.

2. Boston and Providence Railroad, "Rule No. 8," *Instructions for Conductors and Enginemen While Passing Over The Road* (1835), quoted in Jacobs, "Early Rules," 30. Baltimore and Susquehanna Railroad, "Rule No. 13," in *American Railroad Journal* 7 (1841): 166. Baltimore and Ohio Railroad, "Rule No. 9"; I am indebted to James D. Dilts of Baltimore, author of *The Great Road* (Stanford: Stanford University Press, 1993), for this reference. Western Railroad, "Rule No. 10." As early as 1834, the six station clocks along the 136-mile rail link between Charleston and Hamburg were being synchronized; in *Report of the Directors of the South Carolina Canal and Rail Road Company* (Charleston, 1834), 8.

3. The operating rules developed to prevent collisions between trains running in both directions on single-track lines are analyzed in Bartky, "Running on Time" and "Railroad Timekeepers." The central importance of operating rules of all sorts and their British roots are discussed in superb detail by Frederick C. Gamst in the recent translation of Gerstner's *Early American Railroads*, 822–24, n46.

4. British engineers and clockmakers were discussing the need for a single time throughout Great Britain even earlier; however, the first railway industry proposal appears in an 1840 inspection report of a forty-mile section of the Birmingham & Gloucester Railway, where down-trains used Birmingham time while up-trains took Cheltenham time. Although the time difference between the two places was only fifty-six seconds, town clocks varied by ten to fifteen minutes and trains often ran late. To eliminate any chance of a time-related accident on this double-track line (which must have been linked via crossovers), Captain S. C. Melhuish, Royal Engineers, recommended that "all lines of railway leading to London, as well as all branches . . . should be constrained to adopt the London time, under the appellation of *Railway Time*"; Melhuish, "Birmingham and Gloucester Railway," 62.

5. Through-car service began around 1849. For the roads' early years, see Stevens, *Beginnings of the New York Central*; Hungerford, *Men and Iron*; and Harlow, *Road of the Century*.

6. "Railroad Convention," *Daily Albany Argus*, 4 February 1843, 2.

7. According to Stevens, *Beginnings of the New York Central,* 326, many records of the post-1843 conventions have been lost, making it difficult to trace the specific changes. A December 1850 compilation of the Albany-Buffalo route schedules in *American Railway Guide* indicates three operating times—Schenectady, Utica, and Rochester—but printer's errors confound an exact analysis. Some specifics regarding 1855 operating times are in Hungerford, *Men and Iron,* 81–82 and Harlow, *Road of the Century,* 91; however, their quotations are inconsistent with each other. The quotation given here is from "Time Table No. 90," *American Rail Road and Steam Navigation Guide,* published many years after the Central's adoption of Albany time.

8. In contrast to the conventions held in central New York, the minutes of this New England group were carefully preserved and later published; see New England Association of Railroad Superintendents, *Records.* The discussion and quotations given here draw upon this record and on the association-sponsored *Pathfinder Railway Guide,* December 1849 through June 1851.

9. *Pathfinder* guides published after December 1849 simply repeat the initial notice, so it is impossible to analyze the dynamics of acceptance, and ultimate rejection, of this time standard. Another contemporary publication lists fifteen of the association's thirty-one current members as having adopted the standard; New England Association of Railway [*sic*] Superintendents, *Reports,* 46–47.

10. By 1874 the time of this arbitrary meridian—now termed "Providence time" even though the longitude difference between the city and Boston is one minute and twenty-four seconds, not the chosen two minutes—was used by only four minor railroads, while thirty-one companies operated on Boston time. Only with the general adoption in 1883 of Standard Railway Time did these holdouts alter their operating times.

11. Railroad guidebooks of the era indicate that railroads serving New York and Philadelphia adopted the local time of these cities respectively. Early in the 1850s a convention of Ohio railway officials adopted Columbus time, also a rational choice; noted in Colleen A. Dunlavy, *Politics and Industrialization: Early Railroads in the United States and Prussia* (Princeton: Princeton University Press, 1993), 177.

12. In a series of articles, Carlene E. Stephens assigns William C. Bond a pivotal role in the New England Association's decision-making process. Stephens argues that Bond's various timekeeping and astronomical activities relate directly to railroad time policy; see *Harvard Library Bulletin* 35 (1987): 351–84; *Journal for the History of Astronomy* 21 (1990): 21–35; *Technology and Culture* 30 (1989): 1–24, and ibid. 32 (1991): 185–86. I have criticized this conclusion, which is based on less than complete historical records and some technical misunderstandings; see Bartky, idem 32 (1991): 183–85.

13. Simon Willard, Jr., was also a chronometer and watch dealer; unlike those of Wm. Bond & Son, his firm's records have not survived. Willard's astronomical regulator is now part of the Harvard University Collection of Historical Scientific Instruments. The quotations are from Willard, *Simon Willard,* 71–72; however, portions of this 1911 account are suspect; see Joseph E. Brown, "The Willard Clockmakers—an Essay," *NAWCC Bulletin* 36 (August 1994): 464–67. Supporting the view that Willard's firm was a key one in the Boston railroad market is a regulator now on display at the Willard House and Clock Museum, North Grafton, Massachusetts. Dated

ca. 1850, its dial is marked "Boston and Providence Railroad," as well as "Z. A. Willard, Maker," and "Simon Willard and Son." (Zabdiel Adams Willard was a longtime associate and eventual partner in his father's firm.)

14. This is based on a review of Boston city directories for the period. John Ware Willard claimed that for many years Simon Willard, Jr.'s firm "had entire charge of all the public clocks of the city of Boston." This may be an exaggeration, for famous clockmaker Edward Howard's duties while apprenticed (1829–34) to Aaron Willard, Jr., included the weekly winding of the city clocks, demonstrating that yet another Willard was involved in Boston's public timekeeping. William F. Channing, in his 1851 fire-alarm proposal, derived some of his cost estimates from "Mr. F. Kemlo, the skilful and experienced City Clockmaker," implying that this clockmaker had some official role with the municipal government.

15. [Bond] to Jared Sparks, 6 January 1851, HCO-HUA, UAV 630.2, apparently written in response to a university publicity release by President Sparks; see also Bond, "Report . . . for 1850," *HCO Annals* 1 (1856): cxlvi. The Wm. Bond & Son "Daybook," in HUCHSI, merely summarizes the association's vote and the 5 November 1849 start of the firm's duty to provide time to Boston railroad men. The association's rules state that "any person may be chosen an associate member, whose connection with the association will tend to promote its objectives." While railroad timekeeping certainly falls within that category, Bond was never a member.

16. For a summary of this 9 August collision near Old Bridge, New Jersey, see Shaw, *Railroad Accidents*, 31; a pictorial companion to Shaw's important and fundamental study is Reed, *Train Wrecks*.

17. Providence newspapers and the Rhode Island Railroad Commissioners reported fourteen deaths; however, the railroad company reported only thirteen fatalities to Massachusetts officials. See "Record of Accidents . . . for the Year Ending September 30, 1853," *Massachusetts State Legislative Documents* 1854, S. Doc. 2, 198–99. Shaw, *Railroad Accidents*, 31–33; and Reed, *Train Wrecks*, 16, 20–22, each describe this disaster. However, both accounts contain errors. The analysis here extends my earlier one; see Bartky, "Running on Time," 29.

18. In addition to summaries in Boston and New York newspapers, large segments of the inquest testimony were published in the *Providence Journal* and *Providence Daily Post*. These latter accounts are apparently the only remaining sources of information on the conduct of the inquest.

19. Fortunately for the course of American railroad safety, the Valley Falls collision was one of the early rail disasters for which a more dispassionate review was subsequently mounted.

20. The P&W continues to operate to this day as a successful freight railroad. Many of its early records were archived at the Thomas J. Dodd Research Center of the University of Connecticut, Storrs, Connecticut. In addition, I thank Harry A. Snyder, director special projects, Providence and Worcester Railroad Company, Worcester, Mass., for allowing me to examine the directors' annual reports to the stockholders, the stockholders' records, and "Records of the Directors," 1 (1845–62), 181–96.

21. *Providence Daily Post*, 22 August 1853; and *Providence Journal*, 24 August 1853; the public meeting was held on 26 August.

22. The Rhode Island Railroad Commission was established in 1839 and was the first in this country. Yet not until January 1855 did the legislature pass an act making the commissioners responsible for investigating rail accidents. Summaries of the testimony at the commission's hearings were reported in the Providence newspapers, 26–31 August.

23. The commissioners highlighted the fact that the P&W's operating schedule had not been altered since the disaster, so "that under the present arrangement a variation of one minute in the time of the conductor would bring two trains together." The commissioners then opined: "If the president and board of directors permit trains to be run at such times that one minute variation may produce a collision it is . . . incumbent on the president and directors to know that the conductors are provided with watches that run with the highest accuracys."

24. Because of the rush to take legislative steps, the Railroad Commissioners' 20 September report to the General Assembly was not published in the state's formal document series; it can be found at the Rhode Island State Archives in "Report - Railroad Commissioners, 1835–1868." For the legislation, see *Acts, Resolves and Reports of the General Assembly of the State of Rhode Island,* (1853), 257–58; also, *Providence Journal,* 21 and 22 September, and 4 November 1853. With regard to the actions taken by the P&W, by the end of November the first section of its double-track extension was already in use. Presumably, the road's superintendent also adjusted the overly close scheduling of trains and promulgated additional timekeeping measures, but no timetables or rule books from this period have been located.

25. Zerah Colburn, "Safety System of the New York and New Haven Railroad," *American Railroad Journal* 27 (21 January 1854): 37. Jefferson Railroad, *Rules and Regulations,* 17.

26. Lee, *Standard Time,* and Wm. Bond & Son "Daybook," 1 and 7 September 1853, both in HUCHSI. A week later Lee ordered an astronomical timekeeper as a replacement for the road's master standard. Over the next few years, Bond's firm sold astronomical regulators and highest-quality watches to various New England railroads; see Blackwell, "Early Railroad Timekeeping," 460–61; and Stephens, "Most Reliable Time," 16.

27. The effectiveness of timekeeping as a safety measure can be gauged from the fact that only four of the century's more than 175 major accidents were caused by timekeeping errors, an incredibly small number. Besides the two in 1853, there was a 31 December 1855 collision near Pittsburgh, on the Ohio & Pennsylvania Railroad, in which three people died. In that case one train conductor's watch was twenty-five to thirty minutes slow. In 1878, on a road whose operating time came from the Allegheny Observatory, a head-on collision killed eighteen people—another disaster involving an unbelievably slow watch; for the analysis of this disaster, see Bartky, "Running on Time," 30–32.

28. New York and Erie Railroad, "Standard of Time"; the text implies that Bull's service was operating as early as November 1852. Also "Resolution, Regard to Professor Bull, 1854," Office of the Chancellor: 1827–88, New York University Archives. In 1853, Richard H. Bull began a thirty-year career as civil engineering professor at the university, then called the University of the City of New York. A mathematician as

well as an astronomer, Bull was also an important figure in New York's financial world; see *National Cyclopaedia of American Biography* 9 (1907): 472; and various obituaries (d. 1 February 1892). These accounts mention Bull's time services for railroads, but they contain garbled facts, exaggerations—"the first mathematician to put into practice the idea of obtaining the true time by the sun"—and are incorrect with regard to the end of Bull's railroad time service. Dudley, "Railroad Time Service," reported that Bull was still providing time to the New York & Erie in 1881. Another secondary biography—by New York University's E. G. Sihler in *Universities and Their Sons*, ed. Joshua L. Chamberlin (Boston: R. Herndon Co., 1898), 54–55—states that Bull had observatories at both his city residence and his home in New Hamburg, New York; if so, the equipment must have been quite simple (and perhaps portable), and in use much later than the time period discussed here and in ch. 5. A summary history of New York University written in 1964 by Lawrence J. Hollander suggests that Bull was providing time to railroads as early as 1848. If indeed correct, Bull's New York service predates that of Wm. Bond & Son in Boston; however, I have found no evidence to support a date earlier than 1852.

29. The quotation is from the Willard firm's advertisement in *The Boston Directory for the Year 1854* (Boston: Geo. Adams, 1854), 56. "[Alexander] Hall's Telegraph Clock," *Scientific American* 9 (8 April 1854): 233, 236; Hall's patent, No. 11,723, was granted 26 September 1854.

30. Arguing that "an exact knowledge of time is . . . of vital importance to the conductors of all railroad trains," New York University astronomer Elias Loomis wrote: "A small error in a conductor's watch has repeatedly been the occasion of the collision of railroad trains, and the consequent destruction of human life." Loomis, of course, was predisposed to favor accurate time, an astronomical observatory's product; see Loomis, *Recent Progress of Astronomy* (1856), 288. See also "Astronomical Time on Railroads," *New York Daily Tribune*, 22 January 1856, 4.

Chapter 3

1. Bond, "Historical Sketch," *HCO Annals* 1 (1856): xxviii.

2. At the dedication of the Yerkes Observatory; see *Popular Astronomy* 7 (1897): 358–59.

3. Bruce, *Launching of Modern American Science*, 101.

4. The importance of time by telegraph to geodesy and astronomy was still fueling priority-of-invention battles when astronomers became public time-givers; naturally, this controversy affected their accounts of business activities. The secondary literature—works detailing events after 1844—is large. But in my judgment all but two of these studies take too narrow a focus, provide too cursory a review of the extensive primary sources, or tend to take a scientist's published statements at face value. These two exceptions are the first (1850) and third (1856) editions of astronomer Elias Loomis's *The Recent Progress of Astronomy*. They should be read together. That Loomis himself became somewhat entangled in the various priority battles can be inferred from the differences between his two accounts.

5. Quite early, English railways were telegraphing time signals to control the move-

ment of trains. Edward L. Morse, "The District of Columbia's Part in the Early History of the Telegraph," *Records of the Columbia Historical Society* 3 (1930): 173; and "Keeping Time with the Telegraph—Further Examples," *American Telegraph Magazine* 1 (1853): 165, reprinting a Buffalo *Express* account of telegraphing New York's time to Buffalo and Bangor's time to Milwaukee.

6. These mid-1844 trials, however, were not the first transmission of time signals by wire. In 1833, mathematician and observatory director Carl Friedrich Gauss and his physicist colleague Wilhelm Weber used an electric telegraph of their own construction to transmit the time displayed by an astronomical clock at the University of Göttingen's observatory to the physics laboratory, located about a mile away—Gundlfinger, *Hundert Jahre Telegraphie*, 23–25. These signals formed the time base for their magnetic observations, and the Göttingen Observatory meridian became the prime one for a worldwide program of magnetic measurements.

7. Who first suggested telegraphy for determining longitudes, and when, is unknown. Several accounts attribute the idea to Morse himself in 1839—though this is unlikely. Some early Coast Survey histories crown Bache, but these accounts should probably be dismissed as agency puffery. Bache admired François Arago, and Coast Survey documents occasionally credit the French scientist with proposing telegraphic longitude measurements in 1837, although they provide no verifiable citation. Contemporary astronomer/physicists—among them Gauss, Bavaria's Carl A. Steinheil, and Russia's F. G. W. Struve—are all equally likely candidates. The earliest link between electric telegraphy and longitude determinations that I have located is the proposal by physicist Charles Wheatstone of England, reported by astronomer Adolphe Quetelet on 17 October 1840 at the Royal Academy of Sciences meeting in Brussels and published in its *Bulletin* 7 (1840): 133.

8. Wilkes, "Difference of Longitude Determined by Morse's Telegraph," original letter in S. F. B. Morse Papers, LC. (Although listed in the newspaper account as *Captain* Wilkes, the explorer did not attain that rank until the fall of 1855.)

9. Thompson, *Wiring a Continent*, 41. Already the envy of English telegraph engineers, by 1853 cities along the entire Atlantic coast as far east as Halifax, Nova Scotia, were linked to towns lying west of St. Louis and New Orleans—ibid., 91, 202, 240–43; also Highton, *Electric Telegraph*, 144–54, 174–75.

10. Although cited by Coast Survey officials, no copy of this request has been found in the National Archives collection of official correspondence, suggesting that it was in a private letter from Bache to Walker. The quotation is from Walker to George Bancroft, secretary of the navy, 22 May 1845, Bancroft Papers, Massachusetts Historical Society.

11. Perturbing the environment throughout this period was a continuing tug of war between officials of the departments of the Treasury and the Navy, both wanting the Coast Survey in their own organizations; Gustavus A. Weber, *The Coast and Geodetic Survey: Its History, Activities and Organization* (Baltimore: Johns Hopkins University Press, 1923).

12. Bache and Maury had worked closely in 1846 to gain Walker his position at the Naval Observatory. The intensity of Walker's and Maury's subsequent anger to-

ward each other can be gauged by their exchanges in the May through June 1847 *National Intelligencer*. (Walker had just completed his most significant work in astronomy: calculating the orbit of the newly discovered planet Neptune backward in time and establishing an earlier sighting.)

13. Prior to Walker's resignation, Lieutenant Maury received department approval to improve the country's knowledge of the geographical positions of inland cities—to be determined via telegraphy. The superintendent justified his plans and his observatory's central position in the country's geography by citing the programs of two European observatories, those at Greenwich, England, and Pulkova, Russia. (Walker himself used similar arguments with Secretary of the Navy Bancroft during his quest for employment.) After the August collaboration, Maury's plans came into direct conflict with expanded longitude efforts already formulated by Bache and Walker. By law and direction Coast Survey activities were confined to American coastal waters. Nevertheless, Bache, a skilled bureaucrat and administrator, argued his agency's need to include Hudson Observatory observations of its position (near Cleveland) in the agency's transatlantic longitude studies. He also presented the rather weak assertion that Cincinnati Observatory was a necessary link for establishing the precise longitude differences between New Orleans and Atlantic coast cities. Thus the final rupture in relations between these nautical agencies involved inland observatory sites. By early 1848, any possibility of collaboration had vanished; see letter from Maury to Bache, 17 February 1848, USCS-NA, reel 23, #0496. Joint agency efforts did not resume until 1863, after Maury's departure.

14. Loomis, *Recent Progress of Astronomy* (1850), 223.

15. Walker to Bache, 11 August 1848, 12 September 1848, USCS-NA, reel 25, #0790; and ibid., 20 October 1849, USCS-NA, reel 28, #0706. Also important in judging William C. Bond's priority claim is his 17 January 1849 diary entry: "I have decided to make a clock for the magnetic telegraph on my own plan," HCO-HUA, UAV 630.4; and, "I have finished my Electric Clock and it operates perfectly"—Bond to William Mitchell, 15 February 1850, HCO-HUA, UAV 630.2.

16. Walker, "Experience of the Coast Survey," 186, and "Progress of improvement and invention," as reprinted in *HCO Annals*, xxvi. In 1839 astronomer and physicist Carl A. Steinheil received a patent that describes dials driven by, and clocks synchronized to, observatory standard time signals via electrical pulses. Also, in the early 1840s, Scottish inventor Alexander Bain was granted many electrical clock patents. (Most English-language writers view the largely self-taught Bain as the "father of electrical horology," although this claim seems partially biased by romanticism and chauvinism.) Steinheil's patent ("Privilegium") is given in detail in an 1843 issue of the Bavarian *Kunst- und Gewerbe-Blatt.* One of his catalogs of astronomical, geodetic, and physical measuring instruments, including electrical clocks, appeared in an 1847 issue of *Astronomische Nachrichten*, a journal that Sears Walker was receiving. (According to Marc Rothenberg, editor, Joseph Henry Papers, Walker had at least a reading knowledge of German.) I am most grateful to Helmut Franz of the Steinheil optical firm in Munich, and a great-great-grandson of the founder, who supplied much material regarding Steinheil's clock systems. Coast Survey correspondence documents both Bache's and Walker's lack of skills regarding applications of electricity.

17. Walker to Bache, 21 November 1848, USCS-NA, reel 28, #0716, and ibid., reel 25, #0140. Mitchel claimed that Walker's October remarks included the comment that the signal-clock problem had "never been solved," that "the full value of the idea was appreciated at once [by me]," that with his circuit-closing clock causing dots to be printed every two seconds on the paper tape of a Morse register, the "problem [was now] solved," and that "nothing more remained than to elaborate the machinery"—as quoted by Porter, "Ormsby MacKnight Mitchel," 447. Despite these confident words, the crucial step for the development of the recording technology was not reached until 18 November, more than two weeks after Mitchel had left on his railroad survey. Shoemaker, "Stellar Impact," 165–88, credits Mitchel with inventing the recorder, erroneously terming it the "electro-chronograph."

18. Locke to Bache, 29 January 1849, USCS-NA, reel 34, #0460.

19. An edited and incomplete set of this correspondence is in *Astronomical Observations . . . at the National Observatory, Washington* 2 (1851): appendix; elsewhere, Locke gives an erroneous date (13 December) for his 30 December letter to Maury. Throughout the period various names were given to the system: "clock," "telegraphic clock," "magnetic clock," "automatic telegraph clock," with the circuit-breaker itself often termed an "electrotome." Maury coined "electro-magnetic chronograph" ("time-writer") to encompass the entire recording system of astronomical clock, primary circuit-breaker, moving register, and observer's break-circuit key. By March of 1849 Locke was using "electromagnetic chronograph" to describe his system, shortening it later to "electro-chronograph."

20. Like all leaders of organizations, Bache was extremely sensitive to this issue; one example is his strong criticism of Loomis for his August 1847 article on longitudes by electric telegraph, which gave (in Bache's view) scant credit to the Coast Survey. Bache's letter is dated 26 December 1848, and is in *Astronomische Nachrichten* 28, No. 666 (1849). Walker's Coast Survey report is dated 15 December, and was written prior to Locke's granting permission to inform Congress of the invention. In his report Walker garbled citations to Steinheil's and Wheatstone's clocks, and claimed, wrongly, that they performed exactly like Locke's clock attachment. The following January, Henry O'Reilly, battling the Morse interests in an attempt to gain an American patent for Bain's chemical telegraph, claimed priority for the Scottish inventor in the area of electric clocks. This letter to the editor was probably the vehicle by which the Coast Survey first became aware of Bain's work.

21. *National Intelligencer*, 6 January 1849, 4; also Bache to Bond, 29 March 1851, HCO-HUA, UAV 630.2.

22. This is the first of many nineteenth-century dollar amounts mentioned in the work. The question, How much is that in real money? has been "answered" by economic historian John J. McCusker in his study bearing that name, a study which allows the conversion of past prices to comparable values in today's dollars. Following McCusker's procedure (p. 320), Bureau of Labor Statistics annual index numbers, CPI-U for 1992 through 1998 and the semiannual value for the first half of 1999, were spliced into the "Composite Consumer Price Index (Table A-2)." In this particular case, $10,000 in 1849 is equivalent to $214,000 in mid-1999 dollars—certainly a significant award in any era.

McCusker warns, "It should be obvious that there are dangers inherent in any simple cross-temporal . . . comparisons that fail to pay full regard to context," and his further remarks should be read carefully. Nonetheless, using this approach provides an accepted, consistent way to judge dollar values throughout the nineteenth century and to the present day.

23. Mitchel to Bache, 3 February 1849, USCS-NA, reel 34, #0527. Mitchel received financial support from the Coast Survey to build his recording devices, as well as strong encouragement from Benjamin Peirce of Harvard, the agency's technical advisor. Given this alliance, reviews of Mitchel's later right ascension and declination instruments must be read with skepticism, as Sally Gregory Kohlstedt noted in *The Formation of the American Scientific Community*, 126, citing a 3 August 1851 letter from Peirce to Mitchel. That feelings were still running high can also be seen in this letter, for the Harvard mathematician, no stranger to controversy himself, refused to discuss Mitchel's priority claims. Doing so required him to criticize William C. Bond, which "would vex my Boston friends"—from a copy in the possession of the Cincinnati Historical Society Library.

24. Mark Littman, "The Discovery of the Perseid Meteors," *Sky and Telescope* 92 (August 1996): 71.

25. For Walker's purported remarks, see Locke to Bache, 3 February 1849, USCS-NA, reel 34, #0463. That the gag order continued in effect for some time, see Walker to Bache, 21 September 1849, USCS-NA, reel 31, #0802.

26. Bond's diary entries for 2, 10, and 13 January 1849 record his thoughts on the statements that were appearing in the press. The astronomer dismissed them with these words: "It is altogether unfair to me, but I dislike a newspaper controversy and do not intend to notice." Bond's complaint letter appeared in the *Boston Traveller*, 17 March 1849, 2. There are a few additional entries in Bond's diary; however, the observatory's scrapbook (prepared by the Bonds) does not include all the related news stories.

27. Locke's 20 March response was published in several places, including the *New York Herald*, 22 March 1849, 1, and the *Boston Traveller*, 21 March 1849, 2. The priority controversy had an international flavor as well. Early on Locke wrote to Astronomer Royal George Airy and Astronomer-Royal for Scotland Charles Piazzi Smyth, both of whom announced the development with credit to Locke. Somewhat later Mitchel wrote Airy, who gave credit to the two Americans only, thereby angering Bache, the Bonds, and Walker. (Airy subsequently hedged his priority assignments.) Later Urbain Leverrier, director of the Paris Observatory, announced his own electric longitude plans, but much to Bache's annoyance did not mention the Coast Survey's prior work.

28. Bond to William Mitchell, 15 February 1850; and G. P. Bond to Mitchell, 30 December 1850, HCO-HUA, UAV 630.2.

29. For some fifteen years Coast Survey longitude campaigns were linked to Seaton Station, which was not tied telegraphically to the Naval Observatory base meridian until 1863.

30. Walker, "Experience of the Coast Survey." When the *American Journal of Science and Arts* reprinted Walker's AAAS remarks in July 1850, the editors apologized to

their readers for not publishing it simultaneously with Locke's priority-of-invention rebuttal (Locke, "On the Electro-Chronograph"). They footnoted Walker's description with the remark that the "responsibility of published statements must rest with those who make them." The editors themselves, they insisted, are "responsible for no facts or opinions published in these pages, unless uttered in their own individual or collective names"—certainly a proper stance for any journal editor. For subsequent priority statements, see Charles A. Schott, "Appendix 6," *Report of the Superintendent of the U.S. Coast and Geodetic Survey for . . . 1880* (Washington, 1882), 82–83; Schott, "Appendix 11," *Report . . . for 1884* (Washington, 1885), 408; and Schott, "Appendix 2," *Report . . . for 1897* (Washington, 1898), 202.

31. Locke, *Report of Professor John Locke*; Maury, "Electro-Chronographic Clock"; Locke, "Electro-Chronograph Clock of the National Observatory"; and, describing modifications, Locke, "Astronomical Machinery." Simon Newcomb wrote that "[the transmitting clock's] only drawbacks were that it would not keep time and had never . . . served any purpose but that of an ornament," assertions not borne out by the record; *The Reminiscences of an Astronomer* (Boston: Houghton, Mifflin and Co., 1903), 118. Locke's astronomical clock is in the collections of the National Museum of American History, Smithsonian Institution.

32. Bond to William Mitchell, 15 February 1850; Walker to Bond, 20 October 1850; Bond to Walker, 24 October 1850; and G. P. Bond to Mitchell, 30 December 1850, HCO-HUA, UAV 630.2.

33. Bond must have detested the term "electromagnetic chronograph"; using it would be tantamount to acknowledging prior work by others. So the astronomer employed "spring-governor" to encompass the entire time-and-event recorder. He also defined the combined "electric clock and spring-governor" as the apparatus built for the U.S. Coast Survey, and, after winning one of the medals set aside for American inventions at the Great Exhibition in London in 1851, stated that "this invention is known and spoken of in England only as the 'American method,'" thereby obscuring the priority issue even more. Of course Wm. Bond & Son was willing to "unbundle" the apparatus, for the firm sold chronographs separately; Bond, "Report . . . for 1851," *HCO Annals* 1 (1856): clvi–clvii.

34. This connection is discussed further in ch. 4. "Mr. Bond to arrange in regard to transportation . . . or to compare the chronometer[s] at Boston with the Cambridge clock through the medium of the Electric Telegraph." Wm. Bond & Son, "Memorandum of Arrangements," 13 December 1850, HUCHSI.

35. Bond to Bache, 24 March 1851, HCO-HUA, UAV 630.2. Even earlier, Bond alluded to this way for improving precision; see Bond, "Letter to the Secretary." The chronograph (spring-governor) that Bond's son exhibited in England carried the engraved label "Invented and made for the U.S. Coast Survey."

36. Walker to Bache, 28 March 1851, USCS-NA, reel 57, #0564; Bache to Bond, 29 March 1851; Bond to Bache, 2 April 1851; and Bond to Walker, 3 April 1851, HCO-HUA, UAV 630.2.

37. (Walker's letter, received after 9 April, has not been located.) William C. Bond to Richard Bond, 14 April 1851, HCO-HUA, UAV 630.2. Although he now had per-

mission from Walker to do so, Bond did not cancel the instructions to his son regarding no work on the Liverpool-Greenwich telegraphic longitude link.

38. Dated 24 April 1851, Walker's report to Bache covered June 1844 through November 1850, so Bond's equipment was not yet in use. And as one would expect, Walker's report contains scarcely any mention of Locke's 1848–49 efforts. The original copy is in the National Archives, USCS-NA, reel 60; the copy sent by Bache to Bond is in the Harvard University Archives. Garbled sections of Walker's chronology are in *Annals of the Dudley Observatory* 1 (1866): 27–29. Edward S. Holden, mistaken in his belief that this historical account was the work of George P. Bond, adds three paragraphs written by the younger Bond to the end of Walker's chronology; Holden, *Memorials*, 239–44.

39. Walker, "Progress of improvement and invention," *HCO Annals* 1 (1856): xxiv–xxvii; Bond, "Report . . . for 1852," ibid.: clxvii; and Harvard University, Board of Overseers, "Report . . . for 1851–52," ibid.: clxi.

40. During his career at the Coast Survey, Walker became insane, signs of his illness appearing at least as early as August 1851. Committed in January 1852 to an asylum in Baltimore, he was transferred to, and in the fall of that same year released from, one in Trenton. Walker remained in poor health, however, and died on 30 January 1853. Benjamin A. Gould's 1854 eulogy (*Proc. AAAS* 8 (1855): 19–45) should be read in concert with Walker's 14 October 1852 letter to his close friend Alexander Dallas Bache; in USCS-NA, reel 67, #0697.

41. Lamont, *Beschreibung der an der Münchener Sternwarte*, 34–50. Lamont had been developing such equipment for years; see his letter to the Naval Observatory's superintendent, reprinted in the *National Intelligencer*, 9 July 1847. The Bavarian astronomer is credited with the introduction of electrical recording technology into European science; *Dictionary of Scientific Biography* 7 (1973), 608.

42. See, for example, George P. Bond's 1860 certification of astronomer C. H. F. Peters's statement that the probable error of the longitude of Hamilton College Observatory was identical to Harvard College Observatory's; see "Longitude of Hamilton College Observatory," in State of New York, *Report of the Regents of the University on the Determinations of Longitudes in this State*, Accompanying Documents, 32. Senate, 85th sess., 1862, Doc. 95.

43. Bond, "Historical Sketch," *HCO Annals* 1 (1856): xxi. Bond gave the HCO's value for the Cambridge-Greenwich longitude difference, which was not the one used by the U.S. Coast Survey; moreover, the latter's "zero" was Seaton Station. In "Astronomy as Public Utility," Stephens argues that the Harvard College Observatory "served as the *de facto* national observatory" (p. 21), and that "the observatory's longitude work was its greatest public utility" (p. 29). In light of Bache's and Gould's remarks (discussed in ch. 1), I cannot accept this view. Moreover, the Coast Survey did not use the Bonds' chronometer value separately, but averaged results from several techniques, weighting them accordingly.

44. George P. Bond, William C. Bond's successor to the observatory directorship, participated in several of these longitude campaigns. Wm. Bond & Son continued in existence, the firm selling what was then the best chronometric equipment in the

country. With this equipment and a rapid, proven method, a new cottage industry sprang up in the Midwest: astronomers providing longitude differences to states and those in charge of the Survey of the Northern [Great] Lakes. Over the years, many individuals and firms built similar recording devices, among them electrical inventor Moses Farmer, who is highlighted in the next chapter. Also, Mitchel's 1849 chronograph was improved by George Washington Hough during the latter's years at Dudley Observatory. In 1876/77 a "Mitchel" chronograph constructed at Hough's scientific equipment firm in Illinois was purchased by Ripon College; see "The Observatory," *Ripon College Quarterly* 1 (September 1877): 5. I am indebted to college archivist Louise Schang for her efforts in this matter.

45. The determination of longitude differences by electric telegraph continued until 1922, when the subsequently named U.S. Coast & Geodetic Survey introduced radio-transmitted signals into its work; geodetic surveying of the type described here became obsolete in the mid-1980s when the Global Positioning System of satellites became operational; Dracup, "Geodetic Surveying. III," 38–39.

Chapter 4

1. This remark was made five minutes before President Abraham Lincoln was shot; *Trial of John H. Surratt in the Criminal Court for the District of Columbia*, vol. 1 (Washington: Government Printing Office, 1867), 571.

2. Some of the small observatories that sprang up before this period may have distributed true time, but only one example has been found so far: Rittenhouse's observatory in Philadelphia.

3. Bartky and Dick, "First North American Time Ball," for the various citations.

4. *United States Statutes at Large*, 22d Congr., 1st sess., c. 191, p. 1832; this prohibition may have been inserted to prevent Hassler from continuing with a trigonometric survey of the coast. For the depot's early history, see Dick, "How the U.S. Naval Observatory Began"; and Herman, *A Hilltop in Foggy Bottom*. In 1854 the depot became the "United States Naval Observatory and Hydrographic Office."

5. "A ball will be hoisted"—see "Signal for regulating time," *National Intelligencer*, 2 January 1845, 3. "On the top of the dome . . . a ball is hoisted and let fall . . . as is done at the Greenwich Observatory in England"—in "Scientific Apparatus at Washington," ibid., 15 April 1845, 4.

6. Bartky and Dick, "The First Time Balls." Eventually, some two hundred of these navigation aids were erected, with the U.S. Navy installing nearly two dozen at American lake- and seaports.

7. Trelease, in *North Carolina Railroad*, 73, reports that by March 1856 the railroad was using the Naval Observatory's signals to regulate its clocks, Washington time arriving daily at Raleigh via the Washington & New Orleans Telegraph Company wire. The company's operator in Washington could see the time ball on the dome, and sent a signal upon its release.

8. Some astronomers issued time-distance maps so that those only hearing the time gun's discharge could correct for the time lag.

9. In this context, Chicago designated an official timekeeper as early as 1847. The

importance of two generations of Willards in Boston timekeeping was discussed in ch. 2.

10. "Mr. Bond agrees to remove his astronomical . . . apparatus to Cambridge . . . reserving to the said Bond, the exclusive use of said apparatus for any observations necessary for . . . the accurate observations of *time* for the use of his private business, as regulator of chronometers." Dated 29 November 1839, this five-year agreement between the Corporation of Harvard College and William C. Bond is reproduced in the *HCO Annals*; to my knowledge, no historian of the observatory has considered its significance. If Bond had not continued as a businessman, he could not have accepted a nonsalaried position at Harvard. Second, not until mid-1857 did Bond dissolve his partnership in Wm. Bond & Son in favor of his sons, so the observatory's timekeeping endeavors throughout this eighteen-year period are inseparable from the firm's business activities. This agreement also explains why Harvard College Observatory received no income from the time service during the Bonds' directorships. (See also n35 in this chapter.)

11. Synchronicity is fundamental to all timekeeping and -giving. In 1370 Charles V of France supposedly issued a decree to force the tower clocks in Paris to strike the hour bells in unison. However, Dohrn-van Rossum demonstrates convincingly that this famous anecdote is only a legend; see his *History of the Hour*, 135, 217–19.

12. Channing's letter, dated 30 May, was published by the *Boston Daily Advertiser*, 3 June 1845, 2.

13. Some writers claim that between 1845 and 1851 Channing took steps to bring his ideas before the city authorities; however, no evidence has been found. The often-cited 1848 feasibility demonstration of a telegraph alarm system connecting the city's engine houses to transmit the location of a fire to all was the result of a proposal by Mayor Josiah Quincy, Jr., who drew upon ideas and cost estimates prepared by F. O. J. Smith. (Smith, a famous early investor in the Morse telegraph, was president of the Portland Telegraph Company; he also controlled the Morse line between Boston and New York.) Moses Gerrish Farmer, then in Smith's employ, designed the necessary electromagnetic equipment, which was built by the well-known Boston clockmakers Edward Howard and David P. Davis. The testing consisted of an operator stationed in New York pressing his local telegraph key, causing the Boston Courthouse bell to ring; local wags termed the event the "first false alarm." See Boston, *Address of the Mayor*, 8–10; also Boston, *Common Council* 12 (1848–49): 76, 161, 254; and Boston, *Mayor and Aldermen* 26 (1848): 77, 268–69, 561. Locke, whose chronograph was also built by Howard and Davis, mentions the fire-alarm test in passing; Locke, *Report of Professor John Locke*, 43.

14. Channing, *Communication . . . respecting a System of Fire Alarms.* Channing and Farmer are jointly credited with inventing the American fire-alarm telegraph. For the beginnings of their relationship, see Farmer to Channing, 22 February 1873, and Channing to Farmer, 3 March 1873 (not sent); both in the William F. Channing collection, Massachusetts Historical Society.

15. This telegraph connection was needed to fulfill the contract William C. Bond had just signed with the Coast Survey. See J. C. Bond to Father, 21 January 1851,

HCO-HUA, UAV 630.2, with attached artifact: a paper disk showing seconds, recorded via Bain's chemical telegraph. For the route and construction details, see Bond to Walker, 24 October 1850, HCO-HUA, UAV 630.2. Bond in "Report . . . for 1851," *HCO Annals* 1 (1856): cliv, wrote that "we have at Boston [*sic*] a connection with Morse's line." It should have read "at Brighton."

16. "Electro-Magnetic Clocks," *Scientific American* 6 (6 September 1851): 408; Siemens had just founded the firm of Siemens and Halske, which became one of the world's most important electrical companies. Other progress reports on the Boston system are "The [Boston] Municipal Telegraph"; *Scientific American* 7 (27 December 1851): 120, (27 March 1852): 219, (5 April 1852): 227; and "Boston Fire-Alarm Telegraph." A useful historical-technical summary is Bosch, "Historical Sketch of the Fire Alarm Telegraph."

17. Channing, "On the Municipal Electric Telegraph," the quotation is on page 83. Other relevant sources are Channing, "The American Fire-Alarm Telegraph," and Channing to Harvey Jewell (Boston city government), 12 April 1852, draft in the Channing collection, Massachusetts Historical Society. Channing's and Farmer's priority-of-invention claims were upheld by the courts.

18. "Electric Clock," *Scientific American* 7 (1 May 1852): 258. Although they had not seen it, the editors denigrated Farmer's invention, implying that it differed little if any from Bain's electromagnetic clocks of the 1840s. Farmer's "Improvement in Galvanic Clocks," *Patent No. 9279*, was granted 21 September 1852. To my knowledge, this patent was never the subject of litigation.

19. Horsford, *Respecting the Regulation of Timepieces in the City*. Among the subsequent Boston newspaper accounts were *Daily Evening Traveller*, 29 November 1853, 4, and ibid., 14 December 1853, 1, the latter giving Horsford's 23 November plan in detail.

20. Bond's draft, written some time between 1 and 5 December and addressed to the editors of the *Daily Evening Traveller*, either was never sent or not published; in HCO-HUA, UAV 630.2. Bond's commercial interests in time included the family business's maintenance of a railroad time standard and increasing sales of clocks and timepieces to railway companies as a result of the collision on the Providence & Worcester Railroad. Bond noted the "plan of communicating the time from an established Observatory, for commercial and other purposes . . . has, within the last year, been introduced into England, under the direction of the Astronomer Royal," (although he erred somewhat regarding the start of this service); "Report . . . for 1853," *HCO Annals* 1 (1856): clxxi. Adding to the strength of the astronomer's condemnation of Horsford's suggestion was his friendship with Farmer, who had just given the observatory one of his patented batteries. See also the broadside, *Farmer's Improved Sustaining Battery* (Boston, 1853?), a signed copy in Houghton Library, Harvard University.

21. Boston Common Council, Committee on Fire Alarms, *Report . . . on the Regulation of Timepieces*; Farmer's plan is on pages 5–9.

22. Ibid., 13.

23. Horsford's proposal was mentioned in the 7 January *Scientific American*, and

on 8 April came a detailed description of Alexander Hall's system of master and slave electrical clocks for uniform time at railroad stations (see n29 in ch. 2). The *Scientific American*'s editors note Hall's patent application, his sale of some rights to the invention, and a scheduled public demonstration of the time-distribution system in New York. It can hardly have been just a coincidence that Farmer et al.'s incorporation petition was completed exactly one week later, 15 April 1854.

24. Commonwealth of Massachusetts. *Acts and Resolves* (Boston, 1854) chapter 425: 326–27; and the original documents relative to the act, in the Commonwealth of Massachusetts Archives. Two minor mysteries in this routine transaction are the legislature's change of the corporation's name and the deletion of Channing from the list of incorporators.

25. Bond, "Report . . . for 1854," *HCO Annals* 1 (1856): clxxix; the committee's favorable response and appeal to the corporation, dated 23 January 1855, are at clxxv. (At that moment Bond was negotiating the determination of the Fredericton, New Brunswick–Harvard College Observatory longitude difference. Since the electric longitude work for the Coast Survey had ended more than two years earlier, the local circuit from the observatory was not being maintained and the wires were in poor shape; see Thomson, *Beginning of the Long Dash*, 14.)

26. In his resignation letter Farmer wrote: "When I accepted the office of Superintendent of Fire Alarms at the salary of one hundred dollars per month, it was with the mutual understanding that the service would not probably require more than half of my time, & that I was at liberty to use what of my time was not required in the prosecution of my researches, or in any other way which I deemed advisable"; Farmer to mayor and Common Council, 28 July 1855, Common Council Docket Documents, 1855, City of Boston Archives.

27. For constructing the observatory–North Cambridge sector, Bond scavenged the government-owned wires from the Fresh Pond–Brighton link, doing so probably without Superintendent Bache's knowledge or authorization; see Bond to Samuel Hein (Coast Survey disbursing officer), 1 September 1858, HCO-HUA, UAV 630.2.

28. Bond, "Report . . . " (read October 1856), 2; in HCO-HUA, UAV 630.3. In a "Receipts and Expenses" folder, there is a (debit) entry for 31 August 1856: "B[ond] & S[on] paid for Telegraph line to Boston........$200.00"; HCO-HUA, UAV 630.3. Elsewhere Bond estimated the cost of a line of wire at one hundred dollars per mile; since the new link from the observatory to Boston was about four miles long, this debit must be for the section purchased from the telegraph company.

29. Bond used the same phrase when describing an identical situation: "to govern all the Public Clocks in Detroit by a normal pendulum at Ann Arbor"; Bond to Brünnow, 22 December 1858, HCO-HUA, UAV 630.2.

30. Bond, "Report . . . " (read October 1856), 3–5, HCO-HUA, UAV 630.3. Regarding the need for temperature control, see both the Visiting Committee's and Director George P. Bond's "Reports" for 1859, pages 6 and 16, respectively. The normal pendulum was discarded in 1859; see G. P. Bond, "Report . . . [for 1860]," 10 (where he confuses the matter by replacing "normal pendulum" with "normal clock" when referring to his 1859 report).

31. Joseph B. Stearns is listed in Jenkins, *Papers of Thomas A. Edison*, 100–101. Chronometer transports from Wm. Bond & Son to the city office would not have been made prior to Stearns's appointment as superintendent (October 1855), for Farmer would not have been eager to give time away via the fire bells. Most likely such transports began after April 1857, for a lengthy description of the fire-alarm system makes no mention of public time signals; see "Boston Fire Alarm Telegraph," continuing as "The [Boston] Telegraph Fire Alarm," both in the *Firemen's Advocate*; I am indebted to Robert W. Fitz, deputy fire chief (retired), Lebanon, New Hampshire, for these two articles.

32. B[ond], "Boston Bells," 1 July 1858, 3, HCO-HUA, UAV 630.3. Also, Bond to Brünnow, 22 December 1858, HCO-HUA, UAV 630.2; there is a minor error here, Bond's "last year" implying 1857—not 1856—for the completion of the observatory-owned telegraph line. The time signal protocol is in [William C. Bond], "Electric Communication with the Boston Bells at Cambridge," 26 June 1858, HCO-HUA, UAV 630.3. Bond claimed that he could measure any difference in time between the two locations "to the fiftieth of a second." This, of course, is not the same as the average daily error in the observatory's time. Bond also remarked that the striking of the bells automatically from Cambridge was being done "occasionally," suggesting no need to keep as close to the true time of Boston as Bond's error statement implies.

33. Distributing uniform and accurate time via city fire-alarm bells began in New York in January 1857; see ch. 5. (The Naval Observatory was linked to Washington's fire-alarm telegraph in 1864–65.)

34. A description of Wm. Bond & Son's new timekeeping role appeared soon after William C. Bond's death. Announcing its telegraphic connection with the Harvard College Observatory and its own timekeeping equipment in Boston, which together gave the firm the ability to determine time "with the utmost precision," the firm's advertisement continued: "All the public Clocks of the City of Boston, and the Railroads of New England, are regulated by the *Standard time of Messrs. William Bond & Son*"; "Chronometer Manufactory," 1 July 1859, HUCHSI; with emphasis added. (See also n39.)

35. Harvard astronomer Arthur Searle wrote that "previous to 1872, the Observatory had supplied the people of Boston and the vicinity with a standard of time, without receiving any compensation for this service." This statement—*HCO Annals* 8 (1876): 11—is factually correct, but incomplete. It has also been badly misinterpreted; both Jones and Boyd, *Harvard College Observatory*, 159; and Bruce, *Launching of Modern American Science*, 146, read it as a "free" observatory time signal, an anomaly in this era. (See also n10 in this chapter.) Bond was well aware of time's commercial value, writing just before his death that "one of the New York R.R. Companies was willing to pay fifteen hundred dollars *per annum* for such a priviledge [*sic*]"; Bond to Brünnow, 6 January 1859, HCO-HUA, UAV 630.2.

36. Warren, "History and Description of the Boston Fire Alarm Telegraph" (copied from an original); I thank William Greer of Washington for the copy. Except for the time service and an update on the extent of the system, this article scarcely differs from the 1857 description in this weekly (cited in n31, above).

37. In 1860, J. N. Gamewell, who previously purchased all rights to Farmer's and Channing's alarm patents, exhibited a fire- and police-alarm system in New York. Among its enumerated advantages was that its use "establishes uniform time throughout the city, as the hours of 12 M. and 9 or 10 P.M. are daily struck from the central office on all the bells"; "The Fire Alarm Telegraph," *New York Herald,* 25 April 1860, 10. An almost identical statement is in John F. Kennard & Co., *The American Fire Alarm & Police Telegraph* (Boston, 1864), 8; in the William F. Channing collection, Massachusetts Historical Society. (Kennard purchased Gamewell's confiscated patent rights from the federal government.) Also, Gamewell & Co., *The American Fire Alarm and Police Telegraph* (New York, 1872), 9.

38. G. P. Bond, "Report . . . [for 1862]," 12. Between 1856 and 1862 the observatory participated in only three electric-longitude efforts: with Quebec Observatory at the Citadel (1857), Hamilton College (1859), and with Dudley Observatory (1860); none of these determinations used the observatory-to-Boston sector of the telegraph line. The section's sales price was $150, and included a substantial part of William C. Bond's new construction; G. P. Bond to Amos A. Lawrence, 25 January 1862, *Harvard College Papers* 19 (1862).

39. "The Fire-Alarm Telegraph is also employed to designate exact noon, by giving a SINGLE STROKE upon all the alarm bells, and through this agency a uniform and correct time is communicated to our citizens"; in *Russell's Guide,* 26. At some point, time carried from Harvard College Observatory directly to the Fire Alarm Office (and also to Wm. Bond & Son) became the norm; see Augustus MacConnel to Cleveland Abbe, 18 April 1870, Abbe Papers, LC. In the early 1870s, the city of Boston contracted with Harvard College Observatory to have signals telegraphed to the city's alarm office. From there the noon signal was transmitted manually to the electro-magnetic bell strikers.

Chapter 5

1. From an undated summary of costs to transmit time signals to New York City; the date is assigned by context. The document itself is in Gould's hand; in DOA.

2. Before this virtually instantaneous transmission mode even existed, optical telegraphs in England and Prussia were kept in synchrony via manual signaling (and return) of periodic time messages; O. Tuck, "The Old Telegraph," *The Fighting Forces* 1 (1924): 470; and Gerald J. Holtzman and Björn Pehrson, *The Early History of Data Networks* (Los Alamitos, Calif.: IEEE Computer Society Press, 1995), 188. In 1852 the Electric Telegraph Company remarked that "several years" before the start of the Royal Observatory's time service, it "transmit[ted] daily to their more important stations . . . true London time"; *Times* (London), 8 September 1852, 5.

3. Telegraphed time signals from Greenwich to London began in August 1852; Astronomer Royal George Airy had been considering such an activity since 1849. Also in 1852 the director of Munich's Royal Observatory reported that he was providing signals to the city's central telegraph station for the uniform and precise regulation of clocks at railroad stations and telegraph offices; Johann Lamont, *Jahres-Bericht der königlichen Sternwarte bei München* (Munich, 1852), 140. Johann Encke, director of the

Royal Observatory, Berlin, may have transmitted time to the various Siemens-built telegraph systems even earlier. The impetus for these astronomers was, of course, the enormous advantages of electric telegraphy for determining longitude differences.

4. A direct consequence of the U.S. Coast Survey's longitude program, they were (in order of connection) the U.S. Naval Observatory; Philadelphia's Central High School; Lewis Rutherfurd's private observatory in New York; and the observatories at Cambridge; Hudson (near Cleveland); Cincinnati; and Charleston, South Carolina. Over the next decade, three more observatory sites were linked to various commercial lines: Hamilton College's observatory at Clinton, New York; Dudley Observatory in Albany; and the Detroit Observatory at Ann Arbor.

5. Bond to Commander P. F. Shortland, 27 May 1851, HCO-HUA, UAV 630.2, in a discussion of time signals for longitudes. On 30 July 1851, Wm. Bond & Son agreed to pay ten dollars a year for the "use of the line from Boston to Cambridge Observatory 7^m before to 7^m after one P.M. 3 days of the week." This was done so the firm could rate chronometers for the Coast Survey; Wm. Bond & Son, "Daybook," HUCHSI. Bond summarized this direct connection to the chemical telegraph line in his "Report . . . for 1851," *HCO Annals* 1 (1856): cliv. After his Boston firm's move from No. 26 to No. 17 Congress Street, the time-signal code was reiterated; Wm. Bond & Son, "Daybook," 13 May 1852, HUCHSI. The first formal mention of the distribution of time signals from Cambridge appeared in late 1852, when Bond wrote that "the mean time of the Observatory is continued to be transmitted by telegraph to various stations in town and country, on Monday and Thursday of every week"; Bond, "Report . . . for 1852," *HCO Annals* 1 (1856): clxv.

6. Holden, *Memorials*, 248, wrote that "from January 1852, onwards, time signals from a standard clock at Cambridge were *regularly* transmitted to Boston for the convenience of mariners, etc." This incorrect date suggests that Holden may not have consulted William C. Bond's 1851 correspondence or the family firm's records in this regard. Also, he may have been unaware of the typographical error noted earlier— Bond's "we have at Boston [Brighton] a connection with Morse's line"—*HCO Annals* 1 (1856): cliv.

7. Lee, *Standard Time*, details the nontelegraphic transfer of time from the Wm. Bond & Son's timekeeper at Congress Street to the Boston & Providence Railroad station clocks.

8. Sections of the Vermont & Boston Telegraph Company's wires were erected along the routes of the Vermont Central and the Northern Railroads. The Naval Observatory's Washington time, used in the mid-1850s and after by the North Carolina Railroad (Trelease, *North Carolina Railroad*, 73), arrived in Raleigh via the pioneering Washington & New Orleans Telegraph Company, which had been completed in 1848. The South Carolina Railroad probably had the company transmit its operating time, for a section of the wire route was built on the railroad's right-of-way, and from 1834 on, the South Carolina routinely synchronized station clocks along its 136-mile rail link between Charleston and Hamburg.

9. Mott, *Between the Ocean and the Lakes*, 415–20.

10. New York & Erie Railroad, "Rule No. 15," and "Standard of Time of the New

York and Erie Railroad." As already discussed, Bull's astronomical determinations of New York's local time was the basis for the Erie's operating time.

11. The Dudley Observatory's first years ended in great strife, with Gould being ejected from the site in 1859; during his four-plus years as astronomer-in-charge and then as director, he brought none of its planned programs into operation. Gould's near total lack of experience in observatory management made him an extremely risky choice for such a complex undertaking as the inauguration of a major scientific enterprise, so his failure was perhaps inevitable. This stormy era in observatory history was analyzed within a compelling social-history framework by Mary Ann James in *Elites in Conflict*, who draws upon the hundreds of pages of charges, counter-charges, and insults hurled by the protagonists, printed in documents distributed throughout New York and New England and sent to elected officials in Washington. I acknowledge my great debt to Dr. James's research, which was detailed in her Ph.D. dissertation.

12. The naiveté of Henry, Bache, Peirce, and Gould, evidenced at the very start of this unhappy episode, is noteworthy. All concluded that a scientific establishment designed to compete at the state of the art, and simultaneously advance it, could be put into operating condition in the "spare" time of an untried manager who was often not even present on the site.

13. Gould, 3 November 1855, as quoted in Dudley Observatory Trustees, *Statement*, 11; Gould, *Reply*, 229. For the prior art, see Rev. T. R. Robinson, "On the Dependence of a Clock's Rate on the Height of the Barometer," *Memoirs of the RAS* 5 (1833): 125–34; and ibid., *Places of 5,345 Stars . . . at the Armagh Observatory* (Dublin: A. Thom & Sons, 1859), xviii–xxii. This innovative timekeeper, built by Thomas Earnshaw, is shown in A. S. Gunn, "Astronomical Clocks at Armagh Observatory," *Irish Astronomical Journal* 23 (1996): 199–200.

14. For the link to Erastus Corning, see *Proc. AAAS* 9 (1856): 102; Gould to Thomas W. Olcott, 13 October 1855, and Gould to [James H. Armsby], 22 November 1855, DOA. In the early 1860s, Wilhelm Foerster, director of the Berlin Observatory, had F. Tiede build an astronomical clock in an airtight case; the constancy of its rate in this constant-pressure environment was outstanding. For the observatory's eventual sidereal standard, the gift of George W. Blunt, see Gould to Executive Committee of the Observatory, 31 May 1858, DOA. The pendulum and its support were subsequently altered by Wm. Bond & Son; see *Astronomical Notices* No. 15 (24 February 1860): 120.

15. European observatories were already constructing similar systems. Although Moses Farmer's system was installed at Dudley Observatory, it was scarcely used: the chronograph had made subsidiary clocks and dials much less important for observers. Gould probably met Farmer at the AAAS's August 1855 annual meeting in Providence, for the inventor gave a talk there two weeks after his forced resignation as superintendent of Boston's Fire Alarm Telegraph Office. In any event, and fortunately for Gould, Farmer was looking for work at this juncture.

16. Gould to Olcott, 11 September 1855, Olcott Papers, "Dudley Observatory," CS550/Box 3/F-7, Albany Institute of History & Art. Gould, *Reply*, 152, 205–6,

328–29. Farmer's chronograph was altered by Wm. Bond & Son; "Daybook," January–April 1860, HUCHSI; *Astronomical Notices* No. 15 (24 February 1860): 120, notes a modification "of Mr. Bond's construction," which must refer to the Bonds' spring-governor control. Despite the improvement, the chronograph was discarded; subsequent Dudley Observatory annals invariably emphasize Mitchel's more awkward rotating-disk design.

17. Trustee Secretary James H. Armsby, anxious to supply time to New York's shipping, asked Gould to examine the time system used for rating chronometers at Liverpool; Armsby to Gould, 14 September 1855, Bache Papers, LC, reel 1, #0066. The earliest reference located in which Gould mentions observatory mean time and income from its distribution is "The commercial men of N.Y. & the Railroads ought to help us. We can give them their Time, well"; Gould to [Armsby], 22 November 1855, DOA.

18. Walker, "On Controlling Clocks by Electricity." For an appreciation of Airy's impressive horological analysis and design skills, see J. A. Bennett, "George Bidell Airy and Horology," *Annals of Science* 37 (1980): 269–85.

19. Gould to Armsby, 3 December 1855, DOA.

20. [B. A. Gould], n.d., "Estimated expense of time ball"; and Gould to Armsby, [5] and 12 January 1856, DOA. Additional information is in Dudley Observatory Trustees, *Statement*, 12; and Gould, *Reply*, 115–16, 155–56, 191–93.

21. Gould to Olcott, 8 February 1856; and Gould to Armsby, 18 February and 5 March 1856, DOA. The timekeepers' attributes are further detailed in the various controversy documents printed in 1858 and 1859. Note that Gould uses the phrase "automatic regulator" for what we today call a "synchronizer."

22. Joseph Henry to Armsby, 21 January 1856, DOA. The synchronizers were constructed, and two years of testing demonstrated the concept's feasibility; yet the system was never installed. At the time of Gould's removal, two almost complete synchronizer clocks were in storage at the observatory; see Hough, "Time Service."

23. Gould to Airy, 8 January 1856, as summarized in James, "The Dudley Observatory Controversy," 103; and Gould to Armsby, 12 January 1856, DOA.

24. Gould to Armsby, 12 January 1856, DOA, listing the Hudson River, the New York & Erie, the New York & New Haven, and the New Jersey roads, all centered in the New York City area.

25. Signed by Gould, Bache, and Board of Trustees vice president Thomas Olcott, the observatory's letter to the mayor is in the *New York Times*, 14 February 1856, 8. Additional information is in "Time Ball Signals," *Scientific American* 11 (8 March 1856): 204.

26. *New York Daily Tribune*, 1 January 1856, 2; ibid., 7 January 1856, 7; and ibid., 22 January 1856, 4. The railway company's initial investigation, quoted in the *Pittsburg Dispatch* of 2 January 1856, reported the conductor's watch as "twenty-five or thirty minutes wrong."

27. Gould to [Armsby], 26 January 1856, DOA; his citation—*New York Sunday Times*, 21 January 1856—is erroneous. Gould suspected that New York University astronomer Elias Loomis was responsible for the item in the newspaper, and indeed

Loomis had already prepared material on the need for a public observatory that would transmit time to city businesses and railroads; see Loomis, *Recent Progress of Astronomy* (1856), 291.

28. "Resolution, Regard to Professor Bull, [spring] 1854," 4 October 1854, Office of the Chancellor, 1827–88, New York University Archives—an extract from the minutes of the Common Council approving the erection of a dome. Gould's antipathy toward his fellow astronomers is well documented; regarding Loomis he wrote: "The man's [professional] standing is not such as to render him a desirable organ with the public"; Gould to Armsby, 4 March 1856, DOA.

29. See Gould's various letters to Armsby and Olcott, 2 February to 7 July 1856, DOA. Board of Trustees vice president Thomas Olcott requested the marketing document, for which see Bache et al., *Dudley Observatory*.

30. For the details, see Bartky et al., "An Event of No Ordinary Interest."

31. "Astronomical Time for the Port," *New York Times*, 3 September 1856, 4. Gould to Airy, 24 September 1856, as summarized by James, "The Dudley Observatory Controversy," 103.

32. "Observations on Observatories," *New York Times*, 10 December 1856, 4. For Bache's handling of this criticism, see "Science and the Press," *New York Times*, 11 September 1858, 4.

33. Gould to Armsby, 12 and 26 December 1856; and Gould to Olcott, 11 and 31 December 1856, DOA.

34. "To Regulate the Time," *New York Times*, 26 January 1857, 6.

35. In 1857, New York had two municipal telegraph systems operating; apparently both were used in the distribution of time. One system was its Fire Alarm Telegraph, installed in 1852, a simple arrangement of wires connecting the city's eight fire watch towers with City Hall. When someone manning a watch tower saw a fire, signals were immediately sent to the others on duty so that the alarm bells could be struck. The second system was the Police and Fire Telegraph, which connected two dozen station houses with City Hall. In operation since January 1854, the incorporation of a dial telegraph permitted the exchange of police and general information, as well as the transmission of fire alerts. Details on New York's two municipal telegraph systems are contained in competing proposals to the city of Philadelphia; see Philadelphia's Select and Common Council, "Report of the Special Committee . . . in Relation to the Fire Alarm and Police Telegraph," 12 October 1854, *Journal of the Common Council* (1854), Appendix No. 64, 247–69.

36. Gould to Armsby, 17 March 1857, with the version quoted here from a newspaper clipping dated 20 [Jul]y 1858; both in DOA. (After his dismissal as director, Gould savaged several observatory trustees via these "Cock Robin" poems. He denied authorship, but the trustees had this very first one in his own hand.) Gould's ire may have been misplaced; see, for example, university chancellor Isaac Ferris's request for the Dudley Observatory's time signal and his offer of assistance (although Gould was skeptical of his motives); Gould to James Armsby, 18 September 1856, DOA. A more likely suspect is astronomer Charles W. Hackley of Columbia College; Gould regarded him as a possible source for the damaging December editorial in the *New York*

Times. Hackley, a colleague of Lewis Rutherfurd at the latter's observatory, had lobbied the state legislature many times for funds for a public observatory in New York City; C. W. H[ackley] to Matthew Maury, 8 May 1847, Records of the U.S. Naval Observatory, LC. Loomis's discussion shows that the desire for a public observatory continued in 1856; see Loomis, *Recent Progress of Astronomy* (1856), 290.

37. *Albany Atlas & Argus,* 5 January 1860, 3, and 11 January 1860, 3; *Albany Morning Express,* 17 January 1860, 3, and 24 January 1860, 3; *Albany Evening Journal,* 23 January 1860, 3. In 1866, Brünnow, now director of Ireland's Dunsink Observatory (then part of Trinity College, Dublin), supported the introduction of this type of signal device for the city's port area; Wayman, *Dunsink Observatory,* 133; I am indebted to the late Professor Wayman for a copy of this correspondence.

38. "The Time-ball on the Custom-house," *New York Times,* 21 April 1860, Supplement, 2; *New York Herald,* 21 April 1860, 3; *New York Commercial Advertiser,* 21 April 1860, 4; "The Time Ball," in "New Electric Enterprises," *New York Herald,* 26 April 1860, 10. Also, *Albany Atlas & Argus,* 23 April 1860, 3, and 3 May 1860, 3. The signal device was constructed by the New York electrical firm of Chas. T. and T. N. Chester, who also patented and constructed municipal police and fire alarm systems.

39. Months after Mayor Wood's order, observatory trustee Olcott reiterated his wish to have a time ball in New York City as soon as practicable; Olcott to Bache, 23 November 1857, and reply, 27 November 1857, Bache Papers, LC, reel 1, #0386 and #0389. "American Police and Fire Alarm Telegraph," *Scientific American* 2 (5 May 1860): 298. "How America Uses Electricity—The Time Ball and Telegraphic Fire Alarm," *New York Herald,* 25 April 1860, 6; and "The Fire Alarm Telegraph," in "New Electric Enterprises," *New York Herald,* 26 April 1860, 10.

40. *Albany Evening Journal,* 24 January 1861, 3. Mitchel to Olcott, 6 May 1861, Olcott Papers, "Dudley Observatory," CS550/Box 3/F-6, Albany Institute of History & Art.

41. Hough, "Report for 1870," 364, and "Time Service." Hough's remark may not have been an exaggeration, for in 1870 Western Union in New York was getting its operating time from the Dudley Observatory, and would have retransmitted it; B. F. Sands to Abbe, 16 November 1870, Abbe Papers, LC. Jones and Boyd, *Harvard College Observatory,* 159, claim that the Dudley Observatory was receiving compensation for supplying time prior to 1872; however, no evidence has been found. The "Records of the Dudley Observatory Board of Trustees" show no compensation at all through November 1872, a very modest amount from area railroads soon after, and five hundred dollars per year from the city of Albany from 1882 to 1887.

42. Hough left in 1873–74, but observatory time signals continued. Lewis Boss was appointed director in 1876 and began agitating for a transmitting clock to replace Farmer's timekeeper, which he judged to be of "inferior performance"; Boss, "Report . . . for 1877," 8, 15; and "Report [for 1882]," 9–10.

43. "Benjamin Apthorp Gould," *National Cyclopaedia of American Biography* 5 (1907): 108–9.

44. "Determination of the Longitude of Ann Arbor by the Telegraphic Method," *Astronomical Notices* No. 27 (8 October 1861): 17–18; and in *Annual Report of the Sec-*

retary of War, 37th Congr. 2d sess., 1861–62, S. Ex. Doc. 1, pp. 132, 380–84. Also, University of Michigan, *Regents' Proceedings, 1837–1864* (Ann Arbor, 1915), 915, 918, 957, 979; and ibid., *1864–1869*, 46; I acknowledge with pleasure Patricia S. Whitesell, then with the Office of the Vice President for Research, University of Michigan, for providing copies of the university records.

45. All costs of the time service were to be borne by the city of Detroit; influential citizens must have lobbied for these funds to be appropriated; Brünnow to Bond, 30 December 1858; and Bond to Brünnow, 22 December 1858 and 6 January 1859, HCO-HUA, UAV 630.2. Later in Dublin (1873), Brünnow inaugurated two time services. Both involved banks of electric clocks controlled by Dunsink Observatory; Wayman, *Dunsink Observatory*, 96, 132–37.

46. A weekly signal to the Galena & Chicago Union Railroad apparently began in 1864. Later, customers included cities and towns in southern Michigan and northern Ohio and the Michigan Central Railroad; see Adams, "Railroad Watch Inspection," 33–34; and Nourse, "Observatories in the United States," 529. In his May 1879 compilation, Cleveland Abbe, a student of Brünnow's (1859–60), reports that time signals to the Michigan Central began in 1862, but his recollection of specific dates is often faulty; Abbe, *PAMS* 2 (1880): 18. Whitesell, *A Creation of His Own*, 34, 177, writes that the Detroit Observatory's time service "was probably established" in 1861; however, she gives no supporting citations earlier than the last quarter of 1863, well after Watson had been made director. The time service was still operating in 1885; American Railway Association, *Proc. General Time Convention* (Appendix): 725.

Chapter 6

1. In his 1870 compilation (n11, below), Cleveland Abbe suggested that Cleveland's time was being regulated "by a person who obtains it daily from the Observatory at Hudson, Ohio" (in Cincinnati Common Council, *Report on Standard Public Time*, 11). Charles A. Young, who was at Western Reserve College from 1857 to 1866, is credited with inaugurating Cleveland's time service; Edwin B. Frost, "Charles Augustus Young," *Biographical Memoirs of the NAS* 7 (1910), 92. However, Cowles and Company, a jewelry firm on whose transit observations "all the clocks, & rail Road time are dependent . . . for the true time in this city & Vicinity," was paid to regulate the city's clock in 1860 and 1861; "Petition," 27 December 1859 and 7 May 1861, and "Resolution," 7 May 1861, *Records of the City Council of Cleveland*. Perhaps Young provided an occasional time check to some lesser-equipped, subsequent city timekeeper. Similarly, mathematics and astronomy professor William Chauvenet may have provided true time to St. Louis jewelers between 1860 and 1869. However, Washington University's time service did not start until 1878; Woodward, *Washington University Electric Clock System*; "The Erratic Sun," *Republican* (St. Louis), 19 May 1881, 6; and Stephens, "Astronomy at Washington University."

2. Benedict Brothers, *Time Tables*, 26. (In 1883, six years after the start of Western Union's city time service, the firm was still styling itself in this manner; noted in O'Malley, *Keeping Watch*, 126.) For Boston's noon signal see *Russell's Guide*, 26. Also in 1870 Abbe wrote that time telegraphed from Cambridge to Boston "does not con-

tinue at present" (in Cincinnati Common Council, *Report on Standard Public Time*, 10), though he was aware that Harvard College Observatory time came to the Fire Alarm Telegraph Office by the transport of chronometers; see MacConnel to Abbe, 18 April 1870, Abbe Papers, LC.

3. Obendorf, "Samuel P. Langley," 16–17, describes the demanding schedule of the Allegheny Observatory assistant responsible for the time service.

4. Humphreys, "Cleveland Abbe."

5. Abbe, "Letter," *Astronomische Nachrichten* 43 (1868): 43; and "The Resuscitation of the Cincinnati Observatory," *Proc. AAAS* 17 (1869): 172–74, 361.

6. Abbe, *Inaugural Report . . . 1868*, 18, 23; and *Annual Report . . . 1869*, 35–36. Mayor Charles F. Wilstach to Abbe, 15 and 24 February 1869, Abbe Papers, LC.

7. Abbe, *Annual Report . . . 1870*, 16–17. Loomis, *Recent Progress of Astronomy* (1856), 288.

8. *Jahresbericht . . . dem Comité der Nicolai-Hauptsternwarte* (St. Petersburg, 1865), 35–38; and *Jahresbericht* (1866), 39–40. In America only the Naval Observatory was using the Jones system, controlling a clock at the Navy Department. Abbe may have seen this installation, for it was placed in service in 1867–68, the period of his first Washington sojourn.

9. Despite its early promise, using electricity in place of weights or springs to power clocks had not proved terribly successful. Moreover, for many customers, using electrical pulses to advance the hands of subsidiary dials was unacceptable. Time was lost whenever the electric current failed as a result of a broken wire or a discharged battery; and after the current was restored, the again-advancing displays might remain dangerously at variance with the region's true time. Electrical synchronizers, the second means used to ensure uniformity among groups of clocks, had their own problems. Although dedicated lines were not required with this technology, weak batteries could mean that the mechanical devices did not actuate. Using high electric currents to overcome this reliability problem led to spark erosion at the electrical contacts, and could eventually weld them. In Jones's system, clocks still had weights and pendulums, so that a broken wire did not affect timekeeping in dangerous ways. For synchronous timekeeping, the to-and-fro swings of the master clock's pendulum produced pulses in an electrical circuit that included a wire coil mounted on a secondary clock's pendulum. As this latter clock's pendulum swung, the coil passed into the magnetic fields generated by permanent magnets mounted on either side of the pendulum's travel arc. Should the secondary pendulum's swing ever lag the primary clock's, the current-induced magnetic field was attracted to the permanent field of the fixed magnet, thus speeding up the lagging pendulum. Alternatively, should its swing be ahead of the master clock, the same interaction between magnetic fields would retard it. Jones's ingenious design made it easy to achieve clock synchrony to within a few hundredths of a second.

10. John Hartnup at Liverpool Observatory was the first astronomer to use Jones's system, keeping a large tower clock at the Town Hall in synchrony with his observatory's timekeeper. His September 1857 report to the British Association for the Advancement of Science was reprinted: "On Controlling the Movements of Ordinary

Galvanic Clocks," *Horological Journal* 1 (1858–59): 77–78. Hartnup was followed by Astronomer-Royal for Scotland Charles Piazzi Smyth, who allied himself with Jones and with clockmaker James Ritchie. The installation at Edinburgh Observatory included numerous improvements, and Piazzi Smyth's annual reports, which highlighted the observatory's public timekeeping, were probably the basis for the Pulkova Observatory's decision to install James Ritchie & Son's system. As a result of Piazzi Smyth's proselytizing, the University of Glasgow began its own time distribution service to the city and to shipbuilders on the Clyde. By the mid-1860s the university had edged out Edinburgh Observatory's program in Glasgow, as astronomy professor Robert Grant reported. In the same period Astronomer Royal George Airy installed a Jones-type system at Greenwich, and soon after took charge of several time guns in northern England and Scotland that the astronomer-royal for Scotland had been supervising. These conflicts among the observatory directors in Great Britain have been little studied, for the Royal Observatory at Greenwich's accounts have always overshadowed those of the smaller observatories.

11. These details of Abbe's proposal and his comments are found in the Cincinnati Common Council's *Report on Standard Public Time*; additional materials are in Board of Trade president and Common Council member A. T. Goshen's letters throughout this period and into early 1871—in Abbe Papers, LC. Abbe wrote in September 1869 for information on the Royal Observatory's timekeeping and received Ellis's "Lecture on the Greenwich System of Time Signals"; in February 1870 he sent inquiries to Scotland and elsewhere.

Since mid-1869 Abbe was aware that Langley was developing plans for the distribution of time to railroads, and learned of his progress the following February; see Langley to Abbe, 14 June 1869, and 10 and 22 February 1870, Abbe Papers, LC. In his February letter Langley also informed Abbe of Safford's proposal to the city of Chicago; Abbe received more details regarding its status a month later; see Safford to Abbe, 17 March 1870, Abbe Papers, LC. His colleagues' progress in observatory timekeeping may have caused Abbe to intensify his own efforts to gain annual funding from the city of Cincinnati.

12. Charles Piazzi Smyth, "Report to Board of Visitors for 1870," and "Report . . . for 1871," *Astronomical Observations Made at the Royal Observatory, Edinburgh* 13 (1871): R64, R103. The clock for Cincinnati was ordered on 4 January 1871.

13. For the early years of Abbe's meteorological career, see Fleming, *Meteorology in America*, 150–56, 158–59.

14. Ormand Stone, "Cincinnati," *Sidereal Messenger* 1 (1882): 33; Porter, *Observatory of the University of Cincinnati*, 12, 14; and Elliott Smith, "The Scientific Work of the Cincinnati Observatory," *University of Cincinnati Record* 9 (January 1913): 24.

15. Fox, "Dearborn Observatory." Safford to Abbe, 5 May and 12 December 1868, Abbe Papers, LC.

16. In his inaugural report to the CAS's board of directors, Safford described only research in astronomy; Safford, "Report of the Observatory Director for 1866–1868," from a 13 January 1869 article in the *Chicago Times*, as copied by Henry C. Ranney in "Facts Relating to the Chicago Astronomical Society," vol. 1, 192–202. (The early

records of the CAS were destroyed in the Great Fire of October 1871. However, Ranney, the society's secretary for many years, prepared this invaluable compendium, now in the collections of the Chicago Historical Society.)

17. Chicago City Council Proceedings Files, 28 May 1847 and 13 May 1867, Illinois Regional Archives, Northeastern Illinois University.

18. *Chicago Tribune*, 30 May 1869, 4. The identity of "J. C. D." is uncertain. However, at this time insurance company president John C. Dore was the Chicago Board of Trade's president. No records show that Dore was a member of the CAS, suggesting that he was operating independently. In his letter, "J. C. D." cited the New York City time ball and its control by the Dudley Observatory to support the proposal for Chicago, but he was apparently unaware of the visual signal's short life.

19. "Time," *Chicago Tribune*, 7 June 1869, 4.

20. Chicago Common Council, "Proceedings: Petitions and Communications," 12 July 1869, Minute Book, 106, for a summary; details are in "A Prayer for Standard Time in the City," *Chicago Tribune*, 13 July 1868, 4. Also, "Accurate and Uniform Time," *Railroad Gazette* 14 (16 April 1870): 51. The railroad-sponsored timekeeper was J. C. Adams, famous among students of American horology as the "great starter" of watch manufacturing enterprises, among them the Elgin Watch Company.

21. *Chicago Tribune*, 15 and 16 July 1869; I believe that Elias Colbert prepared these editorials.

22. Safford, "Standard City Time," *Chicago Tribune*, 17 July 1869, 2; see also his "Standard Chicago Time," *Evening Post*, 19 July 1869, a copy of which is in the Observatory Records, Northwestern University Archives.

23. Safford's proposal was dated 23 August 1869 and is in Chicago City Council Proceedings Files, 11 May 1870, Illinois Regional Archives.

24. See "Chicago Astronomical Society," *Chicago Tribune*, 18 September 1870, 3, as reported in Ranney, "Facts Relating to the Chicago Astronomical Society," 1:288–89.

25. Chicago City Council Proceedings Files, 22 August 1870, 11 November 1870, and 28 November 1870, Illinois Regional Archives.

26. For newspaper articles printed between July 1870 and July 1871, see Ranney, "Facts Relating to the Chicago Astronomical Society," 1:281–306. In the *Chicago Tribune*, 20 June 1871, 2, the writer noted that "city time [has been] furnished, since the middle of January last." Also, "Common Council: The City Hall Clock," *Chicago Tribune*, 23 June 1871, 3. For the clock's installation by the observatory, see CAS, "Minutes," 15 November 1880, 110–13.

27. Scammon was a remarkable civic leader, and he paid all his debts stemming from the Great Chicago Fire. Some of his history and the esteem Chicagoans held for him are found in "Death of J. Young Scammon," *Chicago Tribune*, 18 March 1890, 3, and the accompanying editorial, "Demise of Mr. Scammon," ibid., 4.

28. CAS, "Minutes," 44–45, 115–20.

29. This summary is based on Obendorf, "Samuel P. Langley"; and Beardsley, "Samuel Pierpont Langley." Beardsley's analysis of the bond between the astronomer and William Thaw is outstanding.

30. Langley wrote to clockmaker Charles Frodsham, astronomer John Hartnup, and Astronomer Royal George Airy in England, as well as to numerous American experts and firms; see the surviving replies, May through December 1869, Langley Papers, AIS.

31. For a description of the firm and its electrical clock systems, see Thomas A. Edison (attrib.), "The Manufacture of Electrical Apparatus in Boston," in Jenkins, *Papers of Thomas A. Edison*, 77–81, and 82, n2. Langley must have also been in contact with this firm in 1868, for Hamblet (employed by Wm. Bond & Son from 1853 to 1862 and thereby aware of its electrical instruments for astronomy) supervised the construction of the Allegheny Observatory's chronograph. Both Langley and Abbe also considered the Kennedy Electric Clock Company of New York; see *NAWCC Bulletin* 31 (February 1989): 47–51, for a description of this company's timekeepers.

32. Edmands and Hamblet to Langley, 6 May 1869, Langley Papers, AIS.

33. Hough to Langley, 13 July and 30 September 1869, Langley Papers, AIS.

34. Edmands and Hamblet, 25 August 1869, Langley Papers, AIS; a copy of their proposal, in Langley's handwriting, is included with this item of correspondence. An undated sketch of the Jones system, based on an 1857 British Patent Office report, is also in the AIS collection.

35. Pennsylvania R.R. (General Superintendent) to Thaw, 23 August and 22 September 1869, Langley Papers, AIS.

36. Beardsley, "Samuel Pierpont Langley," 57, gives 14 September 1869 for the date of this test, which involved the manual transmission of signals, for Langley did not possess a mean-time clock with electrical contacts. See also Langley to [Abbe], 14 September 1870, Abbe Papers, LC, which supports the tentative nature of these 1869 transmissions.

37. Farmer to Langley, 3 November 1869; Hough to Langley, 30 September 1869; William Bond & Son to Langley, 13 and 26 October 1869; and Blunt & Co. to Langley, 9 October 1869; all in Langley Papers, AIS.

38. From the copy in Samuel P. Langley Papers, 1869–1906, Smithsonian Institution Archives, Record Unit No. 7003.

39. Langley to Williams (Pennsylvania R.R.), 14 December 1869, Langley Papers, AIS.

40. E. Howard & Co. to Langley, 27 November 1869, Langley Papers, AIS.

41. E. Howard & Co. to Langley, 20 December 1869, Langley Papers, AIS.

42. Langley was eager to install controlled subsidiary clocks; see his "Uniform Time" articles in the *Pittsburgh Commercial*. The astronomer remained an advocate of the Jones system at least through 1878.

43. Local telegraph operators knew when the daily connections would be made, as well as their own operating time's offset from Pittsburgh (rounded to the nearest minute). For example, one clock-line connection was made at 9:00 A.M. Altoona time. The chief operator in Pittsburgh switched the time circuit into the railroad's east-running main line wire at 8:50 A.M. Pittsburgh time. (At the line's terminus in Philadelphia, the seconds ticks started at 9:20 Philadelphia time.) Similarly, the observatory clock line was connected to the west-running wire at 9:00 A.M. Columbus time (9:13 in Pittsburgh).

44. The switch stopped the observatory's Pittsburgh-based time ticks for ninety seconds, beginning exactly two minutes before the hour. In order to prevent any confusion arising from these periodic intervals of silence, "Altoona time" was taken as ten minutes fast of Pittsburgh, even though its longitude is in fact 6^m24^s east of Pittsburgh. (By 1874 the Pennsylvania Railway was using only two operating times: the local times of Philadelphia and Columbus, Ohio.)

45. Langley, "On the Allegheny System of Electric Time Signals"; also Langley, "Uniform Railway Time" and "Uniformity of Time." Langley was "careless" with regard to the beginnings of his railroad timekeeping system—always citing 1869, the year he published his 1 December pamphlet. But his correspondence reveals that the system was not actually constructed and installed until 1871.

46. In 1874 Langley sent special time signals to Chicago, where Hamblet was testing the performance of an electrical clock system for the Western Electric Manufacturing Company; the test had the completely unintended consequence of restarting the Dearborn Observatory's distribution of time; see E. Howard & Co. to Langley, 26 June 1874, Langley Papers, AIS. Later in the year Langley worked closely with the clock company while it constructed a timekeeping and distribution system for the Philadelphia Local Telegraph Company. After this installation, Philadelphia time transmitted by the Allegheny Observatory arrived at the telegraph company via the Pennsylvania's railroad telegraph wire.

47. Considered as 1875 income, thirty-five hundred dollars was equivalent to about fifty-three thousand dollars in mid-1999 dollars; at the end of Langley's directorship in 1891, the amount was equivalent to over sixty-three thousand dollars.

48. Langley to Holden, 30 January 1885; from the copy in Beardsley, "Samuel Pierpont Langley," 145.

49. Ibid. gives the details of these complex financial arrangements. Payments from the time service continued even after Langley had left Allegheny Observatory for the Smithsonian Institution; S. P. Langley to William Thaw, Jr., 16 December 1889, Langley Papers, AIS.

50. The 1871 quotation is from Langley, "Allegheny System of Electric Time Signals," 378. The 1884 quotation is from Langley, "History of the Allegheny Observatory," 182, 184. These materials form the basis for subsequent priority claims, including one in an official history (1897) of the Smithsonian Institution; also, John A. Brashear's eulogy, "Samuel Pierpont Langley," in *Popular Astronomy* 14 (1906): 269 (in which he alludes to the controversy); and Walcott, "Samuel Pierpont Langley," 248. See also Langley's "Electric Time Service," 101.

51. Jones and Boyd, *Harvard College Observatory*, 159–60, assert that Langley influenced Winlock's decisions regarding Harvard College Observatory's time service, but since similar services were being designed by both Abbe and Safford, a direct causal relationship must be questioned. Winlock purchased a Jones-system clock from James Ritchie & Son around 1872, and unless he read no astronomy journals, the director knew of its use by British and Russian observatories and by the Naval Observatory. Even more compelling, Langley claimed that his influence took place "about 1873," yet the *HCO Annals* 8 (1876): 12, notes income from the new time service starting two years earlier, in 1871.

Chapter 7

1. "Railroad Time," *American Railroad Journal* 25 (21 August 1852): 529–30. For additional details regarding these early suggestions, see Bartky, "Adoption of Standard Time," 25–39, particularly nn10, 11, and 17.

2. *The American Railroader and Universal Weekly Travelers' Guide* 2 (1 February 1868): 8.

3. *Appleton's Railway and Steam Navigation Guide* (New York, 1857). Apparently the first guide combining companies' operating times with cities' local times was prepared in October 1873 by *Appleton's* competitor and eventual successor, the *Travelers' Official Guide.*

4. F. A. Stumm, "A Standard Time for the Whole World," *Scientific American* 18 (6 February 1869): 101–2. Stumm also suggested doing away with A.M. and P.M. designations in favor of a twenty-four-hour notation, an idea later championed by the world-time enthusiast Sandford Fleming.

5. Safford, "Standard Chicago Time," *Evening Post,* 19 August 1869.

6. When the railroads adopted Standard Railway Time in 1883, one resulting zone was 70 minutes wide, another 101 minutes.

7. The quotation is from Langley, "Uniform Time," 12 February. Various editors reported Langley's efforts; see *American Exchange and Review* 17 (March 1870): 69–71; *Railroad Gazette* 11 (9 April 1870): 29; and *Scientific American* 26 (13 January 1872): 34–36.

8. Langley's statement regarding a single—Greenwich—time for railway operations in Great Britain was not correct either. The 1858 court decision remained in force until 1880.

9. After dropping Altoona time, the Pennsylvania Railroad continued with the other two until the industrywide changeover in November 1883; see Langley, "Allegheny System of Electric Time Signals," 386; and Langley, "Uniform Railway Time," 275.

10. "Time for the Continent," *Railroad Gazette* 14 (2 April 1870): 6; "Uniformity of Railroad Time," ibid. (9 April 1870): 29; and "Accurate and Uniform Time," ibid. (16 April 1870): 51. See also X. Sentrick's contribution in the 7 May issue, and the 21 May follow-up comments.

11. Dowd, *System of National Time.*

12. Dowd, "Origin and Early History"; the quotations are from Dowd, *System of National Time,* 5. For an analysis of Dowd's efforts, along with that of a later competitor, see Bartky, "Invention of Railroad Time."

13. Being practical, Dowd didn't "split" a railroad company when its rails crossed a section boundary. Instead, he assigned one operating time, extending it to the rail terminus. A copy of Dowd's map (in color) appears on the cover of the spring 1983 issue of *Railroad History.*

14. Dowd, "Origin and Early History," 93, footnote.

15. Dowd, *Rail Road Time,* attached to Dowd's 20 October 1870 letter to A. J. Cassatt; in Langley Papers, AIS; and *National Rail-Road Time,* shown as plate IV in Charles N. Dowd, *Charles F. Dowd* (New York: Knickerbocker Press, 1930). The astronomers were Elias Loomis and H. A. Newton of Yale College, George Hough of the Dudley Observatory, and W. H. C. Bartlett at West Point.

16. Dowd to Cassatt, 20 October 1870, Langley Papers, AIS; Cassatt to Thaw, 24 October 1870, Langley Papers, AIS; and Obendorf, "Samuel P. Langley," 12–14. Via Thaw, Cassatt invited Langley to address the October convention. See also Dowd, "Origin and Early History," 94.

17. A reprint of the railway association's 13 May 1873 report appears in *PAMS* 4 (1883): 50. Its decision was widely reported, appearing in the *New York Times, American Exchange and Review*, and the *Railroad Gazette*.

18. Abbe's three reports are in the chief signal officer's *Annual Report . . . for 1874*, 383–85; *Annual Report . . . for 1875*, 367–74; and *Annual Report . . . for 1876*, 301–10.

19. Abbe in Signal Officer, *Annual Report . . . for 1876*, 304. Years later he recalled that "the numerous correspondents had used such a great variety of standards of time, many of which could not be identified at all . . . the words 'railroad time,' 'local time,' or 'standard time' seemed to have no definite meaning . . . [and] when several railroads passed near an observer it was really impossible to ascertain what particular railroad time was adopted."; Abbe, "Standard Time in America," 316.

There had been large-scale networks before, but all had been manned by trained observers. The terrestrial magnetic network begun by Gauss—who initiated observatory efforts in 1834 and founded the Magnetic Union of Göttingen to promote magnetic observations—had agreed-on observation times, and each observatory was equipped with a transit and clocks to determine the time directly. A variety of time bases was used in the nineteenth century; for example, Madras Observatory took its magnetic measurements on Göttingen time, its meteorological measurements on Madras (local) time, dropped its time ball on Greenwich (ocean navigator's) time, and distributed civil (local) time.

20. "The time of observations will be the mean time of each Station," with each of three observations being made at "the exact hour, fixed by a well-regulated watch"; Arnold Guyot, *Directions for Meteorological Observations, Adopted by the Smithsonian Institution for the First-Class Observers* (Washington, 1850); reprinted with additions in *Annual Report of the Smithsonian Institution for 1855* (Washington, 1855), 239–40; reprinted in *Smithsonian Miscellaneous Collections* (Washington, 1860), 35–36.

21. See the chief signal officer's *Annual Report* in 1874 and after regarding the data transmission rate. The electric telegraph had made simultaneous observations both possible and useful; before, observations "taken at the same hour of local time" had been adequate, since they could not be utilized (transmitted) at once; Chief Signal Officer, *Annual Report . . . for 1879*, 190–91.

22. "In the new instructions to Signal Service observers no local times will be recognized but only the Washington time, as telegraphed from this office"; Chief Signal Officer, *Annual Report . . . for 1880–81*, 68–69. Years later Abbe noted "the uncertainties in the local standard of time used by our voluntary observers and *in a few cases by our regular observers*"; Abbe, "Meteorological Work of the Signal Service," *Weather Bureau Bulletin No. 11*, 269–71, emphasis added.

23. Abbe in Chief Signal Officer, *Annual Report . . . for 1876*, 317; and in *PAMS* 2 (1881): 237.

24. In existence between 1873 and as late as 1894, this organization is often confused with the American Meteorological Society, founded in 1919. The American

Metrological Society met twice a year in the rooms of F. A. P. Barnard, the president of Columbia College (now Columbia University). Its uniformity goals are set forth in *PAMS* 1 (1880): 5–6. The society served primarily as a means to generate various tracts and reports; more than a third of the *Proceedings'* eight hundred–plus pages is devoted to uniform time. (This journal is one of two major contemporary sources, the other being the *Travelers' Official Guide*, a railroad industry compendium discussed later.) The society acted on Abbe's letter at its 19 May 1875 meeting.

25. Dated 20 May 1879, the "Report" is in *PAMS* 2 (1880): 17–45; copies for distribution became available after 10 March 1880.

26. Abbe credited Harvard's Benjamin Peirce with first proposing the use of Greenwich-based meridians, and in correspondence with *Travelers' Official Guide* editor William F. Allen, Abbe indicated that the mathematician did so in 1873 or 1874. Yale's H. A. Newton, one of those astronomers signing Dowd's 1870 scientific testimonial, was an original member of the society's Committee on Standard Time and also signed Abbe's "Report on Standard Time." Considering the closeness of the community, other astronomers may have been aware of Dowd's hour-difference concept for his proposed railroad zones.

27. While literally true, there had been fewer than ten printed comments by the railroad industry in twenty-seven years. It is amusing to observe that of Abbe's almost three dozen citations in his "Report," all are to astronomers' writings and the annual reports of astronomical and magnetic observatories.

28. Abbe, *Professor Abbe and the Isobars*, 145–48; see also *PAMS* 2 (1881): 124, 166–74.

Chapter 8

1. Abbe, "The Aurora of April 7, 1874," 25 October 1876, in Chief Signal Officer's *Annual Report . . . for 1876*, 310. In the "Report on Standard Time" Abbe claimed that in 1875 the Committee on Standard Time adopted a Greenwich-based system, but so early a date is unlikely. (Abbe's specific dates are often in error.)

2. Admiral Davis served as superintendent of the Naval Observatory in 1865–67 and 1874–77.

3. Time distribution by telegraph companies began very early; for example, by 1856 the Naval Observatory's Washington time was transmitted to the South via the Washington & New Orleans Telegraph Company, an early Morse line absorbed by Western Union. Like other telegraph companies, Western Union transmitted time signals, at no cost, to astronomers in the field, to government longitude parties, and to meteorological observers. These signals were generated by many of the country's observatories; however, the Naval Observatory's time signals were the ones most commonly transmitted.

4. See William Harkness to Superintendent, "Statement respecting early history of the time service of the Observatory," 15 May 1890, USNO-NA, LS (loose in letterbook, this material was prepared to support Captain Pythian's report to the secretary regarding the Naval Observatory's time signal to Western Union; see ch. 15, n46). Other details of the Naval Observatory's post–Civil War time signals appear in the

New York Herald, 21 January 1882, 5, an account probably written by an observatory staff member. See also USNO Superintendent, *Annual Report . . . for 1869,* 56; *Annual Report . . . for 1870,* 44; and *Annual Report . . . for 1871,* 120.

5. Characterized as retransmissions "to nearly every State in the Union"; USNO Superintendent, *Annual Report . . . for 1873,* 94. Winlock died in June 1875, so any reassurance came early in Admiral Davis's second tour at the observatory. The quotation is from Langley to Pickering, 25 March 1877, HCO-HUA, UAV 630.17.7.

6. USNO Superintendent, *Annual Report . . . for 1873,* 94; the telegraph company's actual distribution of observatory time is recorded in Holden's 1876 report to the secretary of the navy (see below).

7. Davis to Harkness, 21 April 1876, USNO-NA, LS-S, in which he credits Harkness with the Philadelphia time ball proposal; also Admiral Davis to Admiral T. A. Jenkins, 26 April 1876, USNO-NA, LS-S. An even earlier proposal for a time ball to be located in New York City is mentioned in USNO Superintendent, *Annual Report . . . for 1872,* 93.

8. An assistant to renowned Naval Observatory astronomer Simon Newcomb, Holden had already prepared other summaries of scientific endeavors. This particular summary carries a 31 October 1876 date. Considered as a trip report, Holden's account is hardly adequate. That portion of the text germane to his official assignment consists of little more than handouts and previously published articles. In particular, Holden fails to mention the only Kensington exhibition event concerned with timekeeping: British railway and electrical engineer Charles V. Walker's lecture, "Galvanic Time Signals." This reinforces my view that Holden's report was primarily a vehicle for Admiral Davis.

9. Highlighting Holden's summary in his own report to Congress, Secretary Robeson also noted with approval the Naval Observatory's system of controlled electrical clocks and its daily time signal to Western Union. He then went on to support the extension of both activities; U.S. Department of the Navy, *Annual Report of the Secretary of the Navy for 1876,* 15. Supt. Davis to Chief, Bureau of Navigation, 4 December 1876, USNO-NA, LS-S.

10. Davis to Green, 5 December 1876, USNO-NA, LS-M.

11. In 1876 and for several years afterward, the Navy Department requested funds to pay for the transmission of time signals to its yards; however, Congress never approved an appropriation. In 1885, Western Union began sending observatory time signals to Navy sites without charge. At once, the number of Navy-operated time balls increased dramatically.

12. According to Dr. Gernot Winkler, retired director of the U.S. Naval Observatory Time Service, time signals telegraphed to Western Union ceased in the early 1970s.

13. To provide a structure for fees, Admiral Davis offered to provide the difference in time between the Washington meridian and those of all American cities with twenty thousand or more inhabitants. After Western Union's request, the Naval Observatory enlisted the services of other government agencies, and this compilation was included in Western Union's announcement document listing its prices for Washington time.

14. Langley to Pickering, 20 March and 9 April 1877, HCO-HUA, UAV 630.17.7.

15. The specter of competition did not initially alarm Dearborn Observatory in Chicago, which was protected by its five-hundred-dollar-per-year time-service contract with the Western Electric Manufacturing Company. However, this five-year agreement with Western Union's affiliate lasted only until 1880.

16. Obendorf, "Samuel P. Langley," 25–26, who highlighted Langley's concerns, mistakenly called the enterprise the "Atlantic and Pacific Tea Co." Langley countered Western Union by publicizing his own efforts, "The Electric Time Service" appearing in *Harper's New Monthly Magazine*. Ignoring the telegraph giant's now-familiar time ball, Langley offered as his example the obscure one dropped briefly in 1860 via Dudley Observatory's signal. Despite gaining national publicity for the Allegheny Observatory's time service, Langley could not prevent another investigation of the prices he was charging the city of Pittsburgh; in "The Other Side," *Pittsburgh Commercial Gazette*, 13 April 1878. In private, Langley denigrated the Naval Observatory's time signals; see his June 1878 report to the Western University of Pennsylvania's Board of Trustees.

17. See Atlantic & Pacific Telegraph Co. to Waldo, 29 September 1877; and J. C. Hinchman (Western Union, New York) to C. F. Wood (Western Union, Boston), 16 October 1877, mentioning the observatory's letters over the past seven months; in HCO-HUA, UAV 630.20.10.

18. In his first report to the Visiting Committee in November, Pickering wrote: "Last spring [1877] . . . it was found that the receipts from the Time Service did not equal the expenditures, [and] its discontinuance was considered." I cannot support this bleak assessment. At that moment the service was generating about twenty-four hundred dollars annually (more than thirty-seven thousand dollars in mid-1999 dollars); indeed, an annuity for Winlock's widow—half the service's net after expenses—was being paid. See Pickering, *Annual Report . . . for 1877*, 9–10; and Searle, "Historical Account," 12.

19. One view regarding motives is that Pickering was quite aware of events in Washington and made a preemptory strike aimed at protecting the New England service area. (Erecting the Harvard College Observatory–sponsored time ball in Boston in 1878 prevented the U.S. Navy from placing one at the Boston Navy Yard.) In any event, Pickering's expansion decision came quite early, for the observatory's Time Service files and letterbook series were initiated in February 1877.

20. The Western Union's circular announcing these Naval Observatory–based time services is dated 2 April 1877. The company also publicized the new services in its 16 April 1877 *Journal of the Telegraph*.

21. This second device must also have failed, for its description in contemporary newspaper accounts is completely at odds with the time ball depicted in *Scientific American* a year later. Regarding the various devices, see *New York Times*, 18 September 1877, 2; *New York World* 18 September 1877, 8; and *Scientific American* 39 (30 November 1878): 335, 337. Constructed as a set of lunes and semicircles cut from copper sheet, the third time ball was designed by the company's manufacturing superintendent, George M. Phelps.

22. Holden to Pickering, 12 November 1877, HCO-HUA, UAV 630.17.7.

23. Despite the fanfare, receipts increased only modestly, Assistant Leonard Waldo estimating them at twenty-seven hundred dollars in 1879. Waldo's Harvard College Observatory time-service materials are extensive, and include *Standard Public Time*, with its cover letter appearing in Abbe, "Report on Standard Time," 39–40.

24. Holden to Acting Supt. C. H. Davis, 1 March 1877—an appendix to Captain R. L. Pythian's 15 May 1890 report to Commodore George Dewey (see ch. 15, n46). A time error, converting New York to Washington time, appears in Western Union's original announcement and was noted by Holden; see the copy in the William Gardner collection, Naval Observatory Library.

25. Western Union's decision may have been based on Admiral Davis's mention of England's Post Office Department's selling the Royal Observatory's time signals to ten London subscribers: "mostly chronometer-makers," according to Astronomer Royal George Airy's 1874 annual report. Although these clock synchronizers can be traced back to Bain's 1840s concepts, they were actually the latest developments of James Ritchie & Son, Edinburgh. I believe that Western Union's seeming misnomer was deliberate and reflected patent concerns.

26. Holden, "On the Distribution of Standard Time"; "Standard Time: The New Western Union Time Ball in Operation," *Journal of the Telegraph* 10 (1 October 1877): 289–90. Holden apparently learned nothing about English technical developments on his 1876 visit. The improved device was the invention of John A. Lund, partner in the London chronometer firm of Barraud & Lunds. Public word of his success appeared quite early; one example is "Greenwich Time," *Times* (London), 15 January 1877, 12; an American example is *Journal of the Franklin Institute*, 3d ser., 72 (April 1877): 227. In October 1877, Lund applied for an American patent, which was granted the following March. (It is worth noting that starting in 1870, Western Union's Office of Electrician was made responsible for reporting on all inventions that might affect the company; see Jenkins, *Papers of Thomas A. Edison*, 55, n12.)

27. One of several ironies here is that in 1876, Swiss inventor Matthaeus Hipp exhibited his already-successful electrical clock system at the International Exhibition in Philadelphia. Seen by American astronomers, it received a rave review from Sir William Thomson (Lord Kelvin), one of the judges for the electrical section. However, an 1887 list of close to one hundred of Hipp's installations worldwide shows that only a single electric clock system was sold in the United States—most likely the one displayed at Philadelphia. See F. A. Walker, ed., *International Exhibition, 1876; Reports and Awards* 7 (Washington, 1880): 481–82, 518; and [Matthaeus Hipp], *Les horloges électriques de M. le Dr. Hipp* (Neuchâtel: H. Wolfrath et cie, 1887), 5–6. Hipp's famous toggle for maintaining the going of a pendulum via electricity was considered satisfactory for public clocks well into the twentieth century.

28. "A New Time Circuit to Be Established," *Journal of the Telegraph* 11 (1 June 1878): 167; the *quid pro quo* was supplying a time signal free to the Fire Department. As before, Western Union described its system as a group of clocks corrected daily by electrical signals that would actuate their synchronizers.

29. Hamblet's services for E. Howard & Co. ended in 1876, when the inventor

joined Stephen Dudley Field's Electrical Construction and Manufacturing Company in San Francisco. He resided in California for a year, coming to New York City in 1877; "James Hamblet," in John B. Taltavall's *Telegraphers of Today* (New York, 1894), 96; "The Death of Mr. James Hamblet," *Electrical World and Engineer* 35 (1900): 56. (A later article, "James A. Hamblet . . . Honored," *Telegraph and Telephone Age* 54 (1 June 1936): 123–24, contains numerous errors.) Hamblet (1824–1900) was not involved in the construction of Western Union's several time balls. Neither was he the only electrical-clock expert available: Vitalis Himmer of New York held several patents, was constructing electrical clocks, and had supplied a pendulum to Harvard College Observatory; see Himmer to Waldo, 28 February 1877, HCO-HUA, UAV 630.377.

30. "Electric Time Service for New York," *Scientific American* 39 (30 November 1878): 335, 337; "Electric Time Service for New York," *Journal of the Telegraph* 12 (1 January 1879): 1–3, 5; and Small, "Electrical Time Services in New York."

31. Hamblet to Waldo, 10 September 1878, HCO-HUA, UAV 630.77. Hamblet was quite successful, as reflected by his subsequent contract: "I have leased the *local wires*, and make my own contracts with subscribers and collect all Bills, and pay the Company a fixed price per mile of wire and an additional sum for each *subscriber* for Battery expenses"; Hamblet to Edmands, 6 March 1882, HCO-HUA, UAV 630.677.

32. The country's fourth time-related accident—at Mingo Junction, Ohio, on 7 August 1878—certainly demonstrates that accurate time does not prevent accidents, for the road's operating time came from Allegheny Observatory; see Bartky, "Running on Time," 30–32.

33. Dudley, "Railway Time Service," a talk given at the AAAS's 1880 annual meeting in Boston; Dudley, "Railway Time Service," *Journal American Electrical Society* 3 (1880): 63–65; Editors, *JC&HR* 12 (July 1881): 139; and Dudley, "Railroad Time and Time-Instruments." For some highlights of this distinguished engineer's industry career, see *National Cyclopaedia of American Biography* 19 (1926): 281–82.

34. The power for the synchronizer came from a local battery; the device reset the clock's minute hand to the start of every hour. Soon after, Dudley modified his synchronizer so that the second hand was zeroed as well, terming this development "a first." No clock patents were granted to Dudley, but as a well-regarded industry insider, he may not have sought them. Railway companies protected themselves from outsiders who demanded large royalties for their inventions by having their own engineers develop products at least as good as the patented ones. For an excellent review of the railroad patent environment, see Usselman, "Patents Purloined."

35. Dudley became a consulting engineer for the New York Central System, so he may have had little time for promoting clock sales.

Chapter 9

1. The five other observatories were those in Albany, Ann Arbor, Cincinnati, Toronto, and Quebec. While Abbe failed to list Chicago's Dearborn Observatory or observatory time services at Washington University in St. Louis, Carleton College in Minnesota, and McGill University in Montreal, only the first and last of these omissions were important at this time.

2. Admiral John Rodgers to T. E. Thorpe, 22 May 1878, USNO-NA, LS-M, not-

ing that the observatory's Washington time was not available at railroad stations on the Great Plains.

3. Hamblet to Waldo, 10 September and 3 December 1878, HCO-HUA, UAV 630.377. Hamblet recorded the Allegheny Observatory's transmissions of Philadelphia time, converting the data to New York local time via the known longitude difference.

4. Hamblet may have made such time-base comparisons earlier, but probably not on a regular basis. In 1874–75 he installed a signal clock at the Philadelphia Local Telegraph Company, where president Henry Bentley had arranged for Langley's Philadelphia time signal to the local offices of the Pennsylvania Railroad to be sent directly to him. Bentley's company was receiving, and continued to receive, the Naval Observatory's noon signal; see his correspondence in Langley Papers, AIS; and Holden, "Report upon the astronomical instruments," 307.

5. Hamblet continued these daily comparisons for many years, eventually also including Yale's erratic time service; *Journal of the Telegraph* 20 (20 April 1887): 50; and Joseph A. Rogers to Charles S. Lyman, 20 January 1886, Yale Astronomical Observatory Records, YUA, 14-E-1. Fewer than two dozen of Hamblet's monthly sheets have been located, the bulk of them in HUA. Some 1881 data are in U.S. Army Signal Office, "Information Relative to Time Balls," 17–18.

6. "Electric Time Service for New York," *Scientific American* 39 (30 November 1878): 337.

7. Hamblet to Waldo, 1 February 1879, HCO-HUA, UAV 630.20.10, which has the comparison for January. Hamblet to Naval Observatory, 1 February 1879, USNO-NA, LR; Rodgers to Hamblet, 4 February 1879, USNO-NA, LS-M. Hamblet's comparisons for March and April 1879 are included in Langley Papers, AIS.

8. Hamblet to Rodgers, 30 January and 2 March 1880; Rodgers to Norvin Green, 3 February 1880, and Rodgers to Hamblet, 3 February 1880, USNO-NA, LS-M.

9. USNO Superintendent, *Annual Report . . . for FY 1880*, 126–27. The Naval Observatory's "get-well" program may have started somewhat earlier, for the superintendent's prior report noted that a change "in the method of transmitting time-signals and of dropping the Washington and New York time-balls is nearly completed, and will probably be in operation by the end of October [1879]."

10. Hamblet to Edmands, 23 May 1881, HCO-HUA, UAV 630.77.

11. Abbe included some of Waldo's Harvard-era writings in his "Report on Standard Time." Then while proofing the printed text, he added a footnote citing the Winchester Observatory's time service, commending both its quality and the efforts of its founder, Leonard Waldo. Since the footnote is dated February 1880 and Yale's time service did not begin until late April, Abbe's praise came rather early!

12. Dorrit Hoffleit, "The Winchester Observatory," in her *Astronomy at Yale*, provides many details of the fee-for-service activities Waldo inaugurated; also, *Fourteenth Annual Report of the Sheffield Scientific School of Yale College, 1879–1880*, New Haven, 1880, 9; and *Transactions of the Astronomical Observatory of Yale University* 1 (1904): v–vi. Also called "Yale College Observatory," in late 1882 Winchester Observatory's official name became "The Observatory in Yale College."

13. The Harvard/Yale longitude expedition is noted in *Fifteenth Annual Report of the Sheffield Scientific School of Yale College, 1880–1881*, New Haven, 1880, 7; however,

the details were not published until 1885 (in a now-missing Yale Ph.D. dissertation), two years after the Horological Bureau's transit instrument had been relocated. (Hoffleit, *Astronomy at Yale*, 71, 214, gives 1882 as the date of the longitude campaign—actually that was the year in which the longitude result was first reported.)

14. *Proc. AAAS* 29 (1881): xv, 279, 747. Abbe claimed that he suggested establishing the committee; *PAMS* 2 (1880): 176. His purpose was to gain AAAS support for his particular position on uniform time for the United States. (For a measure of Abbe's influence, contrast Stone's article "Uniform Time," *Science* 1 (10 July 1880): 13, with his later articles written as committee chairman.) Waldo, who was anxious to continue the use of New York meridian's time, forced the inclusion of a minority statement in the committee's first report; *Proc. AAAS* 30 (1882): 4–6, 385. Disagreements among the astronomers apparently continued, and at the August 1882 meeting association officers shifted timekeeping matters to a new committee; *Proc. AAAS* 31 (1883): 612, 626, 633–34.

15. Waldo presented a talk describing the testing of timepieces and the verification of thermometers. During the meeting he may have spoken with P. H. Dudley, for the latter's analysis of railroad timekeeping in New York City (published the following year) includes unpublished details of Yale's time signals to the New York, New Haven & Hartford Railroad.

16. Waldo was already lobbying Connecticut's railroad commissioners. Thanks to his efforts, in March 1881 the state became the first to adopt uniform time, passing a law that defined Connecticut's official time in terms of New York City's meridian. Until 1885, Yale received two thousand dollars annually from the state for its observatory's daily signals.

17. Waldo's "Distribution of Time" for the *North American Review* was reprinted in *Science*; quotes are from the reprinted version. In the article Waldo also discussed uniform time, citing the American Metrological Society's Greenwich-based standards. He couched his opposition to one time for the United States—Abbe's ultimate choice —in terms of "a single time from the Naval Observatory."

18. Waldo's criticism of the Jones system anticipated a new system of clocks, to be synchronized by electrical signals from Winchester Observatory.

19. Abbe, "Report of Committee on Standard Time," *PAMS* 2 (30 December 1880): 176.

20. Ibid., 176, 181–84. General William B. Hazen, Abbe's superior, took up his duties as chief signal officer also in December, but he probably did not see the drafts at that time.

21. For a positive view of General Hazen's support of science, see Dupree, *Science in the Federal Government*, 189–92. However, the time-services episode described in the next chapter brought down the wrath of the chief signal officer's civilian superiors and raised concerns among several members of Congress—certainly no way to gain appropriations increases.

22. This and subsequent quotations are from U.S. Army Signal Office, "Information Relative to Time Balls."

23. Among the world's timekeeping laboratories today, averaging the values from ensembles of clocks is standard practice. Since this 1881 scheme appears to be the ear-

liest one ever described, identifying the scientist who proposed it is of some interest. Decades later Abbe credited Leonard Waldo's brother Frank, who came to the Signal Service's Washington study room in July 1881 from the Harvard College Observatory Time Service; however, the 3 June date of the inquiry discussed here weakens that attribution. Harvard College Observatory director Edward Pickering credited his time-service assistant, J. Rayner Edmands, but his proposal (*Proc. AAAS* 31 (1883): 612) came a year after that of the Signal Service.

24. Abbe, "Report of Committee on Standard Time," *PAMS* 2 (24 December 1881): 310.

25. Abbe, "Progress Report to Am Met Soc 1882/83 [*sic*]," 6–7, Abbe Papers, MS60, JHU. This nineteen-page manuscript is a summary of events, and includes hitherto-unknown details of Hamblet's multiple-observatory comparisons and his adjustment of Western Union's master clock to give New York City its true time. The report demonstrates that Abbe was quite aware of the lesser accuracy of time balls relative to other modes of time distribution. The manuscript is not in Abbe's handwriting, although its section heads are; it may have been prepared by a Signal Service scribe. Its title is incorrect as well, for it was prepared between September and 24 December 1881 and was never published as a Committee on Standard Time progress report by the American Metrological Society.

26. This clock purchase was later criticized; see House, *Disbursements of Public Moneys*, 21, 34–35, 101, 123.

Chapter 10

1. Supt. John Rodgers to Hon. J. Floyd King, 17 September 1879, USNO-NA, LS-M; King to Rodgers, 15 December 1879, USNO-NA, LR.

2. H.R. 3769, 21 January 1880, 46th Congr., 2d sess. Congressman King's inquiry and his first legislative attempt had little effect on Naval Observatory matters. After the bill's death in 1880, Admiral Rodgers wrote Western Union to inform them that he was about to recommend erecting time balls at Atlantic- and Gulf-coast navy yards. He requested the company's charges for transmitting daily signals from Washington to these sites; twice he received no reply. Save for Hamblet's monthly list of daily comparisons, no letters were exchanged between the two organizations until late in 1881, when the agency conflict erupted.

3. *Congressional Record* (11 December 1881), 108; King introduced ten other bills on that day.

4. In recognition of its new service, the Naval Observatory announced that "the wants of the Army Signal Corps have called for the distribution of daily time signals to be greatly extended; they are now sent to all parts of the country from the Atlantic to the Pacific, and from the great lakes to the Gulf of Mexico"; USNO Superintendent, *Annual Report . . . for 1871*, 119–20.

5. Hazen to Rodgers, 3 June 1881, USNO-NA, LR.

6. Rodgers to Hazen, 11 June 1881, USNO-NA, LS-M. It is likely that Signal Service personnel were aware of King's first bill; moreover, Superintendent Rodgers's remark certainly warned the agency of legislative interest, continuing or otherwise.

7. "Notes," *Nation*, 5 May 1881, 317; Waldo, "Time-Balls and Standard Time."

8. Hazen to Rodgers, 15 October 1881, USNO-NA, LR; Rodgers to Hazen, 18 October 1881, USNO-NA, LS-M.

9. Rodgers to Hazen, 28 May 1881, USNO-NA, LS-M; the superintendent must have erred deliberately when he wrote: "Congratulations on your plan of sending *Washington* time" (emphasis added).

10. Abbe, "Report of Committee on Standard Time," *PAMS* 2 (24 December 1881): 310, urging the society to memorialize Congress.

11. Yale College's Horological Bureau had announced its time-service plan two days earlier in New Haven, the newspaper's reporter writing: "Those to whom this correspondent spoke on the subject say that the Signal Service is at the helm of the movement, it being one of benefit to shipping interests"; "Uniform Time," *New York Herald*, 23 December 1881, 5, a copy of which was attached to Hamblet to Harkness, 23 December 1881, USNO-NA, LR. Also Waldo, "Standard Time," *JC&HR* 13 (February 1882): 18; and Editor, "Proposed Uniform Time Service," ibid., 38.

12. While the performance of Western Union's time ball may have been as poor as Abbe announced, I have located no data that support his statement.

13. "The Standard Time," *New York Times*, 24 December 1881, 1; "Time Ball Service," *New York Herald*, 24 December 1881, 3; "Standard Time," *New York Times*, 1 January 1882, 6.

14. Rodgers to Hon. William H. Hunt, 5 January 1882, USNO-NA, LS-S. For the admiral's letter, coupled with a favorable opinion, see "A Breeze about Standard Time: Clashing Bureaus," *New York Herald*, 7 January 1882, 4.

15. From an untitled (24[?] January 1882) editorial in "Montgomery C. Meigs Scrapbooks," 11 (1870–90): 68, National Museum of American History, Smithsonian Institution.

16. "General Hazen's Reply," *Washington Post*, 20 January 1882, 2; also, "Standard Time: The Controversy between the Signal Service and Naval Observatory," *New York Herald*, 20 January 1882, 4.

17. Robert T. Lincoln to Hon. John D. White, 30 January 1882, National Archives, RG 107, Letters Sent by the Secretary of War Relating to Military Affairs.

18. At entry No. 107 of his National Academy of Sciences bibliography, Abbe refutes General Hazen's disclaimer, writing: "This official interview . . . was written out by Cleveland Abbe at General Hazen's request." (Abbe lists the wrong date for the news conference, a minor error); Humphreys, "Cleveland Abbe," 496. General Hazen's actions prior to the incident certainly support Abbe in this matter.

19. "An Adjusted Trouble," *New York Times*, 24 February 1882, 4.

20. Ironically, a day before H.R. 594 was introduced, Langley voiced similar fears regarding Abbe's drive to have railroads adopt Greenwich-based meridians; Langley to Holden, 12 December 1881, WO-WIS Archives, Series 7/4/2. The quoted remarks are in Langley to Pickering, 22 February 1882, HCO-HUA, UAV 630.17.7. Based on this letter's date, two historians erroneously concluded that Congressman King introduced his bill in 1882; see Obendorf, "Samuel P. Langley," 26, and O'Malley, *Keeping Watch*, 93–94. As a result, O'Malley's analysis of the episode is at variance with the facts. Some observatory directors may have complained to Congress. George Hough

wrote that he had received "a remonstrance against the appropriation of $25,000 to the U.S. Naval Observatory," presumably with a request to write to Congress; Hough to Holden, 16 April 1882, WO-WIS Archives, Series 7/4/2. No other correspondence has been located, even though Commerce Subcommittee chairman White said later that the time ball legislation "was not opposed in the last session of Congress except by two or three little observatories that were then selling their computations of time to certain railroad companies"; *Congressional Record* (10 March 1884), 1761.

21. P. H. Dudley provided the equipment cost figures.

22. H.R. 5009 also included United States navy yards and state capitals—the former probably inserted by the Naval Observatory, and the latter inserted to help ensure passage and to speed the nationwide adoption of Washington time. See House Committee on Commerce, *Meridian Time and Time-Balls on Custom-Houses.* An article favorable to the bill appeared in the trade press; "A Proposal for Standard Time," *JC&HR* 13 (June 1882): 130–31.

23. A. S. Brown to Rodgers, 8 February 1882, USNO-NA, LR; the Mutual Union Telegraph Company was short-lived. Hamblet to Edmands, 6 March 1882, HCO-HUA, UAV 630.377; Hamblet to Harkness, 15 March 1882, USNO-NA, LR. Admiral Rodgers pointed out to Hamblet that the Naval Observatory would only transmit time signals, since the Treasury Department had responsibility for selecting equipment; however, the various cities that were to receive the signals "will of course select their own agents." He suggested that Hamblet prepare a pamphlet or letter to send to these cities "as soon as the bill passes, so as to be foremost in the field"; Rodgers to Hamblet, 23 March 1882, USNO-NA, LS-M.

24. Waldo to Superintendent, 1 May 1882, USNO-NA, LR; Commander W. F. Sampson to Chief, Bureau of Navigation, 3 May 1882, USNO-NA, LS-S. Apparently, the bureau turned down the offer; Acting Superintendent Sampson to Chief, Bureau of Navigation, 11 May 1882, ibid.

25. The writer of the 20 February article in the *New York Times* had added: "Perhaps it would be well for Congress . . . to also define the limits of the . . . [Signal Service's] work." Abbe to Holden, 13 February 1882, WO-WIS Archives.

26. *Congressional Record* (29 March 1882), 2388; and ibid., (11 April 1882), 2793.

27. For Abbe's role, see his commentary with Entry No. 101 in his NAS bibliography; Humphreys, "Cleveland Abbe," 495–96. House Committee on Foreign Affairs, *To Fix a Common Prime Meridian*; and U.S. Senate, Committee on Foreign Relations, *Report on Joint Resolution.*

28. *Congressional Record* (27 July 1882), 6578–80; and House, *Journal,* 1687–88, 1749, 1750.

29. On 10 December 1883, during the first session of the 48th Congress, Chairman White reintroduced the committee bill as H.R. 598; on 24 January 1884, King reintroduced his own bill as H.R. 3965, substituting "standard time" for "Washington meridian time." White then tried but failed to gain approval of his committee's bill as a substituting amendment for the bill making seventy-fifth meridian time the District of Columbia's legal time; see *Congressional Record* 10 (March 1884), 1760–63.

30. *United States Statutes at Large,* 47th Congr., 1st sess., 1882, c. 433.

31. The action was taken just days before General Hazen testified before the so-called Allison Commission, a body established by Congress in 1883 to investigate the overlapping functions of the science agencies. For brief summaries of the Hydrographic Office's time services, see Thomas G. Manning, *U.S. Coast Survey vs. Naval Hydrographic Office* (Tuscaloosa: University of Alabama Press, 1988), 166, n15; and Bartky, "Naval Observatory Time Dissemination before the Wireless."

32. Chief Signal Officer, *Annual Report . . . for 1884*, 20; and *Annual Report . . . for 1885*, 566. This timekeeper is in the collections of the Smithsonian Institution's National Museum of American History.

33. Thomson, *Beginning of the Long Dash*, 28–32, 42. Embodying a decade of Canadian efforts, these 1892 data illustrate the promise of, and the challenge inherent in, distributing accurate time.

34. Abbe, "Meteorological Work of the U.S. Signal Service," 40. In contradiction, see Hazen to Rodgers, 15 October 1881, USNO-NA, LR, in which the chief signal officer refers specifically to Hamblet's comparisons as the model for the agency's program.

35. Abbe, "Standard Time in America," 317; the article is listed as No. 233 in Abbe's NAS bibliography. Not only was Abbe in error regarding the incident's date—January 1882, not 1884—he also ignored the fact that the Naval Observatory began transmitting seventy-fifth meridian time on 18 November 1883.

Chapter 11

1. "I would call attention to the fact that, this [Standard Time] committee two years ago recommended that certain gentlemen be invited to co-operate. . . . [W]e request that they may be invited again"; Abbe, 16 December 1880, in *PAMS* 2 (1880): 177.

2. An earlier analysis of the events leading to Standard Railway Time is Bartky, "Adoption of Standard Time," the research supported by the National Bureau of Standards; a review of accounts by the era's principals and their direct descendants is in n1. A history of events was prepared subsequently by O'Malley, *Keeping Watch*, 100–130.

3. *Travelers' Official Guide of the Railway and Steam Navigation Lines of the United States and Canada* (New York: National Railway Publication Co.); hereinafter the *Travelers' Official Guide*.

4. The two railroad groups merged in 1886 to form the General Time Convention, and in 1891 named itself the American Railway Association, now known as the Association of American Railroads.

5. Allen was elected to the American Metrological Society in May 1880, but apparently he was never informed. See Allen to American Metrological Society, 6 December 1881: "My attention has been called today for the first time to the mention of my name on page twenty-nine of the published Proceedings of the Society issued in 1880"; and Allen to Abbe, 7 December 1881: "Only yesterday I became accidentally aware that my nomination to membership . . . in 1879 was made in the Report of your Committee." Both are in Allen Papers, NYPL. Allen (1846–1915) is listed in the *Dictionary of American Biography*.

6. During the adoption period, April through November 1883, five railroad weeklies printed forty articles; Allen, "History of the Movement," 49.

7. This was the March 1881 law that Leonard Waldo had lobbied for, which required the railroads to use Yale Observatory's New York meridian time. Details regarding the issue are found in the Connecticut Railroad Commissioners' *Annual Report to the State of Connecticut* (Hartford, 1880 through 1885). In 1883 the commissioners, dismayed by the lack of railroad compliance, succeeded in having an amendment passed that fined a road twenty-five dollars if it did not use the state's standard "in their public advertisements and time tables . . . for all stations within this State." Allen emphasized the Connecticut Railroad Commissioners' earlier warning only in his two documents: "*As the neglect . . . to comply causes three kinds of time to be now in use in New London by the railroads centering there . . . it is hoped that . . . the companies will find it for their interest . . . to make, at an early day, such changes . . . as will make them conform to the uniform time established by law.*" He also added: "If we agree that the system . . . here proposed is the one best adapted for practical use on our railway lines . . . whether it conforms to the whims of 'the ruling classes' who run our legislatures." In contrast, the *Railway Age*'s editors, briefly noting Fleming's American Society of Civil Engineers questionnaire, termed the attempt to establish uniform time "timely" and viewed the American Metrological Society's memorial to Congress for the establishment of standard time as a "movement in the right direction"; *Railway Age* 7 (1882): 253, 323.

8. Allen to Michigan Central Railroad, Allen Papers, NYPL. Allen's first remarks at the October 1884 International Meridian Conference (discussed in the next chapter) also expressed concern over government intervention. He urged that no change be made in the Greenwich-based meridians just adopted by the railroads, and presented a strong statement from the General Railway (*sic*) Time Convention. Opposition to government intervention on most matters was common among railroaders; in Bartky, "Adoption of Standard Time," 45.

9. Allen, "Report on Standard Time." Regarding this and subsequent railway time-zone maps, see Bartky, "Adoption of Standard Time," 46, n64.

10. Perhaps the most quoted comment was that of astronomer Simon Newcomb. When asked for his views regarding various uniform time systems for the United States, he responded that he saw "no more reason for considering Europe in the matter than for considering the inhabitants of the planet Mars."

11. Allen's assertion has the ring of truth, for Dowd's earlier partitioning of the Eastern railroads was almost identical, suggesting that natural break points—rivers, mountains, and the like—influenced their boundary selections; see Bartky, "Invention of Railroad Time," 18–21, and n16.

12. Allen, "Report on Standard Time," 691, and "History of the Movement," 34. Averaging to locate a mid-meridian is not unusual. In 1879 Sweden's national meridian for railroads was chosen to bisect the most populous region of the country. And in an editorial in the February 1874 *Travelers' Official Guide*, Allen used the same process.

13. Compare the meridian 75.0° west of Greenwich with Montreal (73.5° W),

New York (74.0° W), Philadelphia (75.1° W), Ottawa (75.7° W), and Washington (77.0° W)—a range six minutes later than the Greenwich-based meridian to eight minutes earlier. Dowd, whose aim was similar to Allen's, chose New York.

14. At that moment, there were an estimated 115,000 miles of trackage in the United States. Of the convention's members, the vote was thirty-three roads totaling 27,781 miles in favor, two roads totaling 1,714 miles against, and two abstentions. A few of the country's Northeastern roads changed to the Eastern standard on 7 October, before the convention met; Allen, "History of the Movement," 44, and the voluminous materials in Allen Papers, NYPL.

15. Schufeldt to Allen, 9 October 1886, quoted in *Proc. General Time Convention*, Appendix, 698.

16. Allen to Schufeldt, 17 October 1883, USNO-NA, LR; Schufeldt to Secretary Chandler, 19 October 1883, USNO-NA, LS-S; Schufeldt to Allen, 19 October 1883, USNO-NA, USNO-NA, LS-M.

17. Allen, "History of the Movement," 26–27. Both protagonists received public recognition early on; both collected and preserved materials supporting their respective priorities. For an analysis of these claims, see Bartky, "Invention of Railroad Time."

18. Allen, "Report on Standard Time," 703. To celebrate the centennial of this event, the Smithsonian Institution's National Museum of American History mounted an exhibition and prepared a brochure; see Stephens, *Inventing Standard Time*.

19. For example, Carlton J. Corliss, *The Day of Two Noons* (Washington: Association of American Railroads, 1953). Fleming, "Time-Reckoning for the Twentieth Century," 357.

20. In Bartky, "Adoption of Standard Time," 49, I mistakenly wrote that Chicago changed to Central time only in December.

21. The attorney general's supporting analysis made clear the fundamental problem associated with the adoption of any new time standard: its legal validity; Benjamin H. Brewster, "Change of Time at Washington," 31 October 1883, *Official Opinions of the Attorneys-General of the United States, Advising the President and Heads of Departments in Relation to Their Official Duties* 17 (Washington, 1890): 619–20; *Washington Post*, 18 November 1883, 1, and ibid., 19 November 1883, 1. A Senate bill making the seventy-fifth meridian the District of Columbia's legal standard of time was introduced on 11 December and signed into law on 13 March 1884.

22. "Standard Time," *Chicago Tribune*, 20 November 1883, 6. Not only is this story apocryphal but it also conflicts with the Baltimore city operating time used by the Baltimore & Ohio Railroad prior to the time change. *New York Times*, 13 November 1883, 4.

23. The name "Dogberry" recalls the stupid public official in Shakespeare's *Much Ado About Nothing*. See *Railway Age* 8 (29 November 1883): 753, and the various editorials and brief articles in the Bangor *Daily Whig and Courier*, 13 through 29 November 1883.

24. "Railroad Time," *Indianapolis Daily Sentinel*, 21 November 1883, 4. Two days earlier in this same paper a humorous discussion of the change, which would cause

everything to be constantly sixteen minutes slow, had appeared. (A note reminding people of the imminent change was printed on 16 November.)

25. The issue here is the one-hour discontinuity at the borders between adjacent time zones. The United States is too broad to be governed by a single time, so the zone boundaries must be placed somewhere. People who work in one time zone but live in the other will often try to "shift" the boundary away from themselves. A number of areas in the country did so unofficially between 1918 until the late 1960s, when more coercive legislation was enacted. Today the only option open to a state that wishes to mitigate the effects arising from time-zone boundaries is to choose not to adopt daylight-saving time. Thus the entire state of Indiana, which straddles two time zones, observes one clock time during the annual daylight-saving time period. Only this exception, and the fact that the state of Arizona remains on Standard Time throughout the year, perpetuates nonuniform timekeeping in the continental United States.

26. For contemporary awareness, see Waldo, "Railroad and Public Time," 606, and Dowd, *System of Time Standards Illustrated with Map.* Dowd noted that "as a rule, places located within about two and a half degrees [ten minutes] either way from this [thirty minute] border line, disregard Standard time for home use, and still retain their old diverse local times." Cincinnati, assigned to the Central railroad zone, was eight minutes away from this halfway line and was adamantly opposed to the change; see "Standard Time," *Electrical Review* 5 (5 November 1884): 5.

27. Allen, "Report on Standard Time," 691. Also, *Travelers' Official Guide* 15 (May 1883): xxxi: "The difference between local and standard time averaging about fifteen minutes, and not in any case to exceed about thirty minutes." Also, *Travelers' Official Guide* 15 (April 1883): xxxii, and (September 1883): xxxi. Amusingly, Allen's son wrote the following over sixty-five years later: "One of the theories which dominated Mr. Allen's plan was that any large section of the population would live contentedly in an area where the time was fixed with a difference of not more than *forty or forty-five minutes* from sun time"; in John S. Allen, *Standard Time in America* (New York: National Railway Publication Company, 1951), with emphasis added.

28. Allen tried to get the state of Georgia to adopt Standard Railway Time, *PAMS* 5 (1888): 159. He was unsuccessful, and we have the 1889 opinion of a Georgia court: "The only standard of time . . . recognized by the laws of Georgia is the meridian of the sun. . . . An arbitrary and artificial standard of time, fixed by persons in a certain line of business, can not be substituted at will in a certain locality for the standard recognized by the law"; see USNO, "Present Status of Standard Time," G15.

29. Editorial comment in *New York Times*, 18 November 1883, 8.

30. Although never "the law of the land," Standard Railway Time remained in general use until World War I. It officially ended at 2:00 A.M. on Sunday, 31 March 1918, when "An act to save daylight and to provide standard time for the United States" became effective. With this legislation Congress provided a legal basis to advance the country's clocks one hour during the next seven months. In many areas, the resulting federal zones were radically different from Allen's railroad time belts, for their designers adhered closely to the hour-wide consequences of Greenwich-based meridians. The official record on civil-time standardization since 1918 is voluminous

and the testimony is sometimes contentious. For an official summary, see U.S. Department of Transportation, *Standard Time in the United States*. Also, Bartky and Harrison, "Standard and Daylight-saving Time."

Chapter 12

1. The other American delegates included a senior U.S. Navy admiral and the observatory's assistant superintendent. (The Naval Observatory's superintendent attended on behalf of Colombia.) When Congress appropriated funds for the conference, it increased the number of American delegates beyond the act's specified three. This action was a skillful way to incorporate the Senate's expressed wishes, for its amendment to the Joint Resolution specifying a maximum of five delegates was dropped in the rush to pass legislation in the last days of the session.

2. Fleming attended the Meridian Conference as a member of the British delegation, listed as "representing the Dominion of Canada." The thrust of Fleming's anecdote—quoted by Burpee, *Sandford Fleming*, 217–18—may be wrong, for he did communicate on the subject at the 1876 meeting in Glasgow; BAAS, *Report of the Forty-sixth meeting of the BAAS* (1877): 182, 244.

3. Most writers identify Fleming as the leader of world time reform; however, the substance of his proposals has never been analyzed. From the first, many of his ideas were ridiculed; in 1884 opposition came from his fellow British delegates to the Meridian Conference. For the details of Fleming's attempts to gain acceptance of his concepts, see Creet, "Sandford Fleming."

4. Established through the efforts of Barnard's friend, the eminent New York jurist David Dudley Field, the organization is today the International Law Association; *Encyclopaedia Britannica*, 1943 ed., s.v. "Field, David Dudley."

5. Barnard, "Uniform System of Coinage and of Time," with the quoted remark following on p. 83. For Barnard's complaint and subsequent report, see his "Uniform System of Time-Reckoning," with the association's resolution following on p. 197."

6. Wheeler, *Third International Geographical Congress*, includes a brief history and an extensive bibliography of time reform, extending through 1885.

7. Hirsch and Oppolzer, "Unification des longitudes."

8. The delegates were also received by President Chester A. Arthur. The leisurely pace of their deliberations was first noted by Malin, "International Prime Meridian Conference," 203.

9. The "Protocols of the Proceedings" is almost 240 pages long, with half in English, half in French; U.S. Department of State, *International Conference Held at Washington*.

10. Ibid., 57–58, 77–83. Several of Allen's examples of Greenwich time for civil purposes were wrong; see Bartky, "Adoption of Standard Time," 48, n69. Allen's twenty-four globe-circling meridians were the same ones advocated by Fleming.

11. Wheeler, *Third International Geographical Congress*, 35. The quotation is from the "Protocols of the Proceedings," 77. In this context, "national time" means a country's civil time, a uniform time based on a meridian within its borders.

12. Howse, *Greenwich time*, 156–57, 160.

13. Malin et al., "Longitude Zero."

14. Executive branch responsibility for the conference was assigned to the Navy Department. For the postconference submissions, see Senate Committee on Foreign Relations, *Report to accompany concurrent resolution*, 1885; and U.S. President, *International Meridian Conference*, 1888.

15. Of course, many in the specialist worlds had already accepted a number of these concepts. For example, while reviewing telegraphic methods of time comparison perfected by thirty years of worldwide experience, the American delegate to the Geographical Congress stated: "The connection of the meridians of the astronomical observatories . . . has . . . been made in many cases with great accuracy . . . and hence . . . [in all nautical and astronomical] tables reference can be made to a single meridian in all the publications"; Wheeler, *Third International Geographical Congress*, 31.

16. Senate Committee on Foreign Relations, *Report on joint resolution*, 1882, 2. The House report reminded members of the existence of Washington time for many government purposes and the (soon-to-be-defeated) time ball bill; House Committee on Foreign Affairs, *To Fix a Common Prime Meridian*, 1882, 2.

17. United Kingdom, *Public General Statutes*, 43 & 44 Vict. 9, c. 14. Only in 1916 did Greenwich mean time become the basis for civil time for the entire British Isles; see United Kingdom, *Public General Statutes*, 6 & 7 Geo. 5, c. 45.

18. Fleming appears to have had no influence on the worldwide adoption process in the two decades following the conference, even though Creet presents a very positive view of his activities. Those writers who praise Fleming tend to condemn Astronomer Royal Airy for opposing Fleming's ideas; see Humphrey M. Smith, "Greenwich Time and the Prime Meridian," *Vistas in Astronomy* 20 (1976): 223; Howse, *Greenwich Time*, 135; and Creet, "Sandford Fleming," 74–75. Yet Airy's mid-July 1881 letter to Barnard, quoted in part here, is a thoughtful approach to gaining general public acceptance of a new system of timekeeping; in *PAMS* 5 (1888): 113–14.

19. Legislation defining the seventy-fifth meridian from Greenwich as the basis for the District of Columbia's time was enacted early in 1884. The few court cases are cited in Frederick Warner Allen, "The Adoption of Standard Time in 1883: An Attempt to Bring Order into a Changing World," Undergraduate Intensive Program (Yale University, 1969), 56–57, 61–63; Orville Butler, "From Local to National: Time Standardization as a Reflection of American Culture," in *Beyond History of Science: Essays in Honor of Robert E. Schofield*, edited by Elizabeth Garber (Bethlehem: Lehigh University Press, 1990), 256–58; and O'Malley, *Keeping Watch*, 139–40. Most legal issues were resolved in March 1918 with the passage of "An act to save daylight and to provide standard time."

20. New Zealand Mean Time, eleven and a half hours in advance of Greenwich time, was in use in the colony starting on 2 November 1868; see New Zealand, Dominion Observatory, "Mean Time and Time Service," which quotes from the *New Zealand Gazette* of 31 October 1868; see also Pawson, "New Zealand times."

21. The slow pace of the European adoption process appears well documented by the era's literature. Howse's summary table of adoption dates (*Greenwich time*, 154–55) is suspect: the first entry should be moved to 1880, and the second and fourth ones

deleted. At least one other table error has been noted—see Pawson, "New Zealand times," 287, n55.

Chapter 13

1. Lund, "Synchronizing the Time Signals." Clock synchronizers had many advantages compared with traditional systems—electrically driven slave dials and the Jones method of clock control—and Barraud & Lunds' system enjoyed great success in England. An 1882 list published in New Haven includes literally hundreds of establishments: public buildings, railways, banks, restaurants and hotels, manufacturers and merchants, scores of taverns, and a few private residences. As we have seen, the 1877 announcement of this important English advance caused the Western Union Telegraph Company to reconsider its initial decisions regarding electrical clocks for New York City.

2. Earlier (October 1878) William Payne established a time service at Carleton College in Northfield, Minnesota. His traditional system included E. Howard & Co.'s mean-time transmitting clock, whose pulses the astronomer periodically sent via a local telegraph company's line to railroads centered in Minneapolis and St. Paul. See Payne's "Time Service of Carleton College Observatory"; and Greene, *A Science Not Earthbound*, 5–8. Also in 1878 Washington University in St. Louis inaugurated a formal time service based on electrically driven slave clocks; Woodward, *Washington University Electric Clock System*.

3. Waldo to Barraud & Lunds, 16 February 1877, HCO-HUA, UAV 630.377.

4. Incorporated on 18 May 1880, Connecticut Telephone became the Southern New England Telephone Company in 1884; "New Haven's Telephone Capitalists," *Electrical World* 6 (7 November 1885): 192; see also J. Leigh Walsh, *Connecticut Pioneers in Telephony* (New Haven: Telephone Pioneers of America, 1950). Morris F. Tyler (1848–1907), SNET's second president, also taught at Yale's Law School, served as the university's treasurer for several years, and was a benefactor of the Yale Library; the lawyer was a close personal friend of the Waldo family.

5. National Telephone Exchange Association, *Proceedings of the Second Meeting*, 74–75; "The Telephone Convention," *Operator* 12 (15 April 1881): 129; and *Journal of the Telegraph* 14 (16 April 1881): 119.

6. Another telephone time service began on 24 June 1880 in Portland, Maine. The exchange operator held a phone transmitter next to a telegraph sounder wired to receive time signals sent from the Boston area. Harvard College Observatory was probably unaware of this novel distribution of its time. (Connecticut Telephone's service may have predated this activity.) See AT&T Archives, "Portland Telephone Exchange, 1879–1884," 33–36.

7. Robert Wheeler Willson (1853–1922) assigned his patents to the Standard Time Company (see below). Willson had been an assistant at Harvard's observatory, as well as a physics tutor and lecture demonstrator there; he and Waldo were friends. His dismissal in 1884 by Yale Observatory's board of managers proved ultimately a boon to astronomy, for the talented scientist and inventor left the United States to earn a Ph.D. in astronomy at the University of Würzberg. Later professor of astronomy at

Harvard, in the 1900s Willson developed the bubble sextant, a significant contribution to early air navigation; Williams, *From Sails to Satellites*, 117, 127; see also H. T. Stetson, "Robert Wheeler Willson," *Popular Astronomy* 31 (1923): 308–13. Waldo's "Distribution of Time" was discussed in ch. 9.

8. Specifically cited in the agreement were Marshall Jewell, Charles L. Mitchell, H. P. Frost, Leonard Waldo, and Robert W. Wilson [*sic*]; U.S. Patents, *Index to Assignments*, L-6 (1879–1883): 229, National Archives, RG 241. The Lunds' patents were "Electro-magnetic Apparatus for Synchronizing Clocks," *Patent No. 201,185*, 1878; and "Electric Clock Regulator," *Patent No. 210,133*, 1878.

9. Connecticut. General Assembly, "Incorporating the Standard Time Company," *Special Acts and Resolutions Passed . . . at the January Session* (Hartford, 1882), 386–88. Other board members were Tyler, the Standard Time Company's president; STC treasurer Mitchel; Jewell; and Frost.

10. Standard Time Company, *Announcement of the Company, Yale College Observatory Standard Time Signals, Barraud and Lunds' Patents*, circulars in Yale Observatory Records, Yale University Archives. Also, *Nation* (8 June 1882): 481; and F. W., "The American Standard Time Company," this last article undoubtedly prepared by Leonard Waldo's brother Frank.

11. Extensive correspondence detailing Holden's inauguration of the Washburn Observatory's time service and its first years of operation are to be found in WO-WIS Archives and LO-UC Archives. Holden saw time-services income as a way for the observatory to supplement by its own efforts the inconsistent funding voted by the state legislature. Writing on 21 August 1881 to former governor and observatory benefactor Cadwallader C. Washburn, Holden promised to "try to earn something from the railroads for time," and that he would "sell it [time] for as much as it will bring"; in WO-WIS Archives.

12. None of the Midwestern railroads used Madison time. According to Holden, "*Chicago* mean time . . . is practically the standard time of the whole state of Wisconsin," but this was an exaggeration; *Publications of the Washburn Observatory of the University of Wisconsin* 1 (1882): 13; also, Professor [Edward] Holden, "Standard Time for Madison," unknown 1881 journal clipping in LC collection, in which the astronomer hedged by writing: "Chicago time is 7 minutes 11.1 seconds faster than Madison time . . . and should be taken, on account of its convenience, as the standard time of Wisconsin." Not surprisingly, Holden was opposed to Abbe's uniform time system, for adopting it would reduce the need for his own time service. For example, in a 10 February 1882 letter to Payne, Holden wrote: "All of us must oppose the Greenwich VI [that is, 90° W] plan"; in LO-UC Archives.

13. *P. H. Dudley's Electrically-Controlled System of Railroad and City Time Service*, with "Clock for" (written in ink) substituted for "System of"; attached to 20 January 1882 letter from Dudley to Holden, WO-WIS Archives, Series 7/4/2. A somewhat later pamphlet is in the files of the Smithsonian Institution's National Museum of American History; see O'Malley, *Keeping Watch*, 347, n10. As noted earlier, Dudley prepared the cost estimate included in the Naval Observatory's March 1882 comments on the time ball bill being considered by the House Commerce Committee.

14. Dudley's ads appeared on the back covers of the May through December 1882 issues. Editor Payne also drew attention to Dudley's timekeeping system; *Sidereal Messenger* 1 (1882): 48.

15. Holden to Standard Time Co., 19 April 1882, LO-UC Archives; Waldo to Holden, 29 April 1882, WO-WIS Archives, Series 7/4/2; Holden to STC Secretary, 4 May 1882, LO-UC Archives.

16. Payne, *Sidereal Messenger* 1 (August 1882): 137. (George Hough at Dearborn Observatory in Chicago was the fourth Midwestern director selling time.) In 1884 Holden bought one or two clocks from the now-collapsed Standard Time Company.

17. In a 22 September 1882 letter marked "Confidential," Lewis Boss asked Edward Pickering to verify Waldo's statements, which, according to Boss, included the unfounded claim that STC was purchasing Harvard College Observatory's time signals. Waldo was urging Boss to invest in the company; in HCO-HUA, UAV 630.17.7. Boss, *Dudley Observatory. Report of the Director [for 1882]*, 9–11. The Dudley Observatory received five hundred dollars per year from the Albany city government, and set its annual charge to prospective customers at twenty-five dollars.

18. *Nation*, 8 June 1882, 481. However, "Trade Gossip," *JC&HR* 13 (February 1882): 32, suggests that STC was a subcontractor, not a principal, in this sale of Howard clocks. Regarding business in Chicago, see Waldo to Comstock, 1 June 1883, WO-WIS Archives, Series 7/4/2.

19. U.S. Patents, *Liber Patent Transfer Files*, vol. H30, (20 September 1883): 287–96, National Archives, RG 241. For a short while, a caretaker operation continued in New York, STC's new secretary using New Haven Clock Company stationery. In New Haven itself, the Standard Time Company was succeeded by the National Time Regulating Company (1884) and the Standard Time and Electric Company (1885–86)—both with ties to SNET officials—and then by the Standard Electric Time Company (1887). Ironically, this last-named company began on a small scale in nearby Derby-Ansonia on 1 March 1882, just eight days before the Standard Time Company's incorporation. Proprietor and electrician Charles D. Warner incorporated his company on 7 February 1887, with headquarters in Waterbury; also in 1887 this company took over SNET's maintenance of Yale Observatory's time service. Like its competitors, the corporation licensed and sold its time systems throughout North America. However, unlike many enterprises of this era, the Standard Electric Time Company flourished, continuing to manufacture electric clock systems well past the mid-twentieth century.

20. STC to President and Fellows, Yale College, 17 September 1883, Yale Astronomical Observatory Records, YUA, 14-E-1; the "company now organizing" did not come into existence for several years. In November Waldo attempted to gain control of the observatory's Horological Bureau; the resulting public row eventually led to still another reduction in his income; the episode is discussed in Hoffleit, *Astronomy at Yale*, 69–71.

21. National Telephone Exchange Association, *Third Meeting*, 79, 85.

22. *Operator* 13 (4 November 1882): 505. Used here is the description in "The Oram Time Indicator," *Electrical Review* 4 (24 May 1884): 1–2; see also Hayes, "The Oram Telephone Time-Indicating System"; and "A New Telephone Time Repeater,"

Electrical Review 7 (10 October 1885): 5. In early 1883 the University of Michigan's observatory director described the use of a nonautomatic telephone-time system invented by C. W. Ruehle; Harrington, "A Telephonic Time-Transmitter."

23. "Standard Time," *Boston Herald*, 19 November 1883, 4. "Standard Time by Telephone," *Electrical World* 2 (1 December 1883): 229.

24. "Time by Telephone," *Scientific American* 50 (19 April 1884): 247. Connecticut, Office of the Secretary of State, *Records of Joint Stock Corporations* 18 (1884): 306–9. SNET's H. P. Frost was one of five incorporators; the others, all Boston telephone officials, included Theodore N. Vail of the Bell interests. Oram received stock for his patents.

25. "At present the subscribers to the telephone are getting our [HCO] time signals in their instruments, and there seems to be no practical method by which we can prevent this [effect]"; Frank Waldo to Pickering, 1 November 1880, HCO-HUA, UAV 630.77. The time-signal confuser, analogous to today's telephone and pay-TV scramblers, was patented in 1884 by one of the company's founders; in all, the company controlled five patents.

26. The International Electrical Exhibition, partly sponsored by the national government, was organized by the Franklin Institute. The AAAS and the United States Electrical Conference met in Philadelphia at the same time. For Oram, see Franklin Institute, *International Electrical Exhibition*, 34, 83; based on a review of city directories, this catalog's New York location for the National Time Regulating Company is incorrect. Also, *Electrical Review* 5 (13 September 1884): 5; and *Electrician and Electrical Engineer* 3 (November 1884): 246.

27. It is interesting to note that all these trials were held in cities with observatories: the Boston region, San Francisco, St. Louis, Cincinnati, Chicago, and Pittsburgh. However, no evidence has been found to suggest that any observatory but those in Boston, San Francisco, and New Haven ever supplied time signals for Oram's time indicator.

28. *Electrical Review* 7 (14 November 1885): 6; and "Oram Time Service, 1885–1886," AT&T Archives. Regarding subscriber queries, see "The Telephone in Honolulu," *Electrical Review* 4 (21 June 1884): 6.

29. *Electrical World* 6 (25 July 1885): 33; some of this telephone time equipment was still operating in 1886. Oram was appointed city of Dallas electrician; *Electrical World* 13 (1889): 180.

30. "Distribution of Time by Compressed Air," *Electrician* 1 (1882): 47, 49; Edmund A. Engler, "Time-Keeping in Paris," *Popular Science Monthly* 20 (1882): 307–12; Joseph Sternfeld, "Airclocks," *NAWCC Bulletin* 3 (1949): 355–57. Pneumatic systems had one advantage: their time displays were not very noisy.

31. "Another Time Telegraph Proposed," *Electrician and Electrical Engineer* 4 (1885): 119; *JC&HR* 35 (3 November 1897): 43; W. Barclay Stephens, "Herman Wenzel and His Air Clock," *California Historical Society Quarterly* 27 (1948): 1–8.

Chapter 14

1. New York, Department of State, "The Time Telegraph Company: Certificate of Incorporation," *Corporation Records*, vol. 76, Document Set 144 (25 October 1882);

Operator 13 (20 October 1882): 486; "Time Furnished by Wire," *Operator* 13 (4 November 1882): 504–5; "Furnishing Time by Telegraph," *New York Times*, 11 November 1882, 3. *The Time Telegraph Company*, (New York, 1882); the LC copy of this prospectus is stamped 25 November 1882. The company's address was Temple Court, 5 Beekman Street, New York.

2. For biographical details, see "Chester H. Pond," *National Cyclopaedia of American Biography* 15 (1916): 160–61; Chauncey N. Pond, "Chester Henry Pond," *Oberlin Alumni Magazine* 9 (1912): 51–53; and *New York Times*, 12 June 1912, 13. What little that has been published regarding Pond's horological efforts is suspect. To my knowledge, the account here is the first that identifies Pond's timekeeping enterprises in existence prior to 1886. The quotation given is from Pond to J. B. T. Marsh, 3 August 1882, Oberlin College Archives.

3. New York clockmaker Vitalis Himmer invented this particular spark-erosion-control device. Mentioned earlier (ch. 8, n29), he is one of those key but forgotten inventors whose efforts contributed to the rapid growth of electrical timekeeping in this decade. Himmer also invented for others; for example, the short-lived Standard Electric Clock Company of New York (1885–87) controlled several of his later patents. In 1890 Himmer was listed as manager of a dry-cell battery manufacturing firm; see *Electrical World* 6 (26 September 1885): i; *Electrical World* 11 (21 January 1888): 33; and *Electrical World* 15 (20 December 1890): xxvi, 442.

4. New York, Department of State, *Corporation Records*, vol. 77, Document Set 166 (28 December 1882); vol. 80, Document Set 47 (19 May 1883); and ibid., vol. 82, Document Set 117 1/2 (18 October 1883).

5. *Electrical World* 2 (13 October 1883): 109.

6. Ibid., (24 November 1883): 225. In December 1883 Senator Hawley introduced S. 616, the bill establishing a legal standard of time for the District of Columbia.

7. Franklin Institute, *International Electrical Exhibition*, lx, 63, 83. The Rhode Island Telephone and Telegraph Company also mounted an extensive display at the exhibition; its large advertisement complemented Time Telegraph's notice. In 1885 and 1886 the Franklin Institute published extensive reports on the exhibition, as supplements to its journal.

8. Among others, F. A. P. Barnard, A. E. Dolbear, Moses Farmer, M. W. Harrington, Samuel P. Langley, T. C. Mendenhall, Simon Newcomb, E. C. Pickering, Frank L. Pope, H. A. Rowland, Charles A. Young; *Electrician and Electrical Engineer* 3 (September 1884): 191–92.

9. At least four of the subcommittee's five participants were astronomers. Leonard Waldo requested that he not review the Time Telegraph Company's exhibit; he had some conflicts of interest forbidden by the "Rules of the Exhibition." After the exhibition closed, a summary of electrical time systems appeared in *Science* 5 (6 February 1885): 120; very likely written by Waldo, its author damned the Time Telegraph Company with very faint praise. (Waldo was annoyed because Time Telegraph had not yet established a railroad-time corporation, although it had agreed to do so when it took over the Standard Time Company of New Haven.)

10. Although unsigned, most of the report was prepared by W. A. Rogers of Harvard College Observatory; "Report of Sub-committee on Time-pieces," 90–96.

11. Pond, "Electro-mechanical clock," *Patent No. 308,521*, filed 30 September 1884, granted 25 November 1884. For a review of this and earlier technical developments, see Terry, "Application of Electricity to Horology." An associate of Pond's well-known patent attorneys, Terry also became famous in this field; see his obituary in the *New York Times*, 20 February 1939, 17.

12. *JC&HR* 18 (May 1887): 142. Power for winding the clock's mainspring came from wet cells mounted within the clock case. (Dry cells did not enter the market until the 1890s.) Early on, Connecticut's competing Standard Electric Time Company began advertising its self-winding master clocks, with movements made by the Self-Winding Clock Company; see SNET, *Connecticut State Directory of Subscribers* (New Haven, 1 July 1888): ii; *Electrical World* 9 (14 May 1887): 233.

13. The Time Telegraph Company was terminated by consolidation in April 1885, one share of stock in the new company being issued for every two shares of the old. "Electrical Time," *Electrical Review* 6 (18 July 1885): 1–7; H. L. Bailey, "The Electric Distribution of Time," *Electrical World* 6 (18 July 1885): 25–26. Bailey was the Telegraphic Time Company's electrician.

14. The Self-Winding Clock Company was incorporated on 18 October 1886 by Pond, Marshall E. Hunter, and John A. Green[e]. Contrary to all accounts published in recent decades, Charles W. Pratt of New York was neither an incorporator nor one of the company's original trustees. Pond to Almeda Gardner Pond, 24 May 1894, Sunflower County Library collection, Sunflower, Mississippi.

15. Also in 1887 Pond's English self-winding clock patent was licensed to a company managed by John A. Lund, whose own patents had initiated a transformation in timekeeping; see *JC&HR* 18 (July 1887): 216; and *Horological Journal* 29 (March 1887): xiii, 102–3. Pond moved to Chicago in late 1887 or early 1888; by mid-1889 he was living in Moorhead, Mississippi; see Marie M. Hemphill, *fevers, floods and faith: A History of Sunflower County, Mississippi, 1844–1976* (Indianola, Miss.: n.p., 1980), 253–61, 264–66, 271–73, 276–77.

16. Self-Winding Clock Co., "History of the Time Service." The 1885 date given is consistent with the incorporation of the first railroad-time companies; however, also described there is the marriage of Gardner's patented, add-on synchronizer to Pond's self-winding clock, a most improbable union.

17. Bartky, "Railroad Timekeepers," 403–4.

18. *Proc. General Time Convention* (1886): 15; ibid. (1887): 65, 119; ibid. (1888): 164.

19. In the context of track miles served, Harvard College Observatory was now a minor purveyor of time. Observatory rankings may have changed dramatically during this period of rapid railway growth, for one 1885 list shows the Naval Observatory as third (in miles of track served), trailing both Allegheny Observatory and the observatory at Washington University in St. Louis. An 1888 tabulation places the Naval Observatory first. However, the data are from quite different sources; see *Proc. General Time Convention* (Appendix): 725; and *Journal of the Franklin Institute*, 3d ser., 126 (1888): 16.

20. In 1868 John Locke's clock, now relegated to a mean-time standard, was fitted with a Jones-type device to control a clock at the Navy Department; USNO Superintendent, *Annual Report . . . for 1868*, 85, and ibid., *Annual Report . . . for 1869*, 56.

Evidence of financial support is in U.S. Patents, *Liber Patent Transfer Files*, vol. Y39, (22 October 1883): 345–46, National Archives, RG 241; Marean's and Royce's assistance was not recorded until mid-April 1889.

21. William Harkness, and to a lesser extent J. R. Eastman, are also credited with these advances in instrumentation. On the other hand, P. H. Dudley, who likely contributed as well, is never mentioned, for contractors to agencies are seldom acknowledged.

22. By 1884, eleven separate wires were in use. Of course several of these must have been redundant ones to guard against breaks in signal transmissions to specific users.

23. Gardner witnessed Dudley's 1882 written estimate of the cost of synchronized electric clocks included in the time ball report sent by Admiral Rodgers to the House committee; Dudley to Rodgers, 1 March 1882, USNO-NA, LR.

24. Gardner, "Time-Controlling System," *Patent No. 287,015*, 23 October 1883, lines 44–45. At this juncture, to suggest that a technical link existed between the two synchronizers is speculation. However, in 1881 the Royce & Marean firm was the agent for Dudley's (unpatented) system of controlled clocks; certainly, the proprietors and Gardner must have examined its workings. Dudley himself claimed that his device was the first to "zero" the second hand at synchronization, and, in 1882 Holden termed Washburn Observatory's public time service "the Dudley system." An 1885 trade journal notice, its statements no doubt supplied by Royce & Marean, suggests a close relationship between the two systems as follows: "W. F. Gardner's system of correcting ordinary clocks, by electricity, . . . is in use at the Washburn Observatory, Madison, Wis."; see *Electrical Review* 6 (4 April 1885): 8.

25. *Proc. General Time Convention* (Appendix): 698–99, 746. The "Captain J. [*sic*] H. MacDonald," who presented details of Gardner's system, was actually I. H. MacDonald, the inventor's patent attorney and sometime business partner.

26. USNO Superintendent, *Annual Report . . . for 1885*, 107. A few years later the observatory reported 350 clocks on the government circuit alone; *Journal of the Franklin Institute*, 3d ser., 96 (1888): 23.

27. Gardner, "Electric Time-Controlling System," *Patent No. 307,287*, filed 23 April 1884, granted 28 October 1884. Commodore S. R. Franklin to Commissioner of Patents, 30 April 1884, USNO-NA, LS; Patent Office to Superintendent, 2 May 1884; and Commissioner of Patents to Superintendent, 5 May 1884, USNO-NA, LR; also, Franklin's 6 May 1882 followup to his immediate superior requesting a letter from the secretary of the navy to the commissioner of patents.

28. Royce & Marean established a two-tier pricing system for Gardner's clocks: twenty-five dollars each for clocks sold to the government, while businesses could only rent them, at forty dollars per clock per year; "How the Correct Hour is Known to the Departments," *Electrical Review* 7 (12 December 1885): 7.

29. As noted earlier, other time balls were located close to branches of the Naval Hydrographic Office that received the daily noon signal from Washington. This particular expansion of the Hydrographic Office's maritime services began in 1884. During that same year the Naval Observatory placed a time-signal station and time ball

at Mare Island Navy Yard, California; beginning in May 1885, Western Union transmitted the station's noon signal to San Francisco, where the branch office used it to drop yet another time ball. Gardner himself patented a time ball, whose design was used for many navy installations.

30. After reviewing timekeeping at European observatories, one naval officer reported that Gardner's system was unsurpassed. Concluding that general knowledge of its superior performance was too little known, he urged that Naval Observatory time services be publicized; Albert G. Winterhalter, "The International Astrophotographic Congress and a Visit to Certain European Observatories and Other Institutions," *Washington Observations for 1885* (Washington, 1889): Appendix II, 322. Also, Brown, "Electrical Distribution of Time"; and his talk at the annual meeting of Railway Telegraph Superintendents, *Electrical World* 12 (21 July 1888): 31.

31. Royce & Marean, *Gardner's System for Correcting Clocks*; in the Gardner collection, Naval Observatory Library. Waldo, "The distribution of time," 120; "The Standard Electrical Time System Company," *American Manufacturer and Iron World* 36 (27 February 1885): 19. D. D. Maxwell to William Thaw, 27 January 1885; Thaw to Langley, 28 January 1885; Pritchett to Langley, 4 April 1887; and Langley to Pritchett, 12 April 1887, Samuel Langley Papers, AIS. Also, Gardner to Professor [Holden], 18 February 1885, WO-WIS Archives, Series 7/4/2.

32. Gardner to Holden, 24 February, 17 April, and 10 March 1887, LO-UC Archives; Gardner was unaware that his franchise rival was no longer known as the Time Telegraph Company. Holden's business dealings at Lick Observatory are discussed in a later chapter.

33. In mid-1999 dollars, $25,000 represents more than $430,000. U.S. Patents, *Liber Patent Transfer Files*, vol. X39, (3 January 1888 and 28 June 1888): 320–28; ibid., vol. J37, (26 March 1888): 339–40, National Archives, RG 241; some of these agreements were not recorded until 16 February 1889. Also "U.S. Standard Railway Time Co.," in the Gardner collection, Naval Observatory Library.

34. U.S. Patents, *Liber Patent Transfer Files*, vol. G22, (31 December 1884): 344, National Archives, RG 241. Seth Thomas Clock Co. to Holden, 28 January and 4 February 1885, WO-WIS Archives, Series 7/4/2. Also *JC&HR* 19 (February 1888): 99, and ibid., (March 1888): 52; *JC&HR* 20 (May 1889): 43, and ibid., (July 1889): 3; "Standard Time Service and Seth Thomas Clocks," *JC&HR* 20 (September 1889): 30, 33; and ibid., (October 1889): 34–36. (Seth Thomas's "clock of precision" was designed in part by Waldo and Yale's Charles Lyman.) Gardner was sent to the Universal Exposition in Paris to install the equipment. He stayed to explain the electrical timekeeping and distribution system to visitors, among whom was President Benjamin Harrison's son. According to the clock-industry trade press, the U.S. Naval Observatory Time Service's exhibit dazzled everyone. Moreover, the International Jury awarded the time service a "Grand Prix" (highest honors), and Gardner received a gold medal as "collaborator"; Rotch to Pythian, 29 September 1889, USNO-NA, LR. During his stay in Paris, Gardner applied for French patents and had a French-English brochure published for the exposition; [W. F. Gardner], *The Gardner System of Observatory Time*, in the Gardner collection, Naval Observatory Library.

35. In one of his diatribes in the *Sidereal Messenger*, William Payne wrote of Self-Winding payments of thirty-thousand dollars—certainly in line with the sum offered by the investors in the ill-fated Gardner Time System Company. The quotation is from *JC&HR* 21 (February 1890): 128.

36. U.S. Patents, *Liber Patent Transfer Files*, vol. T42, (16 October 1890): 155–56, National Archives, RG 241.

37. In 1887 Pond, no longer associated with the Self-Winding Clock Company's management, assigned to the corporation his retained rights to the self-winding clock patent. By March 1889 the trade press was writing that the company "has obtained nearly all the patents relative to transmitting time or synchronizing clocks." Self-Winding's trustees were members of the enormously wealthy and influential Pratt and Bedford families of Brooklyn.

38. Gardner's device may never have actually been used in the company's products; see Pritchett, "Correct Time—How Shall We Maintain It?" Unfortunately, no engineering analysis exists that compares the synchronizers invented between 1882 and 1890 with those used in Self-Winding's various timekeepers.

Chapter 15

1. Reproduced in the *Sidereal Messenger* 9 (August 1890): 331; the official copy is in National Archives, RG 80, General Records of the Navy Department, LR, 1890, File #7586.

2. After relocating its transit in temporary quarters, the Dearborn Observatory tried to retain its other Chicago customers—two banks and one jewelry store—by placing a mean-time clock at the offices of the *Chicago Tribune*. This Chicago time service was canceled in April 1889.

3. Payne studied briefly with Ormand Stone at Cincinnati Observatory; however, his major training was in the law. He used the modest income from Carleton College's time service to help support the small astronomy department that he established. Railroads centered in Minneapolis-St. Paul, and various regional cities and towns, were Carleton's customers; in 1888, this group was paying seven hundred dollars annually for the college observatory's time. Payne's personal income came from the sale of astronomical photographs and subscriptions to the *Sidereal Messenger*—a semipopular astronomy journal that he founded in 1882—plus what the college paid him for teaching.

4. Payne to Comstock, 14 and 21 July 1888, WO-WIS Archives, Series 7/4/2.

5. Hall to Comstock, 19 July 1888, WO-WIS Archives, Series 7/4/2. Hall's letter was penned just after Gardner's failure to establish a railroad time company. It seems unlikely that the astronomer was privy to the instrument-maker's business activities. But, Hall certainly heard rumors, for he wrote that Gardner had patented his system of controlled clocks "which he is trying to extend, and out of which he wants to make some money"; ibid.

6. No doubt the Midwestern rumors resulted from discussions at the seventh annual meeting of the Association of Railway Telegraph Superintendents, held on 10 July in New York. Invited guest Commander A. D. Brown, the Naval Observatory's

assistant superintendent, described Gardner's system and reiterated its great utility in railroad timekeeping. The Chicago, Burlington & Quincy Railroad's superintendent spoke of ongoing tests of self-winding synchronized clocks, of Chester Pond's new time company, and that Pond "proposes to synchronize"; *Electrical World* 12 (21 July 1888): 31.

7. W. Keyes (attorney) to Comstock 27 July 1888; Hall to Comstock, 9 August 1888; and C. S. Jones to Comstock, 19 July 1888, WO-WIS Archives, Series 7/4/2.

8. In 1889 Western Union in Minneapolis proposed the purchase of time from the Washburn Observatory, planning to distribute the signals to individual subscribers, but not "to interfere in any way with the arrangements between the [Washburn] Observatory" and its railroad customers; I. McMichael (Superintendent) to G. E. Bross (Madison Office manager), 25 March 1889. Soon after, Payne began negotiations with Western Union regarding its use of Carleton College Observatory's time signals; Payne to Comstock, 30 April and 4 May 1889. All items are in WO-WIS Archives, Series 7/4/2.

9. "It may be a matter of interest . . . ," *Electrical Review* 5 (17 January 1885), 9; Wabash Railroad Company circulars in Henry S. Pritchett Papers, LC.

10. Pritchett to E. T. Bedford, 12 April 1889; in Self-Winding Clock Company advertisement, *JC&HR* 21 (February 1890): 91.

11. See Frank Jaynes (Western Union) to Holden, 26 March 1889, LO-UC Archives; and telegram, F. H. Tubbs (Western Union) to T. C. Hoag, 1 April 1889, reading in part: "I gave Mr. Hough [a] pamphlet showing we will furnish Synchronized Self-Winding Clocks and correct them daily by Telegraphic Time Signals"; Records of Dearborn Observatory, Northwestern University Archives. These items represent the earliest mention of the joint venture thus far located.

12. *JC&HR* 20 (June 1889): 90, and (September 1889): 106. I have been unable to locate the contract between the two corporations; apparently, annual receipts were divided equally. Self-Winding supplied the clocks, and Western Union's employees installed them and performed the necessary maintenance, including battery replacements. By 1893 Self-Winding was reporting over fifteen thousand clocks in service; *Scientific American* 69 (29 July 1893): 69; and *American Jeweler* 13 (September 1893): 14.

13. One very early customer was the Chicago, Burlington & Quincy Railroad. After a year of field trials, the Burlington selected Self-Winding Clock Company products as its standard timekeepers, and the road's purchasing agent asked for a price quotation on self-winding clocks that could be "connect[ed] with the central clock at division points"; Chicago, Burlington and Quincy R.R. Co. to Self-Winding Clock Co., 3 August 1889, printed in Self-Winding Clock Company's February 1890 trade advertisement, cited above. Soon after, the trade press announced the Burlington's order for "160 self-winding and self-setting clocks" controlled by a master synchronizing clock. See also *Electrical World* 15 (15 February 1890): 108.

14. Pritchett to Pythian, 8 August and 16 September 1889, USNO-NA, LR; Pythian to Pritchett, 27 August 1889, USNO-NA, LS; and Winterhalter [for Pythian] to Pritchett, 19 September 1889, USNO-NA, LS. (The 8 and 27 August and 16 September letters were published in the February 1890 *Sidereal Messenger*.)

15. Pritchett to Comstock, 24 October 1889, WO-WIS Archives, Series 7/4/2.

16. Payne to Comstock, 6 and 10 December 1889, WO-WIS Archives, Series 7/4/2.

17. I believe that this particular letter never existed. Captain Pythian concluded that it came from Pritchett, but both Payne and Pritchett denied that the latter had written it.

18. Payne to Pythian, 6 December 1889, USNO-NA, LR; extracts of this letter appeared in the February 1890 *Sidereal Messenger*.

19. Payne, *Sidereal Messenger* 8 (December 1889): 452–54.

20. Edward T. Bedford was president of the Self-Winding Clock Company. That Payne was unaware of the three-year-old company's actual name until somewhat later indicates that he had little more than astronomers' gossip when he inaugurated his campaign. Indeed, just prior to its start, Payne extolled the performance of Pond's self-winding clock in the pages of the *Sidereal Messenger* 8 (1889): 237, 273–74, 330. Never again would he take such a position; see *Sidereal Messenger* 10 (1891): 151.

21. Payne to Pythian, 20 December 1889 and 11 January 1890, USNO-NA, LR; Pythian to Payne, 12 and 14 December 1889, and 10 January 1890, USNO-NA, LS.

22. Since May 1885, Naval Observatory signals from its time station at Mare Island Navy Yard had been transmitted to San Francisco via a Western Union line; the telegraph company also sent them to many other West Coast cities and railroads. Despite this ongoing service, in 1886 Holden made an agreement with the Southern Pacific while he was president of the University of California. I acknowledge with pleasure a key point made by Dorothy Schaumberg, curator, Lick Observatory Archives, that Holden did not have the authority to commit Lick Observatory to any contract, for the facility was not transferred to the university until 1 June 1888. I conclude that Holden had commercial sales—including timekeepers—firmly in mind, for the 24 March 1886 draft of the observatory contract contains the following condition: "and that you [Southern Pacific] allow me [Edward Holden] to designate one such recipient [of daily time signals via the railroad's telegraph wires] in every town in California of over 5000 inhabitants"; in LO-UC Archives. The Southern Pacific did not accept the requirement, for it is not mentioned in its 12 April 1886 acceptance letter. (Regarding some of Holden's business interests, see Gardner's 1886 letters to him, in LO-UC Archives.)

23. Holden to Payne, 8 January 1890, LO-UC Archives.

24. Holden to Pritchett, 24 March 1890, LO-UC Archives. In this letter the astronomer mistakenly cited his 1877 report to the USNO superintendent regarding Naval Observatory reorganization, instead of his key November 1876 report to the secretary of the navy.

25. Obendorf, "Samuel P. Langley," 27; Langley to William Thaw, Jr., 16 December 1889; and Very to Payne, 20 January 1890; Langley Papers, AIS.

26. By this time Langley had become cautious, for he described his first letter (18 January) as "purely informal and personal" and not "adapted for official reference in connection with any official correspondence that might arise upon the time-service question." This letter and Pickering's 20 December and 20 January letters to Langley

have not been located; Langley's 22 and 24 January ones are in HCO-HUA, UAV 630.17.7.

27. Pickering used his time service's supposed immunity from Western Union competition as an opening gambit, describing it as a continuous service with time ticks every two seconds, "and individuals who have become accustomed to . . . these signals . . . are generally unwilling to exchange them, even at some saving of money, for others given at only long intervals." However well this introduction demonstrated an "unbiased third-party" position in the dispute, including it actually displayed the astronomer's deep misunderstanding of railroad timekeeping needs.

28. Pickering to Pythian, 20 and 29 January 1890, USNO-NA, LR.

29. *Sidereal Messenger* 9 (February 1890): 57–67. Payne's discussion of Naval Observatory personnel attending the railroad industry's General Time Convention sessions and the statements he attributes to them contain many errors. A significant, and perhaps deliberate, one is his 1889 date for these 1888 meetings.

30. Payne continued with this issue long after the controversy ended, writing in the 1900s of "a three-cornered deal in which $30,000 was involved," and that "part of this tripartite agreement involving the sale of the clock patent was that Washington time should be given . . . from the Naval Observatory by means of the Western Union Telegraph Company"; see Payne, "Time and Its Measurement," *Popular Astronomy* 11 (1903): 462; and Payne, "Graft Disturbing Peaceful Astronomers," *Popular Astronomy* 14 (1906): 577.

31. Taylor, "U.S. Government System of Observatory Time"; the Self-Winding Clock Company coupled this article to its full-page advertisement; *JC&HR* 21 (1890): 91. Surely this coupling was "evidence" supporting Payne's claim of a tripartite arrangement, but curiously he never mentioned the advertisement. Payne's drawing attention to Lieutenant Taylor's remarks caused Hough to prepare "Time Service," in which he listed some of the Dudley Observatory's early time services.

32. Hough's statement from Evanston was a bit disingenuous: "Two years ago the Western Union Telegraph Company so interfered with the Observatory service as to practically destroy all revenue. During the past year the service is entirely discontinued"; in *Sidereal Messenger* 9 (March 1890): 139. A more likely reason for Dearborn Observatory's termination is in the next chapter.

33. Pritchett, "Observatory Local Patronage."

34. Pritchett was familiar with the university astronomers' private approach to the Naval Observatory's superintendent. Early on he also discussed the substance of the astronomers' petition with Asaph Hall, his former mentor at the Naval Observatory; see Pritchett to Comstock, 15 April 1890, WO-WIS Archives, Series 7/4/2.

35. Pythian to Pickering, 10 March 1890, USNO-NA, LS.

36. Langley to Pickering, 13 and 20 March 1890, HCO-HUA, UAV 630.17.7.

37. Langley understood the ramifications of synchronizer technology, but he knew almost nothing about the specific clocks in Pittsburgh and Allegheny; see Very to Langley, 25 January 1890, Langley Papers, AIS. Very reported there that "Standard Oil Co. hold patents on the [Self-Winding] clock"—his incorrect statement is actually a reference to another of the Self-Winding trustees' business interests.

38. Langley to William Thaw, Jr., 21 March 1890; and Langley to Thomson, 21 March 1890, Langley Papers, AIS. In both letters Langley claims that "no pecuniary interest of any kind in the direction of the Allegheny Observatory" ever influenced him. Based on Langley's 16 December letter to Thaw (note 25 above), I conclude that, at best, he was engaging in hair-splitting. In addition to his call on the time service's gross, Langley controlled the observatory's cash account, and, further, did not announce his decision to resign as director until 18 April 1891.

39. "What Time is It?" *Chicago Tribune*, 27 March 1890, 1; "Western Union Time-Service," *Chicago Tribune*, 28 March 1890, 5; "Time is Money," *New York Herald*, 30 March 1890, 16; "The Time Pedler," *New York Herald*, 31 March 1890, 8.

40. The error was fifteen seconds. Made extremely nervous by Payne citing him, Langley fired off a special delivery letter to his assistant demanding an explanation. Very quoted his 20 January letter to Payne; Very to Langley, 3 April 1890, Langley Papers, AIS.

41. "Capt. Pythian's Explanation," *Chicago Tribune*, 28 March 1890, 5; "The Western Union Time," *Sidereal Messenger* 9 (May 1890): 232.

42. "Why Dudley Observatory is willing to let the telegraph company have it," *New York Herald*, 31 March 1890, 8.

43. Boss himself avoided these implications, urging instead that "the [mid-]Western observatories, which have relied entirely on the income from the local time service for their support . . . [be given] a chance to adjust themselves to the new [competitive] conditions"; ibid.

44. The list of astronomers signing that was published in the *Sidereal Messenger* is nearly complete; the official document included M. W. Harrington, University of Michigan; J. M. Van Vleck, Weslyan University; F. P. Leavenworth, Haveford College; and C. H. McLeod, McGill College [*sic*], Montreal; also, Charles F. Emerson replaced Edwin B. Frost as director of Shattuck Observatory, Dartmouth.

45. Obviously Dudley Observatory director Lewis Boss did not sign the petition. Neither did Washburn Observatory's George Comstock, who apparently objected to the petition's assertion that the Naval Observatory's signal was essential to Western Union's success; see Pritchett to Comstock, 15 April 1890, WO-WIS Archives, Series 7/4/2.

46. Pickering to Hon. G. F. Hoar, 7 May 1890, National Archives, RG80; this file, cited in n1 above, contains the various documents and letters quoted here. Included is the Naval Observatory's response, Captain R. L. Pythian to Commodore George Dewey, 15 May 1890, comprising twenty pages of text and six appendices; these items are also in USNO-NA, LS (1890), 263–82.

47. The document was drawn up by Pickering, and astronomer H. A. Newton of Yale made some minor changes; see Pritchett's mid-April letter to Comstock, already cited.

48. In later years, the Postal Telegraph Company and the American Telephone & Telegraph Company both brought their wires onto the Naval Observatory grounds.

49. Captain Pythian, who would have consulted with Naval Observatory astronomers, suggested five thousand dollars as "a liberal estimate" for the entire cost for the

equipment (including a suitable building) needed to measure stellar positions and reduce the resulting observations correctly for determining time accurately; further, "two thousand dollars will cover the expense of maintaining it"—minor costs to any telegraph company. These estimates seem rather high to me. Moreover, I suspect that Western Union would have found several private observatories willing to sign a contract to supply a base time signal for the "National Time by Western Union."

50. In contrast to Langley's view regarding responsibility expressed in his 22 January letter to Pickering (n26 above), Stephens argues that the Naval Observatory was "not entirely innocent" in the matter, citing its free time signal and well-publicized 1877 arrangement with Western Union, and also invoking Gardner's sale of his patent rights to the Self-Winding Clock Company; see Stephens, "Impact of the Telegraph," 9.

51. Senator Hoar's return of the documents to the Department of the Navy would be taken as a signal that his interest in the matter was over: a constituent had written, and he had responded.

52. I found no evidence that Congressman Lodge ever sent a copy of the astronomers' petition to the secretary of the navy; Pritchett may have been trying to enlist additional congressional support.

53. Pritchett's attempts to earn a living and acquire funds for his old age via sale of time signals may have influenced him in subtle ways. Later in his distinguished career the sometime astronomer interacted with Pittsburgh philanthropist and industrialist Andrew Carnegie. Their relation led to the formation of the Carnegie Foundation for the Advancement of Teaching and its successor, the Teachers' Insurance and Annuity Association, organizations addressing the retirement income needs of university professors. Pritchett was founding president of both organizations; see Henry S. Pritchett Papers, LC. Captain Pythian left the Naval Observatory soon after completing his report, having been appointed superintendent of the U.S. Naval Academy. Later, as Commodore Pythian, he once more served as the Naval Observatory's superintendent.

Chapter 16

1. Hough, *Annual Report of Director . . . for 1890–1891*, 4; Northwestern University, Board of Trustees, "Minutes" (1889–92): 249; and ibid., (1892–94): 272. As a result of Hough's service, Northwestern canceled its time-service contract with Western Union and declined its synchronized, self-winding timekeepers; Tubbs to Hoag, 1 April 1889, Records of Dearborn Observatory, Northwestern University Archives.

2. Holden, "Clocks and Time-Keeping"; Keeler, "Time Service of the Lick Observatory," and "Experiments with Electrical Contact Apparatus"; William W. Campbell, "Lick Astronomical Department," 1 July 1902, in *President's Biennial Report, University of California* (1902), 92; and S. A. Pope, "'Ask the Conductor; He Has the Right Time,'" *Southern Pacific Bulletin* 12 (August 1923): 6. The end of Lick's time service was apparently never recorded, but it came before 1923. (Western Union used Naval Observatory signals from Mare Island to points along the Pacific coast until 1926, when it shifted its entire service to the Washington signal.)

3. Pickering, *HCO. Time Service*, with date of 26 December 1891—copy in National Archives, RG 37, Naval Hydrographic Office, Entry 57, Correspondence and

Reports File 193.11. This circular was later published without a date; *Science* 19 (1892): 87–89. Sections were also printed in *Astronomy and Astro-Physics* 11 (1892): 344–45, and by the clock-industry trade press.

4. Pickering, *Annual Report . . . for 1885*, 9; and subsequent reports.

5. Edmands to Pickering, 7 November 1891; and Pickering to Charles W. Eliot, 7 November 1891, HCO-HUA, UAV 630.14.

6. *Boston Daily Advertiser*, 29 November 1888 through 7 December 1889. *Electrical World* 14 (7 December 1889): 377, and ibid. (21 December 1889): 408; *Electrical Engineer* 9 (January 1890): 24, and ibid. (February 1890): 81. (Allegheny Observatory and a time-service company in New Haven suffered similar shorted circuits and fires; fortunately, the consequences were less severe.)

7. Pickering to Eliot, 24 November 1891, HCO-HUA, UAV 630.14. The quotation is from the 26 December circular, n3 above.

8. "No More Time Service from Harvard University," *JC&HR* 23 (20 January 1892): 18. How complete the termination was is unknown. As late as 1901 one astronomer wrote that the "Observatory sends out continuous signals to the fire alarm headquarters of the city of Cambridge," which suggests something more than a gown-and-town *quid pro quo*; R. G. Aitken, "The Sources of Standard Time in the United States," *Popular Astronomy* 10 (1902): 13. See also Pickering's announcement of what must have been a quite short-lived public time service for Boston having "marked advantages over the time ball"; in *Annual Report for . . . 1902*, 13–14.

9. Pickering, *Annual Report for . . . 1892*, 10.

10. However, astrophysicist Donald E. Osterbrock documents in his biography of Keeler how magnificently "old" Allegheny contributed in the hands of one of the "new" astronomers.

11. Western University of Pennsylvania, *Catalog . . . for 1893–94*, 111. No evidence has been found to suggest that Keeler was aware of Langley's and Pickering's private efforts with the Naval Observatory's superintendent.

12. Osterbrock, *James E. Keeler*, 144–45. Keeler left in 1898 to become director of Lick Observatory. Subsequently, Allegheny's trustees increased funding significantly, and the observatory moved into new facilities with modern equipment. Under the next director, F. L. O. Wadsworth, the observatory received about nineteen hundred dollars annually from its time service. In the early 1900s, plans were afoot to have the Pittsburgh City Council pass a time ordinance, to increase observatory revenues— shades of Leonard Waldo!—but nothing came of this. In 1920, Allegheny Observatory stopped making stellar observations for time, sold its Riefler clock to Yale, and corrected its transmitting clocks via receipt of the Naval Observatory's radio time signals; from observatory reports and correspondence. I thank David DeVorkin for the use of his microfilm. Allegheny Observatory's time business ended on 3 February 1947; Beardsley, "Samuel Pierpont Langley," 42.

13. Among them Cincinnati, whose city time service was still active in 1912; Elliott Smith, "The Scientific Work of the Cincinnati Observatory," *University of Cincinnati Record* 9 (1913): 24. Yale, whose service ended in 1918, reported a respectable thousand-dollar income at the turn of the century; *Congressional Record*, 33 (11 April 1900): 4025.

Although a Yerkes Observatory time service was discussed, Osterbrock has found no evidence that such a service was ever inaugurated; see Hale, *First Annual Report of the Director*, 7. (I thank Donald Osterbrock for this information.)

14. *JC&HR* 24 (27 September 1893): 8; William Greer, *A History of Alarm Security* (Bethesda, Md.: National Burglar & Fire Alarm Association, 1991), 62–63; Winslow Upton, "Ancient and Modern Observatories," *Sidereal Messenger* 10 (1891): 488–89.

15. Payne retired in 1906, receiving a Carnegie Foundation pension. He designed an astronomical observatory and time-distribution system for the Elgin National Watch Company in Illinois and became the observatory's first director, serving until 1926. His public timekeeping career spanned a half-century; Greene, *A Science Not Earthbound*, 17, 19; and Neidigh, *Elgin Observatory Story*, 1–3.

16. The Washington University Observatory Time Service probably did not continue past 1901, when the university's downtown buildings were sold. The poor seeing at St. Louis, astronomer Pritchett's 1897 removal to Washington to become superintendent of the U.S. Coast and Geodetic Survey, and Western Union's success in the railroad-time market all make any continuation beyond that year unlikely.

17. "Astronomers in Arms: They Object to the Time Signal Service of Western Union," *New York Times*, 6 April 1891, 1.

18. Boss, "An Irrepressible Conflict," *Sidereal Messenger* 10 (1891): 161–68, with author's date of 23 February 1891; the Naval Observatory's copy of the journal is stamped with a 6 April date. Soon after, Payne wrote that Naval Observatory timekeeping and the call for civilian control were not linked; *Sidereal Messenger* 10 (1891): 249. To my mind, Payne's statement was just another feint in his disinformation campaign. For Keeler's mid-July 1891 statement, see Osterbrock, *James E. Keeler*, 144.

19. By questioning the quality of the Naval Observatory work, astronomers finally gained the upper hand, insofar as research directions were concerned. For the story, but without any link to the private observatories' time services, see Howard Plotkin, "Astronomers versus the Navy: The Revolt of American Astronomers . . . 1877–1902," *Proc. American Philosophical Society* 122 (1978): 389–91.

20. From a 12 March 1892 memorandum given in USNO Superintendent, *Report . . . for 1893*, 4; these program limits may have come from the department. During the height of the controversy the Naval Observatory ended its subsidy of controlled Gardner clocks in various government departments, and began charging for their maintenance; Bureau of Equipment to USNO Superintendent, 14 July 1890, USNO-NA, LR.

21. Thomson, *Beginning of the Long Dash*, 30–32.

22. Newcomb, *Improvements in Astronomical Instruments*, 25–27; Safford, "How to Make Good Meridian Observations," *Sidereal Messenger* 10 (1891): 115.

23. Abbe, "Meteorological Work of the U.S. Signal Service," 38–40. The problem of lack of synchronicity among time signals became acute with the advent of wireless time signals from several international sources and the results of comparisons at various stations; see Howse, *Greenwich time*, 164–66. A leader in this area was R. A. Sampson of Edinburgh Observatory, the astronomer-royal for Scotland, whose "minor observatory" became a center for the study of precision timekeeping.

24. USNO Superintendent, *Annual Report . . . for Fiscal Year 1903*, 29.

25. "The Great Exhibit of the Self-Winding Clock Company," *Scientific American* 69 (29 July 1893): 69; and similar articles in the trade press. Janson, "Time Services of the Telegraph Companies," 541. The companies' 1893 receipts are equivalent to $3.3 million in mid-1999 dollars; 1932 receipts are equivalent to $18.2 million. The Western Union–Self-Winding partnership continued until 1963, when the telegraph company bought out its partner. Western Union sold the enterprise in the early 1970s, and the Naval Observatory discontinued its signal to the company's Washington office.

26. Self-Winding Clock Co., *Standard Time Railway Map*, and *What is Standard Time?*; both in the Gardner collection, Naval Observatory Library. For a rather different treatment, see *Electricity at the Columbian Exposition* (Chicago: Donnelley & Sons, 1894), 463–65, by city of Chicago electrician J. P. Barrett, which begins with the details of Gardner's system in the Self-Winding Clock Company section.

27. Rental clocks from Self-Winding eventually carried the label "U.S. Observatory Time Hourly by W. U. Tel. Co.," or (later), "Naval Observatory Time Hourly by Western Union." In actual fact, as Western Union pointed out in its catalogs, the customer's clock was "corrected hourly by signals from high-grade master clocks in the telegraph company's service," and these master clocks were "regulated daily by telegraph signals" from the country's two Naval Observatory sites. On occasion the Naval Observatory complained that since neither the local clocks nor the master clocks were actually receiving its signals directly, the label on the rental clocks was not correct.

28. Portions of Crookes's 1892 article are quoted by Hugh Aitken, *Syntony and Spark: The Origins of Radio* (Princeton: Princeton University Press, 1985), 110–14; Aitken's articulation of its significance is persuasive. L. S. Howeth, *History of Communications—Electronics in the United States Navy* (Washington: Bureau of Ships and Office of Naval History, 1963), 109–10, 520, and, Jayne, "The Naval Observatory Time Service and How to Use Its Radio Time Signals"; the references conflict regarding the inauguration date of the Navy's radio time signals. Details of the jewelers' active interest in this area are in Mathys, "The Right Place at the Right Time," ch. 3.

Bibliography

Archival and Manuscript Collections

Albany Institute of History & Art, New York
 Thomas W. Olcott Papers, "Dudley Observatory"
American Antiquarian Society, Worcester, Massachusetts
 Railway Travel Guides
American Association for the Advancement of Science, Washington, D.C.
American Philosophical Society Library, Philadelphia
 Sears C. Walker Papers
AT&T Corporate Archives, Warren, New Jersey
Boston, Massachusetts
 City Archives and Records
California State Railroad Museum, Sacramento
 Railroad Rule Books
California, University of, Los Angeles
 Moses G. Farmer Papers
California, University of, Santa Cruz
 Mary Lea Shane Archives of the Lick Observatory
Chicago Historical Society
 Chicago Astronomical Society Records, 1862–1903
Cleveland, Ohio
 City Archives
Congress, Library of, Manuscripts Division
 Cleveland Abbe Papers
 Alexander Dallas Bache Papers
 Matthew Fontaine Maury Papers
 Henry S. Pritchett Papers
 U.S. Naval Observatory Records
 Charles Wilkes Papers
Connecticut Department of State, Hartford
 Incorporation Records
Connecticut, University of, Thomas J. Dodd Research Center, Storrs, Connecticut
 Providence & Worcester Railroad Records

Dudley Observatory, Schenectady, New York
　Observatory Archives
　Records of the Board of Trustees: 1852–1943
Harvard University Collection of Historical Scientific Instruments
　Bond Family Papers
　Wm. Bond & Son Records
Harvard University Graduate School of Business Administration (Baker Library)
　Railroad Rule Books
Harvard University (Houghton Library)
　Benjamin Peirce Papers
Harvard University, University Archives (Pusey Library)
　William C. Bond Papers
　Harvard College Observatory Records
　Edward C. Pickering Papers
　Joseph Winlock Papers
Johns Hopkins University (Eisenhower Library)
　Cleveland Abbe Papers
Massachusetts, Commonwealth Archives, Boston
　Incorporation Records, Boston Electric Clock Company
Massachusetts Historical Society, Boston
　George Bancroft Papers
　William F. Channing Papers
National Archives and Records Administration
　RG 23, Coast Survey Records (Microfilm M642)
　RG 78, Naval Observatory Records
　RG 241, U.S. Patents
NAWCC Library, Columbia, Pennsylvania
　NAWCC Bulletin Keyword Index
　JC&HR Keyword Index
　W. Barclay Stephens Materials
New Haven, Connecticut
　- Records of the Board of Aldermen and City Council
New York Department of State, Albany
　Incorporation Records
New York Public Library, New York City
　William F. Allen Papers on Standard Railway Time
　Charles V. Walker Scrapbook (Telegraphy)
New York State Library, Albany
　James Hall Papers
New York University (Bobst Library)
　University Archives
Northeastern Illinois University, Illinois Regional Archives Depository, Chicago
　Chicago City Council Files

Northwestern University, University Archives
  Dearborn Observatory Records
  George W. Hough Papers
  Records of the Board of Trustees
Pittsburgh, University of, (Hillman Library)
  Samuel P. Langley Papers, in AIS (microfilm)
  Western University of Pennsylvania Archives
Providence & Worcester Railroad Company, Worcester, Massachusetts
  Annual Reports to Stockholders
  Records of the Directors
Rhode Island State Archives, Providence
  Railroad Commissioners Records
Smithsonian Institution Archives
  Joseph Henry Papers and Henry Library
  Samuel P. Langley Papers
SNET Archives, New Haven, Connecticut
U.S. Naval Observatory Library, Washington, D.C.
  William F. Gardner Collection
Wisconsin-Madison, University of,
  Washburn Observatory Records
Yale University (Sterling Memorial Library)
  Yale Observatory Records

Selected Technical Journals and Annual Reports

Allegheny Observatory reports, in *Catalogue of the Western University of Pennsylvania*, 1869–1908
American Association for the Advancement of Science, *Proceedings*, 1848–56, 1880–83
American Metrological Society, *Proceedings*, 1873–88
American Railroad Journal, 1837–45, 1850–54
American Railway Association, *Proceedings of the General Time Convention*, 1872–89
Astronomical Notices, 1858–62
Astronomy and Astro-physics, 1892–94
Chicago Astronomical Society, Dearborn Observatory, *Reports*, 1874, 1877, 1880–87
Dudley Observatory annual reports, 1862–70, 1877–80, 1882, 1884, 1887–89
Electrical Review, 1884–86
Electrical World, 1883–90
Electrician and Electrical Engineer, 1882–90
Harvard College Observatory annual reports, 1846–55, 1856–58, 1859–64, 1877–92
Horological Journal, 1858–66
Jewelers' Circular (and Horological Review), 1881–98 [Keyword Index]
Mechanics Magazine, 1836–54
Operator, 1881–82

Popular Astronomy, 1894–1913

Science, 1880–86

Sidereal Messenger, 1882–91

Smithsonian Institution, "Astronomical Observatories," 1879–92

Telegraph, Journal of the, 1877–91

U.S. Coast Survey. Superintendent. *Annual Report*, 1844–61, 1864, 1867, 1870, 1880, 1884, 1897

U.S. Naval Observatory. Superintendent. *Annual Report*, 1867–1905

U.S. Navy Department, Hydrographic Office annual reports, 1885–1906, 1924–26

U.S. War Department. Signal Office. Chief Signal Officer. *Annual Report*, 1870, 1874–77, 1879–90

Washburn Observatory, Publications of, 1882–90

Yale Observatory, Board of Managers annual reports, 1880–1910

Yale Observatory, Horological Bureau annual reports, 1880–86

Sources

Abbe, Cleveland. *Inaugural Report of the Director of the Cincinnati Observatory. 30 June 1868*. Cincinnati: Robert Clarke & Co., 1869.

———. *Annual Report of the Director of the Cincinnati Observatory. 1 May 1869.* Cincinnati: Robert Clarke & Co., 1869.

———. *Annual Report of the Director of the Cincinnati Observatory. 4 June 1870.* Cincinnati: Gazette Co., 1870.

———. "Report on Standard Time." *PAMS* 2 (1880): 17–45.

———. "Progress Report to Am Met Soc 1882/83 [*sic*]." Abbe Papers, JHU.

———. "The Meteorological Work of the U.S. Signal Service, 1870 to 1891." *U.S. Weather Bureau, Bulletin* 11 (1893): 269–71.

———. "The Meteorological Work of the U.S. Signal Service, 1870–1891; C. Standard Time." In *Report of the International Meteorological Congress*, edited by Oliver L. Fassig, 38–40. Washington: Weather Bureau, 1894–96.

[———.] "Standard Time in America." *Science*, new ser., 22 (1905): 315–18.

Abbe, Truman. *Professor Abbe and the Isobars: The Story of Cleveland Abbe, America's First Weatherman.* New York: Vantage Press, 1955.

[Adams, J. C. (?)] "Railroad Watch Inspection and Adams' System of Time Records." *Horological Review* 24 (28 July 1897): 33–34.

Albion, Robert A. *Square-Riggers on Schedule.* Princeton: Princeton University Press, 1938.

———. *The Rise of New York Port.* New York: Charles Scribner's Sons, 1939.

Allen, William F. "Report on Standard Time." *Proc. General Time Convention,* appendix (April 1883): 690–92.

———. "History of the Movement by Which the Adoption of Standard Time Was Consummated." *PAMS* 4 (1884): 25–50.

American Rail Road and Steam Navigation Guide. New York, October 1866.

The American Railroader and Universal Weekly Travelers' Guide. Vol. 2, No. 11. Cincinnati: Allen & Boyd, 1 February, 1868.

American Railway Association. *Proceedings of the General Time Convention And Its Successor The American Railway Association, 1886–1893*. New York, 1893.

American Railway Guide and Pocket Companion. New York, 1851.

Andre, C., and G. Rayet. *L'Astronomie pratique et les Observatoires en Europe et en Amérique*. Troisième Partie. États-Unis d'Amérique. Paris: Gauthier-Villars, 1877.

Andrewes, William J. H., ed. *The Quest for Longitude*. Cambridge: HUCHSI, 1996.

Appleton's Railway and Steam Navigation Guide. New York, 1857.

Bache, Alexander D. "Sears C. Walker's 1847 Report on Telegraph Operations for Longitude." *Astronomische Nachrichten* 27, No. 632 (1848): Cols. 119–26.

———. "Abstract of a Report . . . by Sears C. Walker." *Astronomische Nachrichten* 28, No. 666 (1849): Cols. 273–78.

Bache, Alexander D., Joseph Henry, Benjamin Peirce, and Benjamin A. Gould, Jr. *Dudley Observatory*, 11 August 1856. In A. D. Bache Papers, LC, reel 4, #0419–21.

Bailey, Solon I. *The History and Work of Harvard Observatory, 1839–1927*. New York: McGraw Hill, 1931.

Baker, Daniel W. *History of the Harvard College Observatory during the Period 1840–1890*. Cambridge, 1890.

Baltimore & Ohio Railroad. *Instructions and Time Table for the Government of Conductors and Engine-Men Employed on the Baltimore and Ohio Railroad*. Baltimore, August 1843.

Barnard, F. A. P. "Uniform System of Coinage and of Time." *Report of International Association for the Reform and Codification of the Law of Nations* 9 (1882): 69–83, 236–37.

———. "Uniform System of Time-Reckoning." *Report of International Association for the Reform and Codification of the Law of Nations* 10 (1883): 180–96.

Bartky, Ian R. "Naval Observatory Time Dissemination before the Wireless." In *Sky with Ocean Joined*, edited by Steven J. Dick and LeRoy Doggett, 1–28. Washington: USNO, 1983.

———. "The Invention of Railroad Time." *Railroad History* 148 (spring 1983): 13–22.

———. "A Comment on 'The Standardization of Time' by Zerubavel." *American Journal of Sociology* 89 (1984): 1420–26.

———. "Inventing, Introducing and Objecting to Standard Time." *Vistas in Astronomy* 28 (1985): 105–11.

———. "The Bygone Era of Time Balls." *Sky and Telescope* 73 (1987): 32–35.

———. "Running on Time." *Railroad History* 159 (autumn 1988): 19–38.

———. "The Adoption of Standard Time." *Technology and Culture* 30 (1989): 25–56.

———. "Railroad Timekeepers." *NAWCC Bulletin* 31 (October 1989): 399–411.

———. "Comment on 'The Most Reliable Time.'" *Technology and Culture* 32 (1991): 183–85.

Bartky, Ian R., and Steven J. Dick. "The First Time Balls." *Journal for the History of Astronomy* 12 (1981): 155–64.

———. "The First North American Time Ball." *Journal for the History of*

Astronomy 13 (1982): 50–54. Reprinted with additions in *NAWCC Bulletin* 41 (December 1999): 741–44.

Bartky, Ian R., and Elizabeth Harrison. "Standard and Daylight-saving Time." *Scientific American* 240 (May 1979): 46–53.

Bartky, Ian R., Norman S. Rice, and Christine A. Bain. "'An Event of No Ordinary Interest'—The Inauguration of Albany's Dudley Observatory." *Journal of Astronomical History and Heritage* 2 (1999): 1–20.

Beardsley, Wallace R. "Samuel Pierpont Langley—His Early Academic Years at the Western University of Pennsylvania." Ph.D. diss., University of Pittsburgh, 1978.

Bedini, Silvio. *Thinkers and Tinkers: Early American Men of Science.* New York: Charles Scribner's Sons, 1975.

Benedict Brothers. *New York Railroad and Steamboat Time Tables.* New York, 1866.

Bigourdan, M. G. "Le jour et ses divisions." Bureau des Longitudes. *Annuaire pour l'an 1914*, B.1–50. Paris: Gauthier-Villars, 1914.

Blackwell, Dana. "Early Railroad Timekeeping." *NAWCC Bulletin* 28 (1986): 459–63.

[Bless, Robert Charles.] *Washburn Observatory 1878–1978.* [Madison: University of Wisconsin, 1978].

B[lunt], G. W. "Experiments on Longitude." *New York American,* [?] 1840.

Bond, George P. "Report of the Director [for 1859]." In *Report of the Committee of the Overseers of Harvard College Appointed to Visit the Observatory in the Year 1859,* 10–24. Boston: Geo. C. Rand & Avery, 1860.

———. "Report of the Director [for 1860]." In *Report of the Committee of the Overseers of Harvard College Appointed to Visit the Observatory in the Year 1860,* 9–21. Boston: Geo. C. Rand & Avery, 1861.

———. "Report of the Director [for 1862]." In *Report of the Committee of the Overseers of Harvard College Appointed to Visit the Observatory in the Year 1863,* 10–37. Boston: Geo. C. Rand & Avery, 1863.

Bond, William C. "Letter to the Secretary." *Monthly Notices of the RAS* 9 (1849): 151.

———. "Historical Sketch and Description of the Observatory." *HCO Annals* 1 (1856): xiv–xxx.

———. "Report of the Director . . . for 1850." *HCO Annals* 1 (1856): cxlv–cl.

———. "Report of the Director . . . for 1851." *HCO Annals* 1 (1856): cliii–clvii.

———. "Report of the Director . . . for 1852." *HCO Annals* 1 (1856): clxii–clxvii.

———. "Report of the Director . . . for 1853." *HCO Annals* 1 (1856): clxx–clxxiv.

———. "Report of the Director . . . for 1854." *HCO Annals* 1 (1856): clxxvii–clxxxii.

———. "Report of the Director of the Observatory to the visiting committee appointed by the Board of Overseers of Harvard University," read October 1856. In HCO Records, HUA.

Bosch, Adam. "Historical Sketch of the Fire Alarm Telegraph." *Transactions of the American Institute of Electrical Engineers* 14 (1897): 335–50.

Boss, Lewis. *Dudley Observatory. Annual Report of the Astronomer for 1877.* Albany, N.Y.: Weed, Parsons and Co., 1878.

———. *Dudley Observatory. Report of the Director [for 1882].* Albany, N.Y.: Charles Van Benthuysen & Sons, 1883.

Boston. City Records. *Common Council* 12 (1848–49): 76, 161, 254.

——. City Records. *Mayor and Aldermen* 26 (1848): 77, 268–69, 561.

——. Common Council. *Report of the Committee on Fire Alarms on the Regulation of Timepieces*. City Document 88. Boston, 1853.

——. Mayor. *Address of the Mayor to the City Council of Boston*. City Document 1. Boston, 1848.

"Boston Fire-Alarm Telegraph." *Gleason's Pictorial Drawing-Room Companion* 2 (24 April 1852): 264, 269.

"Boston Fire Alarm Telegraph." *Firemen's Advocate*, 5[2], May 1857, [?].

"The [Boston] Municipal Telegraph." *Commonwealth*, 30 December 1851, 2.

"The [Boston] Telegraph Fire Alarm." *Firemen's Advocate*, 9, May 1857, [?].

British Association for the Advancement of Science. "Transactions of the Sections." *Report of the Ninth Meeting of the BAAS* (1840): 27–28.

Brown, A. D. "The Electrical Distribution of Time." *Journal of the Franklin Institute*, 3d ser., 95 (1888): 462–75, and ibid., 96 (1888): 14–24.

Bruce, Robert V. *The Launching of Modern American Science: 1846–1876*. Ithaca: Cornell University Press, 1988.

Burpee, Lawrence J. *Sandford Fleming: Empire Builder*. England: Oxford University Press, 1915.

Cajori, Florian. *The Chequered Career of Ferdinand Rudolph Hassler*. 1929. Reprint, New York: Arno Press, 1980.

Channing, W. F. *Communication . . . Respecting a System of Fire Alarms*. City Document 20. Boston, 1851.

——. "On the Municipal Electric Telegraph; especially in its application to Fire Alarms." *American Journal of Science and Arts* 13 (1852): 58–83.

——. "The American Fire-Alarm Telegraph." In *Ninth Annual Report of the Board of Regents of the Smithsonian Institution*, 147–55. Washington, 1855.

Chicago Astronomical Society. Board of Directors. "Minutes, 1862–1903."

——. *Full Report of the Secretary and History of the Organization, with the Proceedings of the Society, the "Foundation of Dearborn Observatory" and the Mounting of the Great Telescope of Alvan Clarke [sic] & Sons*. Chicago: Inter-Ocean Steam Book and Job Print, 1874.

Cincinnati. Common Council. *Report of Special Committee of the Common Council of Cincinnati, on Standard Public Time*. Cincinnati: Bloch & Co., 1870.

Connecticut. Railroad Commissioners. *Annual Report of the Railroad Commissioners to the State of Connecticut*. Hartford: Case, Lockwood & Brainard Co., 1880–85.

Creet, Mario. "Sandford Fleming and Universal Time." *Scientia Canadensis* 14 (1990): 66–89.

Dent, Edward J. "Difference of Longitude between Greenwich and New York." *Mechanics Magazine* 32 (1839): 143–44.

Dick, Steven J. "How the U.S. Naval Observatory Began, 1830–1865." In *Sky With Ocean Joined: Proceedings of the Sesquicentennial Symposia of the U.S. Naval Observatory*, edited by Steven J. Dick and LeRoy E. Doggett, 166–81. Washington: USNO, 1983.

Dohrn-van Rossum, Gehard. *History of the Hour: Clocks and Modern Temporal*

Orders. English translation of German ed. Chicago: University of Chicago Press, 1996.

Dowd, Charles F. *System of National Time and Its Application, by Means of Hour and Minute Indexes, to the National Railway Time-Table; also a Railway Time Gazetteer, Containing all the Railways in the United States and Canada, Alphabetically Arranged, with Their Stations Indexed in Form for the National Railway Time-Table.* Albany, N.Y.: Weed, Parsons and Co., 1870.

———. *National Rail-Road Time.* [Albany (?), N.Y., 1870].

———. *Rail Road Time.* [Albany (?), N.Y., September 1870].

———. "Origin and Early History of the New System of National Time." *PAMS* 4 (1884): 90–101.

———. *System of Time Standards Illustrated with Map.* [Saratoga Springs, N.Y., 1885].

Dracup, Joseph F. "History of Geodetic Surveying. I. The Early Years: 1807–1843." *ACSM Bulletin* (March/April 1995): 15–19.

———. "History of Geodetic Surveying. II. Following in Hassler's Footsteps." *ACSM Bulletin* (July/August 1995): 22–27.

———. "History of Geodetic Surveying. III. Triumphs of the Mountain Men." *ACSM Bulletin.* (September/October 1995): 38–43.

Dudley Observatory. Board of Trustees. "Records of the Dudley Observatory Board of Trustees, 1852–1943."

———. Trustees. *Dudley Observatory and the Scientific Council: Statement of the Trustees.* Albany, N.Y.: Van Benthuysen, 1858.

Dudley, P. H. "Railway Time Service." *Journal of the American Electrical Society* 3 (1880): 63–65.

———. "Railway Time Service." *Proc. AAAS* 29 (1881): 274–75.

[———.] "The Railroad Time and Time-Instruments." *Railroad Gazette* 13 (30 September 1881): 333–34.

[———.] *P. H. Dudley's Electrically-Controlled System of Railroad and City Time Service.* [New York (?), 1881 (?)].

Dupree, A. Hunter. *Science in the Federal Government: A History of Policies and Activities.* Baltimore: Johns Hopkins University Press, Paperbacks ed., 1986.

Edmands, J. Rayner. "Cooperation of Observations [Observatories] for Maintaining Accurate Time." *Proc. AAAS* 31 (1883): 612.

Ellis, William. "Lecture on the Greenwich System of Time Signals." *Horological Journal* 7 (1865): 85–91, 97–102, 109–14, 121–24.

Farmer, Moses G. "Improvement in Galvanic Clocks." *Patent No. 9279,* 21 September 1852.

[———.] *Farmer's Improved Sustaining Battery.* [Boston, 1853 (?)].

———. "Plan for Regulating the Public Clocks of Boston." In Boston Common Council. *Report of the Committee on Fire Alarms on the Regulation of Timepieces,* 5–9. City Document 88. Boston, 1853.

Fleming, James Rodger. *Meteorology in America, 1800–1870.* Baltimore: Johns Hopkins University Press, 1990.

Fleming, Sandford. "Time-Reckoning for the Twentieth Century." In *Annual Report of the Board of Regents of the Smithsonian Institution . . . for the Year Ending June 30, 1886*, 345–66. Washington: Government Printing Office, 1889.

Fox, Philip. "General Account of the Dearborn Observatory." *Annals of the Dearborn Observatory of Northwestern University* 1 (1913): 1–20.

Franklin Institute. *Official Catalogue of the International Electrical Exhibition, Philadelphia*. Philadelphia: Barr & McFetridge, 1884.

Gardner, William F. "Time-Controlling System." *Patent No. 287,015*, 23 October 1883.

———. "Electric Time-Controlling System." *Patent No. 307,287*, 28 October 1884.

———. *The Gardner System of Observatory Time*. Paris: Charles Schlaeber, 1889.

Gerstner, Franz A. R. von. *Early American Railroads*. Edited by Frederick C. Gamst. English translation of *Die innern Communicationen* (1842–43). Stanford: Stanford University Press, 1997.

Gould, Benjamin A. "An Address in Commemoration of Sears Cook Walker." *Proc. AAAS* 8 (1855): 19–45.

———. *Reply to the "Statement of the Trustees" of the Dudley Observatory*. Albany, N.Y.: Charles Van Benthuysen, 1859.

———. "On the Longitude between America and Europe from Signals through the Atlantic Cable." In *Report of the Superintendent of the United States Coast Survey Showing the Progress of the Survey during the Year 1867*, Appendix 6, 57–116. Washington: Government Printing Office, 1869.

[Greene, Mark.] *A Science Not Earthbound: A Brief History of Astronomy at Carleton College*. Northfield, Minn.: Carleton College, 1988.

Gundlfinger, Karl. *Hundert Jahre Telegraphie in Frankfurt am Main*. Frankfort am Main: Oberpostdirektion, 1949.

Guyot, Edmond. *Histoire de la détermination des longitudes*. La Chaux-de-Fonds: Chambre suisse de l'horlogerie, [1955].

Hale, George E. *First Annual Report of the Director of the Yerkes Observatory of the University of Chicago*. Chicago: University of Chicago Press, 1899.

Harlow, Alvin F. *The Road of the Century: The Story of the New York Central*. New York: Creative Age Press, 1947.

Harrington, M. W. "A Telephonic Time-Transmitter." *Science* 1 (20 April 1883): 302–3.

Harvard University. Corporation. "Agreement with Mr. Bond," 29 November 1839. In *HCO Annals* 1 (1856): lxxxvi–lxxxvii.

———. Board of Overseers. "Report of the Committee appointed by the Board of Overseers of Harvard University for Visiting the Observatory, for the Academic Year 1851–52." *HCO Annals* 1 (1856): clx–clxii.

———. "Report of the Committee appointed by the Board of Overseers of the University at Cambridge, to examine the Observatory, for the Academic Year 1853–54." *HCO Annals* 1 (1856): clxxiv–clxxvii.

———. "Report of the Committee appointed by the Board of Overseers of the

University at Cambridge, to examine the Observatory, for the Academic Year 1854–55." *HCO Annals* 1 (1856): clxxxiii–clxxxvi.

———. *Report of the Committee of the Overseers of Harvard College, Appointed to Visit the Observatory in the Year 1859.* Boston: Rand & Avery, 1860.

Hayes, Alexander L. "The Oram Telephone Time-Indicating System." *Electrical World* 6 (25 July 1885): 33.

Herman, Jan K. *A Hilltop in Foggy Bottom: Home of the Old Naval Observatory and the Navy Medical Department.* Washington: Navy Bureau of Medicine and Surgery, 1991.

Highton, Edward. *The Electric Telegraph: Its History and Progress.* London: John Weale, 1852.

Hirsch, A., and Th. v. Oppolzer, eds. "Rapport sur l'unification des longitudes par l'adoption d'un méridien initial unique, et l'introduction d'une heure universelle." In *Comptes Rendus de la septième Conférence générale du l'Association de géodésique internationale réunie à Rome, en Octobre 1883.* Bureau central de l'Association géodésique internationale, [1883 (?)].

Hoffleit, Dorrit. *Astronomy at Yale: 1701–1968.* New Haven: Connecticut Academy of Sciences, 1992.

Holden, Edward S. "Report upon the astronomical instruments of the Loan Collection of Scientific Instruments at the South Kensington Museum, 1876." In *Annual Report of the Secretary of the Navy on the Operations of the Department for the Year 1876*, Appendix 12, 289–325. Washington: Government Printing Office, 1876.

———. "On the Distribution of Standard Time in the United States." *Popular Science Monthly* 11 (1877): 175–82.

———. "Standard Time for Madison." Reprint from unknown periodical, n.p., [1881]. In Library of Congress collection.

———. *Publications of the Washburn Observatory of the University of Wisconsin* 1 (1882).

———. "Clocks and Time-Keeping." In *Handbook of the Lick Observatory of the University of California*, 96–103. San Francisco: Bancroft Co., 1888.

———. *Memorials of William Cranch Bond And Of His Son George Phillips Bond.* San Francisco: C. A. Murdock & Co., 1897.

Horsford, Eben N. *Respecting the Regulation of Timepieces in the City.* City Document 75. Boston, 1853.

Hough, G. W. "Report for 1870." *Annals of the Dudley Observatory* 2 (1871): 362–67.

———. "Time Service." *Sidereal Messenger* 9 (1890): 173–76.

———. *Annual Report of the Director of the Dearborn Observatory of Northwestern University. 1890–1891.* [Evanston, 1891].

Howse, Derek. *Greenwich time and the discovery of the longitude.* Oxford: Oxford University Press, 1980.

Humphreys, W. J. "Cleveland Abbe, 1838–1916." *Biographical Memoirs of the NAS* 8 (1919): 469–508.

Hungerford, Edward. *Men and Iron: The History of the New York Central.* New York: Thomas Y. Crowell, 1938.

Jacobs, Warren. "Early Rules and the Standard Code." *Railway and Locomotive Historical Society Bulletin* 50 (October 1939): 29–55.

James, Mary Ann. "The Dudley Observatory Controversy." Ph.D. diss., Rice University, 1980.

———. *Elites in Conflict: The Antebellum Clash over the Dudley Observatory.* New Brunswick: Rutgers University Press, 1987.

Janson, G. W. "Time Services of the Telegraph Companies." *Transactions of the American Institute of Electrical Engineers* 51 (1932): 541–45.

Jayne, J. L. "The Naval Observatory Time Service and How to Use Its Radio Time Signals." *Keystone* 36 (1 September 1913): 129, 131, 133, 135.

Jefferson Railroad. *Rules and Regulations for Running Trains &c.* Louisville, Ky.: Morton & Griswold, 1854.

Jenkins, Reese V., Leonard S. Reich, Paul B. Israel, Toby Appel, Andrew J. Butrica, Robert A. Rosenberg, Keith A. Neir, Melodie Andrews, and Thomas E. Jeffrey, eds. *The Papers of Thomas A. Edison.* Vol. 1. Baltimore: John Hopkins University Press, 1989.

Jesperson, James, and Jane Fitz-Randolph. *From Sundials to Atomic Clocks.* NBS Monograph 155. Washington: NBS, 1977.

Jones, Bessie Z., and Lyle G. Boyd. *The Harvard College Observatory: The First Four Directorships, 1839–1919.* Cambridge: Harvard University Press, Belknap Press, 1971.

Keeler, James E. "The Time Service of the Lick Observatory." *Sidereal Messenger* 6 (1887): 233–48.

———. "Experiments with Electrical Contact Apparatus for Astronomical Clocks." *Sidereal Messenger* 7 (1888): 9–14.

Kelly, John T. *Practical Astronomy during the Seventeenth Century: Almanac-Makers in America and England.* New York: Garland, 1991.

King, W. James. "The Development of Electrical Technology in the 19th Century. Vol. 2. The Telegraph and the Telephone." In *United States National Museum Bulletin* 228. *Contributions from the Museum of History and Technology.* Paper 29, 274–332. Washington: Smithsonian Institution, 1962.

Kohlstedt, Sally Gregory. *The Formation of the American Scientific Community: The American Association for the Advancement of Science 1848–1860.* Urbana: University of Illinois Press, 1976.

Lamont, Johann. *Beschreibung der an der Münchener Sternwarte zu den Beobachtungen verwendeten neuen Instrumente und Apparate.* Munich: J. G. Weiss, 1851.

Landes, David S. *A Revolution in Time: Clocks and the Making of the Modern World.* Cambridge: Harvard University Press, Belknap Press, 1983.

Langley, Samuel P. "Uniform Time." *Pittsburgh Commercial,* 12 February 1870, 2; ibid., 17 February 1870, 2.

———. "On the Allegheny System of Electric Time Signals." *American Journal of Science and Arts,* 3d ser., 4 (1872): 377–86; also extracted in *Journal of the Society of Telegraph Engineers* 1 (1873): 433–41.

———. "Uniform Railway Time." *American Exchange and Review* 24 (1874): 271–76.

———. "Uniformity of Time." *Times* (London), 27 February 1874, 5.

———. "The Electric Time Service." *Harper's New Monthly Magazine* 56 (April 1878): 665–71.

———. "The Other Side." *Pittsburgh Commercial Gazette*, 13 April 1878.

———. "Report of Director." In "Minutes," Western University of Pennsylvania Board of Trustees, 11 June 1878, 161–64.

———. "Electric Time Service." *Journal of the American Electrical Society* 2 (1879): 93–101.

———. "History of the Allegheny Observatory." In Parke, John E., *Recollections of Seventy Years and Historical Gleanings of Allegheny, Pennsylvania*, 179–88. Boston: Rand, Avery & Co., 1886.

Lee, W. Raymond. *Standard Time*, 31 August 1853. Boston: Boston & Providence Railroad, 1853.

Liggett, Barbara. "A History of the Adoption of Standard Time in the United States, 1869–1883." Master's thesis, University of Texas, 1960.

Locke, John. "On the Electro-Chronograph." *American Journal of Science and Arts* 8 (1849): 231–52.

———. "Electro-Chronograph Clock of the National Observatory." *National Intelligencer*, 26 November 1849, 3.

———. *Report of Professor John Locke, of Cincinnati, Ohio; on the Invention and Construction of his Electro-Chronograph: for the National Observatory*. Cincinnati: Wright, Ferris and Co., 1850.

———. "Astronomical Machinery." *National Intelligencer*, 9 August 1851, 2.

Loomis, Elias. *The Recent Progress of Astronomy; Especially in the United States*. New York: Harper & Brothers, 1850.

———. *The Recent Progress of Astronomy; Especially in the United States*. New York: Harper & Brothers, 3d ed., 1856.

Lund, John A. "On a Complete System of Synchronizing by Time Signals, as Now Adopted in London and Elsewhere." *Journal of the Society of Telegraph Engineers and of Electricians* 10 (1881): 381–401.

Mailly, Ed. "Précis de l'historie de l'Astronomie aux États-Unis d'Amérique." In *l'Annuaire de l'observatoire royal de Bruxelles*, pour l'année 1860. Brussels: M. Hayez, 1860.

Malin, Stuart R. "The International Prime Meridian Conference, Washington, October 1984 [*sic*]." *Journal of Navigation* 38 (1985): 203–6.

Malin, Stuart, and Carole Stott. *The Greenwich Meridian*. Southampton, England: Ordnance Survey, 1984.

Malin, Stuart R., Archie E. Roy, and Peter Beer, eds. "Longitude Zero 1884–1984." *Vistas in Astronomy* 28 (1985): 1–407.

Mathys, Joan. "The Right Place at the Right Time: The United States Navy and the Development of Wireless Time Signaling, 1900–1923." Master's thesis, George Washington University, 1991.

[Maury, Matthew.] "The Electro-Chronographic Clock and Its Accompanying Instruments." *National Intelligencer*, 21 November 1849, 3.

McCusker, John J. "How Much Is That in Real Money? A Historical Price Index for Use as a Deflator of Money Values in the Economy of the United States." *Proc. American Antiquarian Society* 101 (October 1991): 297–373. Reprint, 1992.

Melhuish, S. C. "Report of Captain Melhuish on *Birmingham and Gloucester Railway*." In United Kingdom. *Reports to the Committee of Privy Council and Returns, &c. Relative to Railways* 25 (1841): 58–62.

Miller, Howard S. *Dollars for Research: Science and Its Patrons in Nineteenth-Century America*. Seattle: University of Washington Press, 1970.

Mott, Edward. *Between the Ocean and the Lakes: The Story of Erie*. New York: John S. Collins, 1899.

Musto, David F. "Yale Astronomy in the Nineteenth Century." Yale Graduate School. *Ventures* 8 (spring 1968): 7–18.

National Telephone Exchange Association. *Report of the Proceedings of the Second Meeting of the National Telephone Exchange Association*. New Haven: Tuttle, Moorehouse & Taylor, 1881.

———. *The Third Meeting of the National Telephone Exchange Association*. New Haven: Tuttle, Moorehouse & Taylor, 1881.

Neidigh, Ray S. *The Elgin Observatory Story, 1909–1960*. Fredericksburg, Tex.: Printing Press, 1982.

Newcomb, Simon. *Report to the Secretary of the Navy on Recent Improvements in Astronomical Instruments*. 48th Congr., 1st sess., 1884. S. Ex. Doc. 96.

New England Association of Railroad Superintendents. *Records of the New England Association of Railway* [sic] *Superintendents: 1848–1857*. Washington: Gibson Brothers, 1910.

———. *Reports and Other Papers of the New England Association of Railroad Superintendents*. Boston: Stacy, Richardson & Co., 1850.

New York & Erie Rail Road. "Rule No. 15." *Instructions for the Running of Trains, etc. to Go into Effect on Monday, March 31, 1851*. New York: Office of Parker's Journal, 1851.

[———.] "Standard of Time of the New York and Erie Railroad." *American Railroad Journal* 26 (3 September 1853): 567.

New Zealand. Dominion Observatory. "Mean Time and Time Service." *Bulletin* 190. Wellington, 1938.

Nourse, J. E. "Observatories in the United States. I." *Harper's New Monthly Magazine* 48 (March 1874): 526–41.

———. "Observatories in the United States. II." *Harper's New Monthly Magazine* 49 (September 1874): 518–31.

Obendorf, Donald L. "Samuel P. Langley: Solar Scientist, 1867–1891." Ph.D. diss., University of California, Berkeley, 1969.

O'Malley, Michael. *Keeping Watch: A History of American Time*. New York: Viking, 1990.

Osterbrock, Donald E. *James E. Keeler: Pioneer American Astrophysicist*. Cambridge: Cambridge University Press, 1984.

Pathfinder Railway Guide for the New England States. Boston: Snow & Wilder, December 1849–June 1851.

Pawson, Eric. "Local times and standard time in New Zealand." *Journal of Historical Geography* 18 (1992): 278–87.

Payne, William W. "Time Service of Carleton College Observatory." *Science* 2 (17 September 1881): 445.

———. "Time Service of Carleton College Observatory." *Proc. AAAS* 30 (1882): 41–43.

———. "Time Service of Carleton College Observatory." *Sidereal Messenger* 1 (1882): 16–18.

———. "Observatory Local Patronage Threatened." *Sidereal Messenger* 8 (1889): 452–54.

———. "The Western Union Time." *Sidereal Messenger* 9 (1890): 232.

Pickering, Edward C. *Annual Report of the Director of Harvard College Observatory [for 1877].* Cambridge: John Wilson and Son, 1877.

———. *Fortieth Annual Report of the Director of the Astronomical Observatory of Harvard College [for 1885].* Cambridge: Harvard University, 1886.

———. *Harvard College Observatory. Time Service.* [Cambridge: HCO], 26 December 1891.

———. *Forty-Seventh Annual Report of the Director of the Astronomical Observatory of Harvard College [for 1892].* Cambridge: Harvard University, 1892.

———. *Fifty-Seventh Annual Report of the Director of the Astronomical Observatory of Harvard College [for 1902].* Cambridge: Harvard University, 1902.

Pond, Chester H. "Electro-mechanical Clock." *Patent No. 308,521,* 25 November 1884.

Porter, Jermain G. "Ormsby MacKnight Mitchel." *Sidereal Messenger* 8 (1889): 442–47.

———. *Historical Sketch of the Observatory of the University of Cincinnati.* Cincinnati: University of Cincinnati, 1893.

Pritchett, Henry S. "Observatory Local Patronage." *Sidereal Messenger* 9 (1890): 113–16.

———. "Correct Time—How Shall We Maintain It?" *Electrical Engineer* 10 (2 July 1890): 20–21.

Providence and Worcester Railroad Company. "Records of the Directors of the Providence and Worcester Railroad Company." Directors Record No. 1 (1845–62).

———. *Ninth Annual Report of the Directors of the Providence and Worcester Railroad Company to the Stockholders.* Providence: A. Crawford Greene, 1854.

Randall, Anthony G. *The Time Museum Catalogue of Chronometers.* Rockford, Ill.: Time Museum, 1992.

———. "The Timekeeper that Won the Longitude Prize." In Andrewes, *The Quest for Longitude,* 235–54.

Ranney, Henry C. "Facts Relating to the Chicago Astronomical Society." In Chicago Historical Society collection.

Reed, Robert C. *Train Wrecks: A Pictorial History of Accidents on the Main Line.* 1968. Reprint, New York: Bonanza Books, 1982.

Reid, James D. *The Telegraph in America. Its Founders Promoters and Noted Men.* 1879. Reprint, New York: Arno Press, 1974.

Reingold, Nathan. "Alexander Dallas Bache: Science and Technology in the American Idiom." *Technology and Culture* 11 (1970): 163–77.

Rogers, W. A. "Report of Sub-committee on Time-pieces." In "Report of Examiners of Section XXII." *Journal of the Franklin Institute*, 3d ser., 91 (January 1886): Supplement, 90–96.

[Royce & Marean.] *Standard Time. Gardner's System for Correcting Clocks by Electricity and for Furnishing Standard Time to Stores, Hotels, Railroads and Private Residences.* Washington: R. Beresford, 1885.

Russell's Guide to the Fire Alarm Telegraph, bound with *The Stranger's Pathfinder.* Boston: Russell, 1864.

Safford, Truman H. "Report of the [Dearborn] Observatory Director for 1866–1868." In Ranney, "Facts Relating to the Chicago Astronomical Society," 192–202.

Schivelbush, Wolfgang. *The Railway Journey: Trains and Travel in the 19th Century.* 1977. English translation of German ed. New York: Urizen Books, 1979.

Schott, Charles A. "Report on the results of the longitudes of the Coast and Geodetic Survey determined up to the present time by means of the electric telegraph. . . . " In *Report of the Superintendent of the U.S. Coast and Geodetic Survey . . . during the fiscal year ending with June, 1880.* Appendix 6. Washington: Government Printing Office, 1882.

———. "Longitudes deduced in the Coast and Geodetic Survey from determinations by means of the electric telegraph between the years 1846 and 1885. Second adjustment." In *Report of the Superintendent of the U.S. Coast and Geodetic Survey . . . during the fiscal year ending with June, 1884.* Appendix 11. Washington: Government Printing Office, 1885.

———. "The telegraphic longitude net of the United States and its connection with that of Europe, 1866–1896." In *Report of the Superintendent of the U.S. Coast and Geodetic Survey . . . [for] 1897.* Appendix 2. Washington: Government Printing Office, 1898.

[Searle, Arthur.] "Historical Account of the Observatory from October 1855 to October 1876." *HCO Annals* 8 (1876): 1–65.

Self-Winding Clock Co. "Advertisement." *JC&HR* 21 (February 1890): 91.

———. *Standard Time Railway Map.* [New York (?), 1893].

———. *What is Standard Time?* [New York (?), 1893].

———. "A History of the Time Service of the Western Union Telegraph Co.," 29 August 1934. Reprinted in *Journal of the Electrical Horological Society* 7 (December 1981): 1–3.

Shaw, Robert B. *A History of Railroad Accidents, Safety Precautions and Operating Practices.* N.p.: Vail-Ballou Press, 1978.

Shoemaker, Philip S. "Stellar Impact: Ormsby Macknight *[sic]* Mitchel and Astronomy in Antebellum America." Ph.D. diss., University of Wisconsin, Madison, 1991.

Slotten, Hugh R. *Patronage, Practice, and the Culture of American Science.* Cambridge: Cambridge University Press, 1994.

Small, Charles S. H. "Electrical Time Services in New York." *Operator* 13 (23 December 1882): 681–84.

Smith, Humphrey M. "Greenwich Time and the Prime Meridian." *Vistas in Astronomy* 20 (1976): 219–29.

Sobel, Dava. *Longitude.* New York: Walker, 1995.

Sobel, Dava, and William J. H. Andrewes. *The Illustrated Longitude.* New York: Walker, 1998.

Standard Time Company. *Announcement of the Standard Time Company.* New Haven [1882]. In YUA.

———. *Yale College Observatory Standard Time Signals. Circular No. 2.* New Haven [1882]. In YUA.

———. *Barraud and Lunds' (English) Patents for the Automatic Setting of Public, Office and Railroad Clocks by Telegraphic Observatory Time Signals. Circular No. 3.* New Haven. May 1882. In YUA.

Steinheil, Carl August von. "Privilegium," 2 October 1839. In *Kunst- und Gewerbe-Blatt des polytechnischen Vereins für das Königreich Bayern* 29 (February 1843): 127–42.

———. "Galvanische Uhren." In *Jahrbuch für 1844*, edited by H. C. Schumacher, 41–48. Stuttgart and Tübingen: J. G. Cotta, 1844.

———. "Verzeichness der astronomischen, geodätischen und physikalischen Mess-Instrumente." *Astronomische Nachrichten* 26, No. 609 (26 August 1847): Cols. 133–42.

Stephens, Carlene E. *Inventing Standard Time.* Washington: Smithsonian Institution, 1983.

———. "Partners in Time: William Bond & Son of Boston and the Harvard College Observatory." *Harvard Library Bulletin* 35 (1987): 351–84.

———. "'The Most Reliable Time': William Bond, the New England Railroads, and Time Awareness in 19th-Century America." *Technology and Culture* 30 (1989): 1–24.

———. "The Impact of the Telegraph on Public Time in the United States, 1844–1893." *IEEE Technology and Society Magazine* 8 (March 1989): 4–10.

———. "Astronomy as Public Utility: The Bond Years at the Harvard College Observatory." *Journal for the History of Astronomy* 21 (1990): 21–35.

———. "Response." *Technology and Culture* 32 (1991): 185–86.

Stephens, Eugene. "Astronomy at Washington University through the Years, 1857–1954." St. Louis: Washington University, n.d.

Stevens, Frank W. *The Beginnings of the New York Central Railroad.* New York: G. P. Putnam's Sons, 1926.

Symonds, R. W. *Thomas Tompion: His Life and Work.* London: B. T. Batsford, 1951.

Taylor, Hiero S. "U.S. Government System of Observatory Time." *JC&HR* 21 (February 1890): 82, 85, 86, 88.

Terry, Charles A. "Application of Electricity to Horology." *Electrician and Electrical Engineer* 4 (1885): 89–92, 169–71, 213–15.

Thompson, Robert Luther. *Wiring a Continent: The History of the Telegraph Industry in the United States, 1832–1866.* Princeton: Princeton University Press, 1947.

Thomson, Malcolm M. *The Beginning of the Long Dash: A History of Timekeeping in Canada.* Toronto: University of Toronto Press, 1978.

Time Telegraph Co. *The Time Telegraph Company.* New York, 1882.

Tobler, A. *Die Elektrischen Uhren und die Elektrische Feuerwehr-Telegraphie.* Vienna: A. Hartleben's Publishing House, 1883.

Travelers' Official Guide of the Railway and Steam Navigation Lines of the United States and Canada. New York: National Railway Publication Co., 1874, 1880–85.

Trelease, Allen W. *The North Carolina Railroad, 1849–1871, and the Modernization of North Carolina.* Chapel Hill: University of North Carolina Press, 1991.

United States. President. Message. *International Meridian Conference.* 50th Congr., 1st sess., 1888. H. Ex. Doc. 61.

U.S. Coast Survey. *Laws Relating to the Survey of the Coast of the United States; with The Plan of Reorganization of 1843, and Regulations by the Treasury Department.* Washington: Public Printer, 1858.

U.S. Congress. House. Committee on Commerce. *Meridian Time and Time-Balls on Custom-Houses.* 47th Congr., 1st sess., 1882. H. Rept. 681.

———. Committee on Foreign Affairs. *To Fix a Common Prime Meridian.* 47th Congr., 1st sess., 1882. H. Rept. 1519.

———. Committee on Naval Affairs. *American Prime Meridian.* 31st Congr., 1st sess., 1850. H. Rept. 286.

———. *Disbursements of Public Moneys of the Chief Signal Officer, United States Army.* 49th Congr., 1st sess., 1886. H. Misc. Doc. 255.

———. *Journal,* 1687–1688, 1749, 1750. 47th Congr., 1st sess., 1882.

U.S. Congress. Senate. Committee on Foreign Relations. *Report on joint resolution . . . to call an international conference to fix on . . . a common prime meridian. . . .* 47th Congr., 1st sess., 1882. S. Rept. 840.

———. Committee on Foreign Relations. *Report to accompany concurrent resolution . . . to communicate . . . the resolutions adopted by the International [Meridian] Conference. . . .* 48th Congr., 2d sess., 1885. S. Rept. 1188.

U.S. Department of Commerce and Labor. Bureau of the Census. *Municipal Electric Fire Alarm and Police Patrol Systems.* Washington: Government Printing Office, 1904.

U.S. Department of the Navy. Secretary. *Annual Report of the Secretary of the Navy on the Operations of the Department for the Year 1876.* Washington: Government Printing Office, 1876.

U.S. Department of State. *International Conference Held at Washington for the Purpose of Fixing a Prime Meridian and a Universal Day.* 48th Congr., 2d sess., 1884–85. H. Ex. Doc. 14.

U.S. Department of Transportation. Office of the Secretary. *Standard Time in the United States: A History of Standard and Daylight Saving Time . . . and an Analysis of the Related Laws.* Washington, 1970.

U.S. Department of War. Army Signal Office. "Information Relative to the

Construction and Maintenance of Time Balls." *Professional Papers of the Signal Service.* No. 5. Washington, 1881.

U.S. Naval Observatory. "The Present Status of the Use of Standard Time." *Publications of the United States Naval Observatory* 4 (1905): Appendix IV.

Usselman, Steven W. "Patents Purloined: Railroads, Inventors, and the Diffusion of Innovation in 19th-Century America." *Technology and Culture* 32 (1991): 1045–75.

Walcott, Charles D. "Samuel Pierpont Langley." *Biographical Memoirs of the NAS* 7 (1913): 247–68.

W[aldo], F[rank]. "The American Standard Time Company." *Sidereal Messenger* 1 (May 1882): 17–19.

Waldo, Leonard. *Standard Public Time.* Cambridge: HCO, 1877.

———. "Report upon the Time Service [for 1877–1878]." In HCO Records, HUA.

———. "Appendix C," 20 November 1877. In Pickering, *Annual Report of the Director [1877],* 28–36.

———. "After an examination of the records of railway disasters. . . . " *Circular letter of inquiry to the superintendents of the New England railroads,* 6 November 1878, 2p. In HCO Records, HUA.

———. "Telling the Time." *Bulletin of the Essex Institute* 10 (1878): 40–52.

———. "On the Longitude of Waltham, Mass." *Proc. American Academy of Arts and Sciences* 5 (1878): 175–82.

———. "Report relating to our Time Service for the year ending Sept. 1st 1879." In HCO Records, HUA.

———. "Progress of the Time Service [for 1879–80]," 1 November 1880. In HCO Records, HUA.

———. "The Distribution of Time." *North American Review* 131 (December 1880): 528–36. Reprinted in *Science* 1 (1880): 277–80.

———. "Time-Balls and Standard Time." *Nation,* 12 May 1881, 332.

[———. (?)] "Standard Time." *JC&HR* 13 (February 1882): 18.

———. "Railroad and Public Time." *North American Review* 137 (December 1883): 606–9.

[———. (?)] "The distribution of time on a commercial basis . . . " *Science* 5 (6 February 1885): 120.

Wales, William, and William Bayly, *The Original Astronomical Observations, Made In The Course Of A Voyage towards the South Pole, And Round The World, In his Majesty's Ships the Resolution and Adventure. . . .* London: W. and A. Strahan, 1777.

Walker, Charles V. "On Controlling Clocks by Electricity." *Monthly Notices of the RAS* 21 (1861): 72–76.

W[alker, Sears C.] "Difference of Longitude by Telegraph." *Littell's Living Age* 15 (October 1847): 186–87.

———. *Report on an application of the galvanic circuit to an astronomical clock and telegraph register in determining local differences of longitude . . . ,* 15 December 1848. 30th Congr., 2d sess., 1849. H. Ex. Doc. 21.

———. "Report of the Experience of the Coast Survey in regard to telegraph operations." *Proc. AAAS* 2 (1850): 182–92.

———. "Brief abstract of the progress of improvement and invention in the art of determining longitudes by the Electric Telegraph," 24 April 1851. Reprinted in *HCO Annals* 1 (1856): xxiv–xxviii.

[Waltham Observatory.] "Standard Time at a Modern Watch Factory." *Scientific American* 92 (1905): 300–301.

Warner, Deborah Jean. "Astronomy in Antebellum America." In *The Sciences in the American Context: New Perspectives*, edited by Nathan Reingold, 55–75. Washington: Smithsonian Institution Press, 1979.

Warren, Daniel. "History and Description of the Boston Fire Alarm Telegraph." *Firemen's Advocate*, 18 December 1858, [?].

Wayman, Patrick A. *Dunsink Observatory: 1785–1985*. Dublin: Royal Dublin Society, 1987.

Western Railroad Regulations, 4 December 1843. [Boston (?), 1843].

Western University of Pennsylvania, Catalogue [for 1893–94]. Pittsburgh, 1893.

Wheeler, George M. *Report on the Third International Geographical Congress and Exhibition at Venice, Italy, 1881*. 48th Congr., 2d sess., 1885. H. Ex. Doc. 270.

Whitesell, Patricia S. *A Creation of His Own: Tappan's Detroit Observatory*. Ann Arbor: Bentley Historical Library, University of Michigan, 1998.

Whitney, Marvin E. *The Ship's Chronometer*. Cincinnati: American Watchmakers Institute Press, 1985.

Wilkes, Charles. "Difference of Longitude Determined by Morse's Telegraph." *National Intelligencer*, 21 June 1844, 3.

Willard, John Ware. *Simon Willard and His Clocks*. 1911. Reprint, New York: Dover Publications, 1968.

Williams, J. E. D. *From Sails to Satellites: The Origin and Development of Navigational Science*. Oxford: Oxford University Press, 1992.

Withington, Sidney. "Standardization of Time in Connecticut." *Railway and Locomotive Historical Society Bulletin* 46 (April 1938): 14–16.

———. "Marking Time in 1883: A Contribution Made by the Railroads to Our National Welfare in Standardizing Time throughout the Country." Connecticut Society of Civil Engineers. *Sixty-Seventh Annual Report* (1951): 120–33.

Woodward, C. M. *Circular Letter Concerning the Washington University Electric Clock System*, 1 May 1879. [St. Louis: N.p., 1879].

Index

New York Central, 21–22, 65, 67, 73; New York Central & Hudson River, 114; New York, New Haven & Hartford, 119, 159, 254n15; North Carolina, 229n7, 235n8; Northern, 235n8; Ohio & Pennsylvania, 67, 221n27, 237n26; Pennsylvania (see Pennsylvania Railroad); Philadelphia, Wilmington & Baltimore, 114; Providence & Worcester (see Valley Falls disaster); South Carolina (see Charleston & Hamburg above); South Eastern (England), 64; Southern Pacific, 187, 197, 274n22; Vermont Central, 235n8; Wabash, 183; Worcester & Nashua, 22, 23

Railroads: accidents, 24–25, 30, 67, 221n27, 222n30, 237n26, 252n3,(see also Valley Falls disaster); associations and conventions, 21, 97, 100, 138, 219n11, 258n4 (see also General Time Convention; New England Association of Railroad Superintendents); operating rules, 30, 60, 172–73, 218n3 (see also under Railroad timekeeping)

Railroad Superintendents, New England Association of, 22–23, 219nn8,9

Railroad timekeeping, 19–20; accuracy and consistency, 20, 30, 53, 65, 67, 113–14, 191, 222n30, 252n32; and multiple operating times, 21–22, 22–23, 93–96, 97–99, 139, 219nn9–11; and observatory time, 30, 143, 173, 175, 191, 221–22n28, 269n19; operating rules for, 20, 25, 27–28, 30, 60, 172–73, 218n3; and Standard Railway Time, 139–40, 141, 142, 143, 146, 219n10, 260n14, 261–62n30; transferring time, 20, 28–29, 60–61, 88, 113–14, 235n7, 244n43, 245n44; and watch inspection, 30, 172, 183. See also Western Union Telegraph Co.

Railway Telegraph Superintendents, Association of, 272n6

Rating chronometers. See Chronometers, marine; and under Observatories

Regulator. See Clock, setting of, to time manually; Synchronizer, clock

"Report on Standard Time," 102–3, 119, 148, 248n1, 253n11

Rhode Island Railroad Commission. See Valley Falls disaster

Rhode Island Telephone & Electric Co., 168–69

Riefler, Sigmund, 202–4

Ritchie, James, & Son (Edinburgh), 77, 86, 87, 242n10, 245n51, 251n25

Rittenhouse, David, 8, 229n2

Rodgers, John: and time ball legislation, 116–18, 127–28, 130, 134, 135, 255nn2,6; and U.S. Signal Service, 128–29, 130, 131–32, 255n6, 256n8. See also U.S. Naval Observatory

Rome Conference. See International Geodetic Association

Royce & Marean (Washington), 174, 176, 270nn24,28

Rutherfurd, Lewis, 70, 147, 239n36

Safford, Truman H., 78, 79, 80, 82, 202; and railroad time zone, 95–96

Sampson, Ralph A., 279n23

Savannah (Central Railway) time, 145–46

Scammon, J. Young, 82, 243n27

Scrambler, signal, 164, 267n25

Seaton Station, 39, 43–44, 73, 226n29, 228n43. See also Longitude, zero of

Self-Winding Clock Co. (New York), 170–72, 177–78, 183, 186, 269nn12,14, 272n37, 274n20; and clock synchronizers, 178, 183, 184, 269n16, 272n37, 273n6, 273n13; and Western Union, 114, 178, 183–84, 185, 189, 193, 202, 273nn11–13, 275n37, 280nn25,27; and World's Columbian Exposition, 201, 202. See also Pond, Chester H.

Seth Thomas Clock Co. (Thomaston, Conn.), 176; and "clock of precision," 176–77

Sidereal Messenger. See Payne, W. W.

Sidereal time. See under Time

Siemens, Werner, 51, 231n16

Smith, F. O. J., 230n13

Smyth, Charles Piazza, 77, 226n27, 242n10

SNET (Southern New England Telephone Co.), 163, 164, 264n4, 266n19. See also Connecticut Telephone Co.

Spellier, Louis, 166

Spellier Electric Time Co. (Philadelphia), 166

Standard Electrical Time System Co., 175

Standard Electric Clock Co., 268n3

Standard Electric Time Co., 266n19, 269n12

Standard Railway Time. See under Railroad timekeeping; Time

Standard Time. See under Time

Standard Time, Committee on. See under AAAS; American Metrological Society

Standard Time and Electric Co., 266n19

STC (Standard Time Co.) and clock synchronizer, 134, 160, 161, 162–63, 168, 264n7, 265nn9,10, 266n16, 268n9. See also Yale College Observatory

Steinheil, Carl A., 39, 223n7, 224n16, 225n20

Stone, Ormond, 119, 137–38

Struve, Friedrich G. W., 16, 217n21, 218n26, 223n7

Synchronizer, clock, 157, 237n21, 241n9, 264n1; Bain's, 110, 251n25; Farmer's, 65, 237nn21,22; Hamblet's, 85; Lund's, 110, 112, 134, 157–58, 160, 162, 172, 174; Ritchie's, 251n25; Steinheil's, 85, 224n16; Willson's, 159. See also